Luftwaffe Night Fighter Victory Claims

1939-1945

This book is dedicated to the memory of the aircrews of Germany and the British Commonwealth who fought and died in the offensive aerial campaign over northwest Europe 1939-45

Luftwaffe Nightfighter Victory Claims

The individual combat victory claims of the Luftwaffe nightfighter pilots 1939-45

by

John Foreman

Johannes Matthews

Simon Parry

Cover illustration:
The tail of Oberleutnant Wilhelm Johnen's Me110G from 5./NJG 5.

Rear cover:
The ***horrific close up of the heavily damaged rear turret of Stirling, BF517 'O' No.75 Squadron, after a raid on Duisburg, 26 April 1943. (ww2images.com)***

First published 2004 by
Red Kite
PO Box 223, Walton-on-Thames,
Surrey, KT12 3YQ
England

www.redkitebooks.com
E-mail: contact@redkitebooks.com

Typeset in Great Britain by
A.C.E.Services,
Radlett, Herts, WD7 8LU

Printed in Malta by
Interprint

ISBN 0-9538061-4-6

CONTENTS

Introduction

During the course of the Second World War, there was no campaign longer, more intense and bloodier than that fought in the night skies over Germany and the German-occupied territories of continental Europe between the crews of Royal Air Force Bomber Command and the *Luftwaffe* nightfighter force. Both the British and the Germans began the war with little more than a token night air defence system but, within a few scant years, the nightfighter forces of both sides had developed into fearsome killing machines. The technical air war has already been ably described before, particularly by Dr. Alfred Price in 'Instruments of Darkness' and by Gebhart Aders' 'History of the Luftwaffe Nightfighter Forces', the losses of Bomber Command by Bill Chorley, the losses of the Luftwaffe nightfighter force by Michael Balss. This volume relates the individual combat successes of the *Luftwaffe* aircrew and gives perhaps an insight into how the Luftwaffe night air defence system, known colloquially as the *Nachtjagd*, developed.

Sources

The principal sources for this book are the recently discovered microfilm records of the *Oberkommando der Luftwaffe* and the *Abschusse Kommission*, the organisation responsible for victory confirmations. The listings are, at times, supplemented by nightfighter unit victory boards, rudder markings, personal accounts, log books and diaries and – very occasionally – by previously published sources. These latter have only been used to confirm and correct details such as names, since the microfilms themselves are notoriously bad in quality and are at times difficult, if not impossible, to decipher. Additions to the microfilm listings are not normally inserted unless they are confirmed by two separate sources or by trusted sources as noted below. Where we have made such insertions, the victory numbers are annotated '*'. In cases where names are unreadable, they have been left blank. Some names are inscribed in quotes i.e. 'Schmidt'; this is where a name appears to be so, but the authors are unsure. The victory numbers may be confusing; as this is a record of the *Nachtjagd*, only those victories achieved by aircrew in dedicated nightfighter units, and those by day fighter pilots operating at night, are included and numbered from '1' onwards. Daylight victories credited to pilots who later transferred to the *Nachtjagd* have been ignored, as are those daylight victories by *Wilde Sau* pilots. Day victories by night units have been included. For example, Walter Borchers, who claimed ten kills as a daylight Bf110 pilot, will have the first ten ignored and his listing as victory No.1 will be his first night claim.

One of the authors was privileged to know Herr Hans Ring, former documentation expert of the *Gemeinschaft der Jadgflieger* – the *Luftwaffe* Fighter Pilots Association – and received many *Experten* victory listings from him. Also the late Emil Nonnenmacher, whose collection of original material was truly amazing but has, alas, now been scattered far and wide. Gerrie Zwanenburg MBE, the noted Dutch historian supplied more original documentation, as did several surviving pilots. These separate listings have been used to supplement the microfilms and are asterisked as noted above.

The *Experten*

The careers of *Luftwaffe Experten*, particularly the Ritterkreuzträger, have been widely published; however, anomalies frequently occur. Usually this is where the microfilm confirmations do not 'match up' to the published totals. Usually the published figures are less than the totals accepted, although occasionally they exceed them. This is in itself confusing. For example, if a pilot 'officially' credited with fifty kills has perhaps forty listed, was the accepted total an overstated guess that has become accepted as fact? Were the additional ten claims rejected? Were they correct but unrecorded? We shall never know. In the cases where there are more victories shown in the microfilms than in the accepted totals, were the 'extra' ones rejected and if so which – or did the pilot truly have extra, uncounted kills to his credit? Again, this is a matter for conjecture. The best we can do is to lay the listings before the reader, as always in the form of a research tool. We believe that the list contained in these pages to be the most accurate possible from the available information. However, we would welcome informed additions and corrections to this listing

Within these pages lies the stark outline of the strategic night air offensive from the perspective of those who strove to halt it; hundreds if not thousands of brave young *Luftwaffe* aircrew who took to the night skies to hunt down their adversaries. As one *Luftwaffe* radar operator said, *"We never hated the Tommies, we were there to defend our homeland, our cities and our families."* They did that professionally, fearlessly, and with an awesome dedication and skill – but failed to halt the storm of destruction that befell the *Third Reich*. Very many fell and, as military historians, we salute them.

John Foreman
Radlett

Johannes Matthews
Southampton

Simon Parry
Walton on Thames

Chapter One

Groping in the Dark

September 1939 - December 1940

At the opening of hostilities in 1939, the Germans had already appreciated that some form of night air defence would be a good thing. However, due to the technical restrictions of finding and destroying enemy aircraft at night, their first attempt at fighter interception was to form *Dämmerungstaffeln* (twilight squadrons), equipped with the Bf109D single-seat fighter, to operate just before dawn and just after dusk. These units attached to *Jagdgeschwader 2* and *26*.

The early British un-escorted bomber incursions had sustained more than acceptable losses to fighters and, following the disastrous raid on 18th December 1939, when nearly half of a Wellington force had been slaughtered over the Heligoland Bight, it became clear to both sides that the British bomber offensive would either fail, or take the less dangerous but inaccurate path of night bombing, which would require a completely different approach from the defenders.

The *Dämmerungstaffeln* achieved three successes in the spring of 1940, clearly not enough, and so on 22nd June 1940 a completely new unit was formed under *Oberst* Wolfgang Falck of *Zerstörergeschwader 76*. This was to be named *Nachtjagdgeschwader 1* and was equipped and deployed as follows:

Stab./NJG 1	*Oberst* Falck	Deelen	Bf110C
I./NJG 1	*Hptm* Radusch	Mönchengladbach	Bf110C
II./NJG 1	*Hptm* von Bothmer	Schipol	Bf109D

This deployment did not last long, for on 1st July there were several changes. *II Gruppe* of *./NJG 1* was renamed *III./NJG 1* and a new *I Gruppe* was formed from other units. Thus the nightfighter units were now:-

Stab./NJG 1	*Oberst* Falck	Deelen	Bf110C
I./NJG 1	*Hptm* Radusch	Mönchengladbach	Bf110C
II./NJG 1	*Hptm* Heyse	Güterslohe	Ju88C
III./NJG 1	*Hptm* von Bothmer	Schipol	Bf109D

Before the year was out, there were to be further changes. On 1st September *II./NJG 1*, having relinquished its Bf109s in favour of Bf110s, was renamed *I./NJG 2*, and moved next to Gilze Rijn to commence intruder missions against the United Kingdom. *II./NJG 1* was reformed for the second time, at Deelen under *Hauptmann* Graf von Stillfried und Rattonitz at Deelen with Bf110s. A month later, *I./NJG 3* was formed at Vechta with Bf110s under *Hauptmann* Günther Radusch, with *Hauptmann* Werner Streib becoming *Gruppenkommandeur* of *I./NJG 1*. The formation of an intruder *Gruppe* employing the long-range Ju88C fighters was to prove a thorn in the flesh of British of many months to come. Meanwhile, until airborne radar became available, the defensive *Gruppen* would have to learn to work closely with ground control, *Flak* and searchlights in the hope of stemming the increasing weight of British bomber attacks.

1940

20-21 April 1940

IV./JG2	Ofw Willi Schmale	Battle	Nr. Wismar	00.45	1*	W

24-25 April 1940

IV./JG2	Ofw Fritz Förster	Whitley	Hörnum		1	W

8-9 May 1940

IV/JG2	Ofw Fritz Förster	Whitley	20km N Heligoland	02.50	1	W

8-9 July 1940

St./KG 4	Unknown crew	E/A	England	0330-37	1	W

19-20 July 1940

2./NJG1	Oblt Werner Streib	Whitley	Saerbeck	02.15	1	W

20-21 July 1940

3./NJG1	Oblt Walter Ehle	Wellington	12km N-W Munster	01.38	1	W
StI/NJG1	Oblt Siegfried Wandam	Wellington	8km N Borken	02.23	1	W

21-22 July 1940

2./NJG1	Oblt Werner Streib	Whitley	10km N Münster	01.22	2	W

22-23 July 1940

5./NJG2	Fw Otto Wiese	Wellington	100km W Texel	22.30	1	W
4./NJG1	Fw Georg Schramm	Wellington	N Sea	23.55	1	W

17-18 August 1940

4./NJG1	Fw Peter Laufs	Hurricane	Lowestoft	22.45	1	W

27-28 August 1940

St./NJG2	Ofw Merbach	Hurricane	Grimsby	21.30	1	W

30-31 August 1940

2./NJG1	Oblt Werner Streib	Wellington	20km N-N-E Emmerich	23.24	3	W
2/NJG1	Oblt Werner Streib	Hampden	Arnheim	00.32	4	W

2-3 September 1940

3./NJG1	Fw Paul Gildner	Hampden	9km S Sittard	00.45	3	W

6-7 September 1940

1./NJG2	Lt Kurt Hermann	Blenheim	Nr. Waddington	04.05	1	W

Werner Streib, who achieved the first 'official' victory of the nightfighter force on 19th/20th July 1940 by bringing down a 9 Squadron Wellington. He would bring his score to seven night victories by the end of 1940, adding to his earlier victory as a daylight Bf110 pilot with 2./ZG 1. By the end of the war he would rise to the rank of Oberst, having been awarded the Ritterkreuz mit Eichenlaub und Schwerter.

9-10 September 1940

2./NJG2	Fw Hermann Sommer	Blenheim	Nr. Waddington	23.15	1	W

18-19 September 1940

3./NJG1	Fw Paul Gildner	Whitley	3km S-E Groenlo	23.25	2	W
3./NJG1	Fw Paul Gildner	Whitley	Zieuwent	00.28	3	W

19-20 September 1940

St./KG53	Unknown crew	E/A	England	03.37		W

29-30 September 1940

I./NJG1	Oblt Heinrich Griese	Wellington	Malgarten	00.34	1	W

30 September-1 October 1940

2./NJG1	Oblt Werner Streib	Wellington	Bersenbruck	22.49	5	W
2./NJG1	Oblt Werner Streib	Hampden	Badbergen	23.19	6	W
2./NJG1	Oblt Werner Streib	Wellington	Menslage	23.35	7	W
I./NJG1	Oblt Heinrich Griese	Wellington	Löningen	00.38	2	W

7-8 October 1940

2./NJG1	Lt Hans-Georg Mangelsdorf	Hampden		00.10	1	W

12-13 October 1940

2./NJG1	Lt Hans-Georg Mangelsdorf	Hampden		03.02	2	W

14-15 October 1940

StI/NJG1	Hptm Werner Streib	Hampden	NE Calbe	03.05	8	W
I./NJG 1	Ofw Gerhard Herzog	Wellington	Gardelegen	00.55	1*	W

16-17 October 1940

4./NJG1	Lt Ludwig Becker	Wellington	Oosterwolde	21.25	1	W

The main opponent for the British bomber forces for the whole of the night air war was to be the Messerschmitt Bf110. In 1940, the standard model was the Bf110C, which was soon to be supplanted by the Bf110D (left) and Bf110E models. (right) Ground crew refuelling a Bf110E of II./NJG 1.

20-21 October 1940

3./NJG2	Hptm Karl Hülshoff	Hereford	Nr. Dishforth	21.30	1	W

24-25 October 1940

1./NJG2	Lt Kurt Hermann	Blenheim	Nr. Hemswell	21.50	2	W
1./NJG2	Lt Kurt Hermann	Blenheim	Nr. Hemswell	22.10	3	W
3./NJG2	Lt Hans Hahn	Wellington	Nr. Linton-on-Ouse	23.02	1	W

27-28 October 1940

2./NJG2	Lt Heinz Völker	Hampden	E Skegness	00.30	1	W

16-17 November 1940

4./NJG1	Oblt E. Lippe-Weissenfeld	Wellington	10km W Medemblik	02.05	1	W

23-24 November 1940

3./NJG2	Hptm Karl Hülshoff	Wellington	80km W Scheveningen	18.30	2	W
1./NJG2	Fw Heinz Strüning	Wellington	50km W Scheveningen	18.40	1	W
1./NJG2	Lt Kurt Hermann	Hereford	over N Sea	18.45	4	W

6-7 December 1940

3./NJG2	Fw Wilhelm Beier	Hurricane	5km E Spilsby	06.36	1	W

15-16 December 1940

4./NJG1	Fw Hans Rasper	Wellington	10km NW Petten	23.23	1	W

16-17 December 1940

4./NJG2	Oblt Paul Böhn	Hampden	Scampton	04.00	1	W

20-21 December 1940

1./NJG2	Uffz Gerhard Blum	Blenheim	over N Sea	09.00	1	W

Intruder pilots of I./NJG 2: (right) Karl Hülshoff, the Gruppenkommandeur would rise to the rank of Oberstleutnant and is believed to have survived the war with eleven credited victories. (left) Oberleutnant Kurt Hermann claimed a total of nine kills, seven with NJG 2, but was shot down by a Hurricane in March 1941 to become a prisoner. Both claimed victories over England on the night of 23rd November 1940.

21-22 December 1940

2./NJG2	Lt Heinz Völker	Blenheim	50km E Aldeburgh	21.40	2	W

Chapter Two

Gaining Experience

January - July 1941

By 1941, the *Luftwaffe* nightfighter crews were mastering the business of hunting down the British night raiders. The Bf110, found to be so sadly lacking in day fighter qualities, was proving to be a more than adequate night fighter and the Ju88C was also an excellent choice, particularly when used as a night intruder. During this period, men like Werner Streib, Paul Gildner, Ludwig Becker, Heinz Strüning and Walter Ehle were taking the first steps on a path that would lead to renown and high honours in the defence of the Reich, while Hans Hahn and Kurt Hermann of *I./NJG 2* were fast gaining a reputation as long-range hunters. The one major problem was the fact that the German fighters had no airborne radar equipment, but this was to rectified in August.

In the following seven months, only one new unit was to be formed, when *Stab./NJG 4* came into existence at Metz under command of Major Rudolf Stoltenhoff. However the establishment of this Geschwader in its entirety would take much longer.

1-2 January 1941

1./NJG2	Lt Rudolf Stradner	Wellington	75km E Lowestoft	18.40	2500	1	W

2-3 January 1941

1./NJG2	Uffz Helmut Arnold	Wellington	50km E Haisborough	18.45	3600	1	W
3./NJG2	Lt Hans Hahn	Whitworth	50km E Withernsea	19.00		2	W

3-4 January 1941

3./NJG2	Lt Gerhard Böhme	Whitley	SE Flamborough Hd	18.50		1	W

An aircraft to fear: The Bf110 flown by the deadly Oberleutnant Helmut Lent of II./NJG 1.

9-10 January 1941

6./NJG1	Oblt Reinhold Eckardt	Whitley	Erlekem, S Nijmegen	23.18	1	W

15-16 January 1941

4./NJG1	Oblt E. Lippe-Weissenfeld	Whitley	5km N Petten	22.46	2	W
2./NJG2	Oblt Albert Schulz	Blenheim	Church Fenton	02.30	1	W
2./NJG2	Oblt Albert Schulz	Blenheim	Church Fenton	02.45	2	W

9-10 February 1941

StI/NJG1	Hptm Walter Ehle	Wellington	15km W Nunspeet	23.33	3	W

10-11 February 1941

2./NJG2	Oblt Albert Schulz	Blenheim	Nr. West Raynham	01.10	3	W
2./NJG2	Hptm Rolf Jung	Wellington	Nr. West Raynham	02.30	1	W
5./NJG1	Lt Leopold Fellerer	Hampden	10km N Alkmaar	03.50	1	W
3./NJG2	Oblt Paul Semrau	Blenheim	Nr. Feltwell	04.15	1	W
3./NJG2	Oblt Paul Semrau	Blenheim	Nr. Feltwell	04.20	2	W
1./NJG2	Oblt Kurt Hermann	Hereford	Nr. Waddington	06.49	5	W
1./NJG2	Oblt Kurt Hermann	Hereford	Nr. Waddington	06.58	6	W

14-15 February 1941

I./NJG3	Oblt Jusgen	Whitley	8km S Nijmegen	23.19	1	W
1./NJG2	Fw Heinz Strüning	Hudson	75km E Gt Yarmouth	09.15	2	W

15-16 February 1941

1./NJG2	Fw Heinz Strüning	Wellington	65km ENE Southend	19.58	3	W
6./NJG1	Fw Ernst Kalinowski	Wellington	10km west Zutphen	23.05	1	W
1./NJG2	Oblt Herbert Bönsch	Blenheim	Nr. Waddington	01.25	1	W

25-26 February 1941

2./NJG2	Fw Ernst Ziebarth	Wellington	Nr. Waddington	23.44	1	W

26-27 February 1941

1./NJG2	Oblt Herbert Bönsch	Blenheim	Nr. Scampton	00.15	2	W
1./NJG2	Oblt Kurt Hermann	Blenheim	Nr. Waddington	01.40	7	W

28 February-1 March 1941

4./NJG1	Ofw Paul Gildner	Blenheim	Nr. Groningen	02.58	4	W

1-2 March 1941

1./NJG2	Oblt Herbert Bönsch	Blenheim	Nr. Scampton	02.55	3	W

8-9 March 1941

2./NJG1	Lt Reinhold Knacke	Wellington		05.50	1	W

10-11 March 1941

StI/NJG1	Hptm Werner Streib	Hampden	15km S-W Venlo	22.18	9	W
2./NJG1	Lt Reinhold Knacke	Stirling		23.42	2	W

12-13 March 1941

4./NJG1	Fw Hans Rasper	Wellington	4km S Medemblik	21.46	2	W
2./NJG2	Fw Heinz Mittelstädt	Wellington	Nr. Wells	22.10	1	W

Paul Gildner of 4./NJG 1 was a fast rising star in the spring of 1941. He would win the Ritterkreuz on 9th July and, after claiming a total of 44 victories, would be posthumously awarded the Eichenlaub after being killed in an accident on 24/25th February 1943. On 13th March 1941 he brought down Flight Lieutenant Dickinson's 110 Squadron Blenheim. He is seen here inspecting the wreckage.

III./NJG1	Oblt Heinrich Wohlers	Hampden	Nordhorn	00.05	1	W
3./NJG3	Uffz Karl Kupfer	Wellington	5km W Meppen	00.55	1	W
3./NJG2	Lt Hans Hahn	Hudson	Nr. Leeming	01.15	3	W

13-14 March 1941

3./NJG2	Lt Hans Hahn	Blenheim	Nr. Waddington	22.00	4	W
4./NJG1	Ofw Paul Gildner	Blenheim	Tolbert	22.48	5	W
2./NJG1	Lt Reinhold Knacke	Wellington		22.50	3	W
4./NJG1	Ofw Paul Gildner	Wellington	Nr. Burlanger	23.10	6	W
I./NJG3	Lt Ernst Schmidt	Wellington	Cloppenburg	00.54	1	W
4./NJG1	Fw Hans Rasper	Wellington	10km N-W Petten	03.20	3	W

14-15 March 1941

StI/NJG1	Hptm Werner Streib	Wellington	Helenaveen	22.32	10	W

15-16 March 1941

1./NJG3	Fw Georg Klever	Hurricane		08.00	1	W

17-18 March 1941

1./NJG2	Fw Peter Laufs	Wellington	Nr. Cromer	06.54	2	W
1./NJG2	Lt Rudolf Pfeiffer	Wellington	Nr. Mildenhall	07.10	1	W

27-28 March 1941

2./NJG1	Ofw Gerhard Herzog	Whitley	10km N Meijel	23.05	2	W
2./NJG1	Ofw Gerhard Herzog	Wellington	10km S-E Helmond	23.30	3	W
I./NJG1	Oblt Walter Fenske	Manchester	1km W Hertogenbosch	23.50	1	W

31 March- 1 April 1941

7./NJG1	Fw Karl-Heinz Scherfling	Wellington	S-E Groningen	22.36	1	W

3-4 April 1941

3./NJG2	Lt Heinz Völker	Wellington	Nr. West Raynham	01.20	3	W

7-8 April 1941

StI/NJG2	Unknown crew	E/A			1	W
4./NJG1	Ofw Paul Gildner	Whitley	8km S-E Gröningen	00.27	7	W
2./NJG2	Lt Hans Feuerbaum	Hudson	Nr. Wells	01.20	1	W
2./NJG2	Lt Hans Feuerbaum	Hereford	Nr. Wells	01.25	2	W
3./NJG2	Lt Hans Hahn	Hereford	Wellingore	01.43	5	W

8-9 April 1941

StI/NJG2	Hptm Karl Hülshoff	Wellington	50km E Bridlington	22.50	3	W
StI/NJG2	Hptm Karl Hülshoff	Hampden	30km N Scarborough	23.00	4	W
3./NJG2	Lt Hans Hahn	Wellington	nr Upwood	00.40	6	W

9-10 April 1941

4./NJG2	Uffz Hans Berschwinger	Wellington	Nr. Marham	23.20	1	W
StIII/NJG1	Oblt Fritz Gutezeit	Wellington	15km N-W Nordhorn	23.27	1	W
7./NJG1	Fw Karl-Heinz Scherfling	Stirling	S-W Lingen	23.35	2	W
3./NJG3	Uffz Karl Kupfer	Wellington	10km S-E Leer	00.14	2	W
4./NJG1	Oblt E. Lippe-Weissenfeld	Wellington	S Den Helder	00.59	3	W
4./NJG1	Ofw Paul Gildner	Wellington	Ijsselmeer	01.07	8	W
1./NJG2	Fw Peter Laufs	Wellington	Nr. Wells	01.30	3	W
2./NJG2	Oblt Albert Schulz	Wellington	Nr. Cambridge	01.30	4	W
3./NJG2	Ofw Wilhelm Beier	Whitworth	Chelmsford	02.35	2	W
5./KG53	Unknown crew	E/A		02.40		W
StI/NJG3	Hptm Gunther Radusch	Wellington	2km S-W Papenburg	03.00	1	W
3./NJG2	Uffz Alfons Köster	Defiant	Upwood	05.17	1	W

10-11 April 1941

2./NJG1	Lt Hermann Reese	Hampden	S-W Beesel	22.34	1	W
StI/NJG1	Hptm Werner Streib	Hampden	W Roggel	22.49	11	W
2./NJG1	Ofw Gerhard Herzog	Hampden	6km N Maastricht	22.55	4	W
StI/NJG1	Hptm Werner Streib	Hampden	Ittervoort	23.02	12	W
StI/NJG1	Lt Hans-Dieter Frank	Hampden	2km S-W Neeroeteren	23.32	1	W

16-17 April 1941

3./NJG2	Lt Heinz Völker	Hereford	Nr. Finningley	23.05	4	W
3./NJG2	Lt Hans Hahn	Hampden	SE. Waddington	01.22	7	W

17-18 April 1941

StI/NJG1	Hptm Werner Streib	Wellington	10km E Weert	23.39	13	W

20-21 April 1941

3./NJG2	Lt Hans Hahn	Hereford	Nr. Digby	03.30	8	W

23-24 April 1941

2./NJG2	Fw Vinzenz Giessübel	Blenheim	Nr. Marham	01.55	1	W

24-25 April 1941

3./NJG2	Lt Heinz Völker	Blenheim	Nr. Lindholme	00.30	5	W

G9+BC of II./NJG 3 was crash-landed at Schleswig on 10/11th April. It was flown by Leutnant Uellenbeck, a former Bf110 pilot with I./ZG 76, who had fought in the 'Battle of the Heligoland Bight' in 1939. (S.Hall)

3./NJG2	Lt Heinz Völker	Blenheim	Nr. Lindholme	00.37	6	W
3./NJG2	Lt Heinz Völker	Blenheim	Nr. Lindholme	00.44	7	W
3./NJG2	Fw Herbert Biehne*	Blenheim	Nr. Lindholme	00.50	1	W

* *Bordschütze Völker*

25-26 April 1941

4./NJG2	Lt Rudolf Pfeiffer	Defiant	Nr. Hemswell	03.10	2	W

29-30 April 1941

2./NJG2	Ofw Hermann Sommer	Blenheim	Nr. Tollerton	00.15	2	W
2./NJG2	Ofw Hermann Sommer	Blenheim	Nr. Tollerton	00.20	3	W
2./NJG2	Ofw Hermann Sommer	Blenheim	Nr. Huckhall	00.50	4	W
2./NJG2	Ofw Hermann Sommer	Beaufort	Nr. Bircham Newton	01.30	5	W

Albert Schulz of I./NJG 2 was not destined to become a 'high scorer', but claimed several victories over England during 1940 - 1941. The intuders were very active in the spring of 1941 and were arguably more successful than their home-based comrades. The effect that their activities produced was far greater than their numbers would suggest; many RAF pilots, whether on operations or training sorties, crashed in hurried landings due to the threat of an intruder in the area. (H.Scholl)

2-3 May 1941

3./NJG2	Lt Hans Hahn	Stirling	Oakington	01.20	9	W
2./NJG2	Lt Hans Feuerbaum	Hereford	Nr. Church Fenton	03.30	3	W
2./NJG1	Lt Reinhold Knacke	Wellington	4km W Weert	04.13	4	W

3-4 May 1941

3./NJG2	Lt Hans Hahn	Fulmar	Stoke Holy Cross	03.10	10	W

4-5 May 1941

3./NJG2	Lt Hans Hahn	Blenheim	Nr. Feltwell	01.12	11	W

6-7 May 1941

1./NJG2	Fw Heinz Strüning	Wellington	Peterborough	02.08	4	W

7-8 May 1941

3./NJG2	Uffz Alfons Köster	Wellington	5km W Southend	00.22	2	W
3./NJG2	Oblt Paul Semrau	Wellington	Nr. Nottingham	01.20	3	W
3./NJG2	Ofw Wilhelm Beier	Wellington	Nr. Wells	02.50	3	W

8-9 May 1941

3./NJG2	Uffz Alfons Köster	Blenheim	E Mablethorpe	23.58	3	W
4./NJG1	Ofw Hans Rasper	Wellington	10km E Medemblik	00.48	4	W
6./NJG1	Oblt Reinhold Eckardt	Blenheim	25km W St.Peter	02.13	2	W
I./NJG3	Uffz Schmolinsky	Wellington	2km N Plantlünne	02.14	1	W
I./NJG3	Fw Heinz Säwert	Wellington	Leer	02.24	3	W
5./NJG1	Lt Rolf Bokemeyer	Whitley		02.35	1	W
4./NJG1	Ofw Paul Gildner	Whitley	E Enkhuizen	02.47	9	W
4./NJG1	Oblt E. Lippe-Weissenfeld	Wellington	Anna Paulowna	02.48	4	W
6./NJG1	Ofw Johann Schönherr	Wellington	10km N Helgoland	05.15	1	W

9-10 May 1941

1./NJG2	Fw Heinz Strüning	Wellington	over N Sea	23.30	5	W
2./NJG1	Lt Reinhold Knacke	Whitley	8km W Weert	03.05	5	W

10-11 May 1941

1./NJG2	Oblt Heinz Harmstorf	Wellington	Sea.N-E Cromer	23.19	1	W
4./NJG1	Oblt E. Lippe-Weissenfeld	Stirling	10km S-W Medemblik	00.20	5	W
I./NJG3	Fw Heinz Säwert	Stirling	5km S-E Cloppenburg	00.36	4	W
StI/NJG1	Hptm Walter Ehle	Wellington	5km W Westerhever	00.57	3	W
5./NJG1	Oblt Karl-Heinz Hollborn	Whitley	5km W Meldorf	01.15	1	W
I./NJG3	Oblt Walter Milius	Manchester	15km WSW Vestrup	01.15	1	W
6./NJG1	Lt E-W von Bonin	Wellington	Tönning	02.20	1	W
6./NJG1	Oblt Reinhold Eckardt	Whitley	6km S Husum	02.48	3	W
6./NJG1	Ofw Johann Schönherr	Blenheim	W Wesselburen	03.02	2	W
3./NJG2	Ofw Wilhelm Beier	Blenheim	50km E Scarborough	03.46	4	W

11-12 May 1941

6./NJG1	Lt Lothar Linke	Wellington	SE Estenfeld	01.16	1	W
6./NJG1	Oblt Helmut Lent	Wellington	SW Suderstapel	01.40	1	W
6./NJG1	Lt E-W von Bonin	Blenheim	3km W St.Peter	01.42	2	W
4./NJG1	Ofw Hans Rasper	Whitley	10km N-W Enkhuizen	02.13	5	W

Feldwebel Hans Rasper (left) and Unteroffizier Erich Schreiber of 4./NJG1. Rasper scored slowly but steadily and, by the end of 1943, had seven credited kills. (S.Parry)

6./NJG1	Oblt Helmut Lent	Wellington	SW Nordstrand	02.49	2	W
4./NJG1	Oblt Helmut Woltersdorf	Hampden	8km SW Enkhuizen	02.57	1	W
2./NJG3	Lt Rudolf Altendorf	Wellington	Cloppenburg	03.20	1	W

16-17 May 1941

2./NJG2	Ofw Hermann Sommer	Wellington	Nr. Wells	23.30	6	W

17-18 May 1941

2./NJG1	Lt Kurt Loos	Whitley	SE Megelsum	01.25	1	W
3./NJG2	Oblt Paul Semrau	Blenheim	5km S Grantham	02.15	4	W

3-4 June 1941

3./NJG2	Ofw Wilhelm Beier	Blenheim	20km SE Lowestoft	01.35	5	W

11-12 June 1941

2./NJG1	Lt Kurt Loos	Wellington	Megelsum	01.34	2	W
3./NJG2	Uffz Alfons Köster	Hampden	N-W London	02.07	4	W
StI/NJG1	Lt Hans-Dieter Frank	Whitley	Nr. Eindhoven	02.13	2	W
4./NJG1	Uffz Ludwig Röll	Hudson	100km W Texel	03.47	1	W
4./NJG2	Oblt Paul Bohn	Wellington	Sea.N-E Bacton	03.55	2	W
4./NJG2	Oblt Paul Bohn	Wellington	Sea.N-E Bacton	04.10	3	W

12-13 June 1941

4./NJG1	Ofw Hans Rasper	Wellington	N-W Enkhuisen	00.37	6	W
3./NJG2	Ofw Wilhelm Beier	Defiant	Nr. Thornby	00.44	6	W
4./NJG1	Oblt E. Lippe-Weissenfeld	Whitley	2km N Medemblik	01.10	6	W
3./NJG2	Oblt Paul Semrau	Halifax	Nr. Finningley	01.40	5	W
2./NJG1	Lt Reinhold Knacke	Whitley	Aalst	03.16	6	W

(Right) Unteroffiziers Heinz Schulz(Bs) Helmut Bahner (Ff) and Jakob Reid of I./NJG 2 were caught by Pilot Officer J.B.Thompson of 25 Squadron and were shot down at Wingland Marsh. (Left) Bahner in front of R4+EM at Gilze Rijn.(S.Hall)

16-17 June 1941

3./NJG2	Ofw Rolf Bussmann	Wellington	N Cromer	00.00	1	W
3./NJG2	Ofw Rolf Bussmann	Wellington	NW Cromer	00.07	2	W
1./NJG1	Fw Reinhard Kollak	Whitley	Nr. Hasselt	02.26	1	W
2./NJG1	Lt Reinhold Knacke	Wellington	10km NW Weert	02.31	7	W
1./NJG2	Oblt Herbert Bönsch	Wellington	120km E Harwich	03.12	4	W
1./NJG2	Fw Helmut Arnold	Whitley	50km N Estend	03.45	2	W
2./NJG2	Ofw Hermann Sommer	Wellington	30km SE Lowestoft	03.55	7	W
2./NJG2	Fw Otto Glas	Wellington	over N Sea	04.05	1	W
2./NJG2	Ofw Hermann Sommer	Wellington	20km SE Lowestoft	04.10	8	W

17-18 June 1941

1./NJG2	Oblt Herbert Bönsch	Wellington	Sea.115km E Cromer	00.08	5	W
2./NJG1	Oblt Wolfgang Thimmig	Whitley		02.41	1*	W

18-19 June 1941

4./NJG1	Oblt E. Lippe-Weissenfeld	Wellington	W Enkhuizen	00.53	7	W
4./NJG1	Ofw Paul Gildner	Wellington	Sea.Nr. Ameland	01.38	10	W
4./NJG1	Ofw Paul Gildner	Wellington	57km N Ameland	02.34	11	W
2./NJG3	Lt Rudolf Altendorf	Whitley	SW Hilkenbroek	02.48	2	W
4./NJG1	Ofw Paul Gildner	Whitley	47km N Ameland	02.57	12	W
3./NJG2	Oblt Paul Semrau	Whitley	90km NE Gt Yarmouth	04.23	6	W

19-20 June 1941

4./NJG1	Uffz Ludwig Röll	Whitley	NW Terschelling	03.17	2	W

20-21 June 1941

4./NJG2	Oblt Paul Bohn	Wellington	200km E Spurn Head	00.20	4	W

22-23 June 1941

1./NJG2	Oblt Herbert Bönsch	Wellington	75km NE Gt Yarmouth	00.15	6	W
4./NJG1	Oblt E. Lippe-Weissenfeld	Wellington	Insinghuizen	02.13	8	W

Oberleutnant Paul Semrau of 3./NJG 2 claimed his sixth victory on 18/19th June 1941 when he caught a Whitley off Great Yarmouth. He was to rise to command NJG 2, having achieved 46 night combat victories. On 8th February 1945 he unwisely flew a daylight test flight and was caught by a Canadian Spitfire pilot as he approached Twente airfield. He and his crew were killed in action and he was posthumously awarded the Eichenlaub to the Ritterkreuz, which he had received in October 1942.

23-24 June 1941

II./NJG1	Oblt Reinhold Eckardt	Halifax	Buxtehude	02.32	4	W

24-25 June 1941

4./NJG1	Oblt Helmut Woltersdorf	Wellington	N Schiermonnikoog	01.40	2	W
4./NJG1	Uffz Ludwig Röll	Wellington	10km SW Petten	04.11	3	W

25-26 June 1941

4./NJG2	Oblt Paul Bohn	Whitley	over N Sea	00.15	5	W
I./NJG1	Oblt Walter Fenske	Whitley	3km W Wittbrock	00.53	2	W
4./NJG2	Ofw Jung	Wellington	Sea.75km E Harwich	01.35	1	W

26-27 June 1941

I./NJG1	Oblt Walter Fenske	Manchester	Adolf Hitler-Koog	01.31	3	W
I./NJG1	Oblt Walter Fenske	Manchester	NW St.Peter	01.46	4	W

27-28 June 1941

3./NJG2	Ofw Robert Lüddeke	Wellington	30km SE Eastbourne	23.45	1	W
2./NJG2	Ofw Hermann Sommer	Wellington	70km E Gt Yarmouth	23.55	9	W
I./NJG3	Uffz Benning	Wellington	S-E Ahaus	00.38	1	W
II./NJG1	Oblt Reinhold Eckardt	Whitley	Zuiderzee	01.36	5	W
II./NJG1	Oblt Reinhold Eckardt	Wellington	Hamburg	01.42	6	W
5./NJG1	Uffz Armin Behrend	Wellington	NE Hamburg	01.52	1	W
II./NJG1	Oblt Reinhold Eckardt	Whitley	10km SE Stade	01.52	7	W
6./NJG1	Oblt Helmut Lent	Whitley	15km W Bremervorde	01.58	3	W
II./NJG1	Oblt Reinhold Eckardt	Wellington	Hanburger Hafen	02.22	8	W

29-30 June 1941

6./NJG1	Oblt Helmut Lent	Stirling	10km S Wesermünde	01.40	4	W
I./NJG3	Oblt Walter Milius	Wellington	40km N-E Bremen	01.45	2	W
StII/NJG1	Hptm Walter Ehle	Wellington	40km NE Bremen	01.52	4	W
2./NJG3	Lt Rudolf Altendorf	Hampden	3km SE Eldenburg	01.55	3	W
6./NJG1	Oblt Helmut Lent	Stirling	20km SW Bremervörde	02.05	5	W
StII/NJG1	Hptm Walter Ehle	Stirling	2km NW Ellerbeck	02.15	5	W
4./NJG1	Ofw Paul Gildner	Whitley	10km NNW Ameland	02.43	13	W
5./NJG1	Uffz Ludwig	Wellington	N Deinstadt	02.50	1	W
4./NJG1	Uffz Ludwig Röll	Wellington		02.50	4*	W
II./NJG1	Oblt Reinhold Eckardt	Hampden	S Uetersen	02.55	9	W

2./NJG1	Ofw Gerhard Herzog	Stirling	3km W Pellworm	03.15		5	W
1./NJG2	Fw Helmut Arnold	Wellington	60km E Bacton	03.50	800	3	W
4./NJG2	Ofw Jung	Wellington	200km ESE Grimsby	04.14		2	W

30 June 1941

I./NJG1	Oblt Walter Fenske	Halifax	S Garding	13.20		5	W
I./NJG1	Oblt Walter Fenske	Halifax	SW Husum	13.30		6	W
I./NJG1	Lt Jakob Bender	Halifax	E Tönnnig	13.31		1	W
2./NJG1	Ofw Gerhard Herzog	Halifax	SW Husum	13.35		6	W

30 June-1 July 1941

StI/NJG1	Hptm Werner Streib	Whitley	2km NW Waldfeucht	01.19		14	W

2-3 July 1941

2./NJG1	Lt Reinhold Knacke	Whitley	8km SW Maastricht	01.03		8	W
2./NJG1	Lt Reinhold Knacke	Hampden	6km NE Maastricht	01.12		9	W

3-4 July 1941

1./NJG3	Gfr Rudolf Frank	Wellington	10km N Grafeld	00.36		1	W
4./NJG1	Oblt Helmut Lent	Wellington	23km SE Assen	00.43	4600	6	W
StI/NJG1	Hptm Werner Streib	Whitley	3km E Asten	02.33		15	W
4./NJG1	Oblt Helmut Lent	E/A		03.41		7	W

4-5 July 1941

1./NJG2	Ofw Peter Laufs	Battle	Nr. Wells	01.00		4	W
I./NJG1	Oblt Heinrich Griese	Hampden	3km NNE Heeze	01.09		3	W
1./NJG2	Ofw Heinz Strüning	Wellington	Nr. Bircham Newton	01.57		6	W

5-6 July 1941

4./NJG1	Oblt Helmut Lent	Whitley	8km W Coeverden	00.56	3000	8	W
3./NJG2	Ofw Wilhelm Beier	Wellington	110km NNW Texel	03.08		7	W
3./NJG2	Ofw Wilhelm Beier	Blenheim	100km W Texel	03.36		8	W

6-7 July 1941

3./NJG2	Ofw Wilhelm Beier	Whitley	60km NE Bacton	23.54		9	W
3./NJG2	Ofw Wilhelm Beier	Whitley	60km NE Bacton	23.58		10	W
3./NJG2	Fw Heinrich Barth	Blenheim	30km E Gt Yarmouth	00.24		1	W
3./NJG2	Oblt Paul Semrau	Blenheim	Nr. Wells	01.25		7	W
4./NJG1	Oblt Helmut Woltersdorf	Wellington	2km W Texel	03.43		3	W

7-8 July 1941

4./NJG2	Fw Hans Berschwinger	Whitley	Sea.NW Sheffield	00.12		2	W
4./NJG1	Oblt Helmut Lent	Whitley	20km SE Assen	00.55	3600	9	W
2./NJG1	Lt Reinhold Knacke	Hampden	Heuthem	01.17		10	W
2./NJG1	Lt Reinhold Knacke	Wellington	Maastricht	01.25		11	W
I./NJG1	Oblt Heinrich Griese	Hampden	Heeze	02.18		4	W

8-9 July 1941

III./NJG1	Oblt Hans-Karl Kamp	Whitley	3km SW Dreierwalde	21.19		1	W
4./NJG1	Ofw Paul Gildner	Hampden	N Gröningen	00.58		14	W
4./NJG1	Lt Rudolf Schoenert	Whitley	60km NW Vlieland	02.51		1	W
2./NJG2	Oblt Albert Schulz	Hampden	5km NE Feltwell	03.01		5	W

Waiting for dusk. A Ju88C of I./NJG 2 at Gilze Rijn. Note the 'solid' nose, containing the heavy cannon and machine-gun armament, and the all-black camouflage.

4./NJG1	Lt Rudolf Schoenert	Whitley	55km NW Vlieland	03.40		2	W

9-10 July 1941

III./NJG1	Lt August Geiger	Halifax	10km S Nijmegen	23.32		1	W
I./NJG1	Oblt Heinrich Griese	Wellington	1km SE Dilsen	01.05		5	W
III./NJG1	Oblt Heinrich Wohlers	Wellington	4km S Meppen	01.31		2	W
4./NJG1	Oblt Helmut Lent	Wellington	10km NW Meppen	02.20	4400	10	W

10-11 July 1941

2./NJG1	Oblt Wolfgang Thimmig	Wellington	Arendonk	02.40		2	W

12-13 July 1941

4./NJG1	Oblt Helmut Lent	Hampden	25km SE Gröningen	00.55		11	W
7./NJG1	Lt Heinz-Martin Hadeball	Hampden	2km E Wettrup	01.05		1	W

13-14 July 1941

4./NJG1	Oblt E. Lippe-Weissenfeld	Wellington	S Medemblik	00.28		9	W

14-15 July 1941

4./NJG1	Oblt Helmut Lent	Wellington	25km SE Gröningen	00.49	3000	12	W
I./NJG3	Uffz Benning	Wellington	5km NW Zeven	01.21		2	W
3./NJG2	Fw Alfons Köster	Blenheim	Nr. Wyton	01.25		5	W
3./NJG2	Fw Alfons Köster	Blenheim	Nr. Wyton	01.30		6	W
6./NJG1	Lt E-W von Bonin	Wellington	E Zwolle	01.40		3	W

15-16 July 1941

III./NJG1	Oblt Hans-Karl Kamp	Wellington	W Osnabrück	22.10		4	W
StI/NJG1	Hptm Werner Streib	Wellington	Someren	00.54		16	W
2./NJG1	Oblt Wolfgang Thimmig	Wellington	SSE Hamont-Achtel	01.31		3	W
StI/NJG1	Hptm Werner Streib	Wellington	Geyspers	01.45		17	W

16-17 July 1941

4./NJG1	Lt Rudolf Schoenert	Wellington	5km W Lemmer	00.50	4000	3	W
4./NJG1	Ofw Paul Gildner	Wellington	SW Harlingen	03.27	4400	15	W

17-18 July 1941

2./NJG1	Oblt Wolfgang Thimmig	Whitley		02.41		4	W
3./NJG2	Oblt Paul Semrau	Blenheim	Nr. Digby	02.50		8	W
3./NJG2	Ofw Wilhelm Beier	Blenheim	120km NE Gt Yarmouth	03.02		11	W

18-19 July 1941

4./NJG2	Oblt Hans-Georg Schütze	Wellington	Nr. Wyton	00.40		1	W
4./NJG2	Oblt Hans-Georg Schütze	Wellington	Nr. Wyton	00.51		2	W

19-20 July 1941

I./NJG3	Uffz Alfred Brackmann	Whitley	Osnabrück	01.25		1	W

21-22 July 1941

2./NJG2	Lt Heinz Völker	Wellington	Ashwell	01.19		8	W

24-25 July 1941

4./NJG1	Oblt E. Lippe-Weissenfeld	Wellington	3km SW Medemblik	02.23	4700	10	W
4./NJG1	Oblt Helmut Lent	Wellington	13km SSW Leeuwarden	03.54	4500	13	W

25-26 July 1941

4./NJG1	Lt Lothar Linke	Hampden	Schiermonnikoog	23.57	4200	2	W
1./KG1	Unknown crew	E/A	Russia	03.10			E
2./KG28	Gfr	Rata	Russia	03.10			E
2./KG28	Gfr	Rata	Russia	03.10			E
4./NJG1	Oblt E. Lippe-Weissenfeld	Whitley	11km W De Kooi	03.20	2200	11	W
1./NJG1	Fw Reinhard Kollak	Whitley	1km W Courcelles	04.22		2	W
4./SKG210	Ogfr Josef Kociok	SB-2	Russia			1	E

27-28 July 1941

I./NJG2	Lt Lothar Bisang	Oxford	Raum "C" England	03.46		1	W

29-30 July 1941

1./NJG2	Fw Helmut Arnold	Whitley	Nr. Hemswell	00.45		4	W

Chapter Three

The Kammhuber Line

August 1941 - May 1942

In mid-1940, *Oberst* Josef Kammhuber had been appointed to command the growing nightfighter force and, in October, had begun the task of creating a co-ordinated defensive structure using ground radar and searchlights. He began with three nightfighter zones, each of which was equipped with a pair of short-range *Würzburg* radar sets. One of these was employed to track the nightfighter and the other to track the bomber and co-ordinate the illumination of the target with the searchlight units. Known as *Helle Nachtjagd* (illuminated nightfighting), this was quickly supplanted by *Himmetbett* (four-poster bed), which retained the twin *Würzburg* radars for fighter and target tracking and utilized the long-range *Freya* radar for overall surveillance. Such was the accuracy of the radar equipment that skilled controllers were usually able to bring the fighter crew to visual acquisition of a target. By mid-1941, the chain of nightfighter zones or 'boxes' extended from west of the Ruhr as far north as the Danish coast. All that was required was the provision of airborne radar to make the system almost perfect.

This defensive barrier, to become known as the *Kammhuber Line*, had one serious drawback. A control centre could only intercept one aircraft at a time in any one box. The British crews had obliged the defences since attacks on Germany had commenced in 1940 by flying individually to their targets, rather than as a cohesive stream. When the British discovered this, the Germans could well be swamped by many bombers flying through the same box simultaneously. The good news for the *Nachtjager* was that a new airborne radar had indeed been designed and built, known as *Lichtenstein SN-2*. The effectiveness of this was demonstrated on the night of 9/10th August 1941 when, flying a radar-equipped Do215B, *Oberleutnant* Ludwig Becker of *4./NJG 1* shot down a Wellington east of Gröningen. However, this equipment was not to come into general use until early 1942, when British tactics were to change dramatically; the determined and implacable Air Marshal Arthur Harris took command of the British bomber forces and began to wage war in more deadly fashion.

At the beginning of August 1941, the Order of Battle was as follows:

Stab./NJG 1	*Oberst* Falck	Deelen	Bf110
I./NJG 1	*Maj* Streib	Venlo	Bf110
II./NJG 1	*Maj* Ehle	St. Trond	Bf110/Do215B
III./NJG 1	*Maj* von Graeve	Twente	Bf110
I.NJG 2	*Maj* Hulshoff	Gilze-Rijn	Ju88C/Do17Z
I./NJG 3[1.]	Vechta	Hptm Radusch	Bf110

1. Less 1 Staffel, which had moved to the Middle East in February and would remain there until October.

Between August 1941 and May 1942, the nightfighter force was to undergo a slow expansion to match the growing weight of Bomber Command attacks. On 29th September, *Stab./NJG 3* was established at Stade. Commanded by *Major* Johannes Schalck, it would be equipped with Bf110s. *II./NJG 3* joined it in the north on 3rd October at Schleswig, commanded by *Hauptmann* Gunther Radusch and similarly equipped with Bf110s. *Stab./NJG 2* was formed at Gilze Rijn on 1st November under command of *Oberstleutnant* Karl Hülshoff and at the same time *II./NJG 2* came into being under *Major* Helmut Lent, also at Gilze Rijn and both equipped with Ju88Cs. On the same date *III./NJG 3* came into being at Stade with Bf110s, under command of *Major* Nacke. Later in the month *I./NJG 2* would be withdrawn from intruder duties and sent to the Middle East. It would not return until late 1942. Two new units were formed in April 1942, III./NJG 2 at Gilze Rijn, equipped with Ju88Cs and commanded by *Hauptmann* Herbert Bönsch, while *II./NJG 4* was formed at Leipheim with Bf110s. The commander of this unit is not known. Finally, *III./NJG 4* was established at Mainz Finthen on 1st May, commanded by *Major* Kurt Holler and equipped with Do217s.

2-3 August 1941

4./NJG1	Fw Siegfried Ney	Wellington	1km W Staveren	00.34		1	W
I./NJG3	Fw Alfred Warrelmann	Stirling	SW Werder	01.48		1	W
4./NJG1	Fw Siegfried Ney	Wellington	30km S Harlingen	04.35		2	W

3-4 August 1941

4./NJG1	Lt Rudolf Schoenert	Wellington	SW Stavoren	00.14	4700	4	W
4./NJG1	Fw Siegfried Ney	Wellington	40km SW Den Helder	02.55	4300	3	W

5-6 August 1941

1./NJG1	Fw Reinhard Kollak	Wellington	2km NE Maastricht	00.07		3	W
1./NJG1	Fw Reinhard Kollak	Wellington	5km NW Marche	00.40		4	W
1./NJG1	Fw Reinhard Kollak	Whitley	5km NE Liege	01.35		5	W
I./NJG1	Lt Hans-Joachim Redlich	Wellington	3km N Tirlemont	02.34		1	W
I./NJG1	Lt Hans-Joachim Redlich	Halifax	5km NE St.Truiden	03.14		2	W

6-7 August 1941

StI/NJG1	Hptm Werner Streib	Whitley	10km NE Eindhoven	02.19		18	W
1./NJG1	Fw Reinhard Kollak	Wellington	15km SE Wavre	03.38		6	W

7-8 August 1941

3./NJG2	Fw Alfons Köster	Whitley	Marham	01.21		7	W
2./NJG2	Oblt Albert Schulz	Wellington	over N Sea	01.26		6	W
2./NJG1	Lt Kurt Loos	Stirling	10km SSW Herwijnen	02.33		3	W
3./NJG2	Ofw Wilhelm Beier	Blenheim	East Dereham	03.20		12	W
3./NJG2	Ofw Wilhelm Beier	Halifax	200km E Withernsea	03.44		13	W
3./NJG2	Ofw Wilhelm Beier	Wellington	200km E Flambor'o Hd	03.48		14	W

8-9 August 1941

4./NJG1	Oblt Ludwig Becker	Wellington	25km E Gröningen	00.25		2	W
I./NJG1	Oblt Walter Fenske	Wellington	3km N Heligoland	03.22		7	W

12-13 August 1941

6./NJG1	Fw Ernst Kalinowski	Wellington	1km SW Grafeld	00.49		2	W
3./NJG2	Ofw Rolf Bussmann	Wellington	Oxford	01.10		3	W
3./NJG2	Ofw Rolf Bussmann	Wellington	Oxford	01.15		4	W
3./NJG2	Ofw Rolf Bussmann	Wellington	Oxford	01.23		5	W
4./NJG1	Oblt Ludwig Becker	Manchester	13km ESE Gröningen	01.25		3	W
4./NJG1	Oblt Ludwig Becker	Manchester	25km E Gröningen	01.25		4	W
1./NJG2	Ofw Peter Laufs	Wellington	Coltishall	03.15		5	W
6./NJG1	Lt Hans Autenrieth	Halifax	1km E Wittstedt	03.16		1	W

Ludwig Becker participated in the 1940 Blitzkrieg as a Bf110 pilot with 3./ZG 1 and proceeded to become a nightfighter pilot in June 1940, claiming his first victory in October. On 9/10th August 1941 he scored the first victory using airborne radar and would go on to win the Ritterkreuz in 1942, having becoming one of the top Experten. On the morning of 26th February 1943 he received the Eichenlaub and then took off on a daylight mission against American heavy bombers. The interception was made over the North Sea, near Schiermonnikoog. He failed to return from this mission. He is credited with 44 night victories.

14-15 August 1941

7./NJG1	Lt Heinz-Martin Hadeball	Halifax	15km ENE Lingen	01.10		2	W
4./NJG1	Oblt Ludwig Becker	Whitley	25km SE Leeuwarden	01.17	4200	5	W
I./NJG3	Uffz Benning	Wellington	W Sogel	01.36		3	W
4./NJG1	Oblt Helmut Lent	Whitley	N Ameland	03.20	4400	14	W
4./NJG1	Ofw Paul Gildner	Whitley	8km WSW Terschel'g	04.13	2500	16	W
3./NJG2	Ofw Robert Lüddeke	Wellington	Humber Estuary	05.20		2	W
3./NJG2	Ofw Robert Lüddeke	Whitworth	Humber Estuary	05.37		3	W

16-17 August 1941

3./NJG2	Lt Hans Hahn	Wellington	Nr. Scunthorpe	00.53		12	W
2./NJG1	Oblt Wolfgang Thimmig	Hampden	Winterswijk	01.41		5	W
I./NJG1	Oblt Heinrich Griese	Whitley	2km N Roermond	01.44		6	W
2./NJG1	Oblt Hermann Reese	Manchester	10km W Maastricht	01.52		2	W
StI/NJG1	Hptm Werner Streib	Manchester	10km NE Sittard	02.05		19	W
StI/NJG1	Lt Hans-Dieter Frank	Wellington	10km NW Roermond	02.15		3	W
2./NJG1	Oblt Wolfgang Thimmig	Whitley	20km W Venlo	02.30		6	W
StI/NJG1	Lt Hans-Dieter Frank	Whitley	2km N Venlo	02.40		4	W
StI/NJG1	Hptm Werner Streib	Whitley	2km N Roermond	02.52		20	W
StI/NJG1	Oblt Wilhelm Dimter	Whitley	5km N Roermond	03.49		1	W
StI/NJG1	Oblt Wilhelm Dimter	Whitley	W Breda	04.35		2	W

17-18 August 1941

4./NJG1	Oblt Ludwig Becker	Hampden	2km S Gröningen	01.44		6	W

18-19 August 1941

1./NJG2	Ofw Heinz Strüning	Blenheim	Nr. Grantham	01.20	20	7	W
1./NJG2	Ofw Heinz Strüning	Blenheim	Nr. Grantham	02.00		8	W
4./NJG1	Fw Siegfried Ney	Wellington	15km N Medemblik	02.07	3000	4	W
I./NJG1	Oblt Heinrich Griese	Wellington	6km S Maastricht	02.19		7	W
2./NJG1	Ofw Gerhard Herzog	Wellington	8km SW Roermond	02.23		7	W
StI/NJG1	Oblt Wilhelm Dimter	Whitley	7km S Nijmegen	02.28		3	W
2./NJG1	Ofw Gerhard Herzog	Wellington	3km N-N-W Roermond	02.30		7	W
I./NJG1	Oblt Heinrich Griese	Wellington	10km N Maastricht	02.36		8	W
2./NJG1	Oblt Hermann Resse	Whitley	10km SE Hertogenb'h	03.03		3	W

19-20 August 1941

3./NJG2	Fw Alfons Köster	Blenheim	20km W Lincolm	04.55		8	W

24-25 August 1941

StI/NJG1	Lt Hans-Dieter Frank	Whitley	3km E Weert	01.15		5	W
1./NJG1	Oblt Rudolf Sigmund	Whitley	5km N Wavre	01.20		1	W
I./NJG1	Oblt Heinrich Griese	Halifax	10km S Ath	01.30		9	W

28-29 August 1941

4./NJG1	Oblt Helmut Lent	Hampden	S Ameland	03.40	3400	15	W

31 August-1 September 1941

StI/NJG1	Oblt Wilhelm Dimter	Wellington	8km NE Maastricht	23.37		4	W
2./NJG1	Oblt Wolfgang Thimmig	Hampden	15km W M-Gladbach	23.39		7	W
StI/NJG1	Oblt Wilhelm Dimter	Hampden	10km NE M-Gladbach	23.47		5	W
2./NJG1	Uffz Heinz Pähler	Whitley	3km E Asten	00.12		1	W
2./NJG1	Fw Richard Launer	Hampden	10km NE Helmond	00.17		1	W
2./NJG1	Uffz Heinz Pähler	Hampden	9km NW Roermond	00.31		2	W
3./NJG2	Fw Alfons Köster	Wellington	Oxford	02.53		9	W
3./NJG2	Fw Alfons Köster	Halifax	Nr. Hatfield	03.42		10	W

1-2 September 1941

3./NJG3	Uffz Karl Kupfer	Halifax	Kreuzbruch	01.28		3	W

2-3 September 1941

4./NJG1	Lt Rudolf Schoenert	Stirling	15km NE Wangerooge	23.34	4400	5	W

6-7 September 1941

4./NJG1	Oblt Ludwig Becker	Whitley	ENE Harderwijk	23.00	2500	7	W
6./NJG1	Oblt Hans-Karl Kamp	Whitley	2km SE Borculo	23.15		3	W
StI/NJG1	Lt Herbert Lütje	Whitley	Metelen	23.22		1	W
7./NJG1	Oblt Helmut Woltersdorf	Wellington	Ruurlo	00.11		4	W
III./NJG1	Fw Richard Philipp	Whitley	Haselunne	00.56		1	W
III./NJG1	Ofw Willi Schmale	Whitley	3km S Boxmeer	01.07		2	W
4./NJG1	Oblt Helmut Lent	Whitley	Bergen	01.25	3200	16	W
4./NJG1	Uffz Heinz Grimm	Whitley	35km N Terschelling	02.00	1200	1	W

7-8 September 1941

7./NJG1	Lt Heinz-Martin Hadeball	Wellington	6km SE Haaksbergen	23.08		3	W
2./NJG3	Lt Rudolf Altendorf	Whitley	N Berlin	00.32		4	W
I./NJG1	Lt Hans Schmitz	Whitley	NE Grevenbroich	00.47		1	W
2./NJG3	Lt Rudolf Altendorf	Whitley	N Berlin	01.00		5	W
7./NJG1	Oblt Helmut Woltersdorf	Wellington	2km NE Nienborg	01.22		5	W
3./NJG3	Uffz Wolfgang Brendel	Whitley	Leissnitz	01.36		1	W
4./NJG1	Oblt Helmut Lent	Wellington	NE Franeker	04.04	4800	17	W
4./NJG1	Oblt Helmut Lent	Wellington	SE Leeuwarden	04.59	2900	18	W
I./NJG3	Fw Heinz Säwert	Hampden	Nieuwendam	05.00		5	W

15-16 September 1941

II./NJG1	Oblt Walter Barte	Hampden	Hamburg/Wilhelmsburg	22.51		1	W
1./NJG3	Gfr Rudolf Frank	Wellington	NW Meppen	23.30		2	W

Hans Hahn of 3./NJG 2 was the most successful intruder pilot during 1940-1941. Flying with 3./NJG 2, he was awarded the Ritterkreuz on 9th July 1941, one of the first nightfighter pilots to receive the award. After being credited with twelve victories over England, he set out on the night of 11th October to hunt near Grantham. He misjudged an attack on an Oxford trainer and collided with it, crashing to his death. Almost immediately, I./NJG 2 was withdrawn from intruder missions and sent instead to the Middle East. This removal of the intruder threat to British crews must surely be regarded as a major blunder on the part of the Oberkommando der Luftwaffe.

II./NJG1	Oblt Walter Barte	Hampden	10km SW Zeven	23.54	2	W
II./NJG1	Oblt Reinhold Eckardt	Wellington	Hamburg/Bramstedt	00.33	10	W
II./NJG1	Oblt Reinhold Eckardt	Wellington	Hamburg/Barsbüttel	00.40	11	W
6./NJG1	Uffz Franz Geidlinger	Stirling	5km E Rotenburg	01.08	1	W

20-21 September 1941

StIII/NJG1	Lt Herbert Lütje	Wellington	1km W Bentheim	22.18	2	W
7./NJG1	Oblt Helmut Woltersdorf	Wellington	5km W Hengelo	22.46	6	W
3./NJG2	Oblt Paul Semrau	Hampden	7km NE Upper Heyford	23.42	9	W

28-19 September 1941

1./NJG1	Fw Reinhard Kollak	Wellington	2km NE Malvoisin	22.45	7	W

29-30 September 1941

I./NJG1	Lt Hans Schmitz	Stirling	S-W Haderslev	22.47	2	W
4./NJG1	Oblt Ludwig Becker	Wellington	35km ESE Gröningen	22.53	8	W
6./NJG1	Fw Ernst Kalinowski	Wellington	5km S-E Hittfeld	00.26	3	W

30 September-1 October 1941

3./NJG2	Ofw Rolf Bussmann	Blenheim	Nr. Winthorpe	23.00	6	W

3-4 October 1941

3./NJG2	Fw Alfons Köster	Stirling	Cambridge	23.52	11	W

10-11 October 1941

4./NJG1	Fw Siegfried Ney	Hampden	Harlingen	05.00	5	W

12-13 October 1941

3./NJG2	Lt Hans Hahn	Oxford	Nr. Grantham		13	W
1./NJG1	Fw Reinhard Kollak	Stirling	35km S Charleroi	21.44	8	W
4./NJG1	Fw Siegfried Ney	Wellington	Blokzijl	21.48	6	W
4./NJG1	Lt Leopold Fellerer	Halifax	3km S Wons	22.17	2	W
I./NJG1	Oblt Heinrich Griese	Wellington	9km S Beaumont	22.59	10	W
I./NJG1	Oblt Heinrich Griese	Wellington	3km S Dinant	23.46	11	W
4./NJG1	Oblt Helmut Lent	Wellington	NE Leeuwarden	00.06	19	W
4./NJG1	Oblt Helmut Lent	Hampden	Zuiderzee	00.33	20	W
I./NJG1	Oblt Heinrich Griese	Whitley	Philippevill	02.44	12	W
4./NJG1	Ofw Paul Gildner	Whitley	10km S-W Roosendaal	04.05	17	W

13-14 October 1941

I./NJG1	Oblt Roderer	Hampden	Waasmunster	22.15	4500	1	W
1./NJG2	Ofw Heinz Struning	B-17	Nr. Upwood	22.20		9	W
I./NJG1	Oblt Heinrich Griese	Manchester	4km E Comblain	04.44		13	W
7./NJG1	Gfr Erhard Brühnke	Manchester	2km S Gorssel	05.30		1	W
2./NJG1	Oblt Reinhold Knacke	Whitley	Lommel	06.00		12	W

15-16 October 1941

2./NJG1	Fw Heinz Meier	Wellington	Grevenbicht/Limburg	21.17		1	W
2./NJG1	Oblt Reinhold Knacke	Wellington		21.37		13	W

20-22 October 1941

2./NJG3	Lt Rudolf Altendorf	Il-4	Berlin	22.15		6	E

26-27 October 1941

I./NJG3	Uffz Alfred Brackmann	Manchester	SSE Bremervörde	22.56		2	W

31 October-1 November 1941

4./NJG1	Ofw Paul Gildner	Wellington	Schiermonnikoog	21.30		18	W
4./NJG1	Oblt Rudolf Schoenert	Halifax	N Wangerooge	21.37		6	W
4./NJG1	Oblt Rudolf Schoenert	Halifax	15km N Langeoog	22.05		7	W
4./NJG1	Ofw Siegfried Ney	Wellington	W Callantsoog	23.28	3900	7	W
4./NJG1	Ofw Paul Gildner	Whitley	E Texel	23.50		19	W
5./NJG2	Uffz Heinz Grimm	Whitley	N Norderney	00.50		2	W

1-2 November 1941

I./NJG1	Lt Hans Schmitz	Whitley		22.35		3	W
5./NJG2	Oblt Rudolf Schoenert	Hampden				8*	

5-6 November 1941

5./NJG2	Oblt Rudolf Schoenert	Hampden				9*	W
5./NJG2	Uffz Heinz Grimm	Hampden	Spiekeroog	04.52		3	W

7-8 November 1941

5./NJG2	Lt Leopold Fellerer	Stirling	8km SW Bergen	22.35	5300	3	W
7./NJG1	Lt Heinz-Martin Hadeball	Wellington	N Cloppenburg			4	W
StIII/NJG1	Oblt Herbert Lütje	Halifax	4km E Terlet	00.04		3	W
5./NJG2	Oblt E. Lippe-Weissenfeld	Whitley	E Medemblik	00.41		12	W
7./NJG1	Ofw Wilhelm Engel	Wellington	5km W Bocholt	00.47		1	W
5./NJG2	Oblt E. Lippe-Weissenfeld	Wellington	W Alkmaar	01.20	500	13	W
StII/NJG2	Hptm Helmut Lent	Wellington	N Akkum	01.21		21	W
6./NJG2	Oblt Ludwig Becker	Whitley	13km E Staveren	06.30		9	W

8-9 November 1941

7./NJG1	Lt Fritz Carstens	Whitley	2km N-E Wassenberg	20.39		1	W
6./NJG2	Ofw Siegfried Ney	Hampden	20km E Alkmaar	21.07		8	W
StI/NJG1	Oblt Wilhelm Dimter	Hampden	30km NW Venlo	21.30		6	W
7./NJG1	Lt Werner Rowlin	Wellington	2km N Doetinchem	21.36		1	W
StI/NJG1	Oblt Wilhelm Dimter	Wellington	25km NE Venlo	21.45		7	W
5./NJG2	Oblt E. Lippe-Weissenfeld	Whitley	18km N Alkmaar	23.03	3000	14	W

On the night of 31st October 1941, Warrant Officer Snowdon headed his 115 Squadron Wellington towards Bremen. His crew would neither reach Bremen nor return to base. En route he was caught by Oberfeldwebel Paul Gildner of 4./NJG 1 and was forced to belly-land X9873 on the shore of Schiermonnikoog Island, where he and his crew were captured.

18-19 November 1941

7./NJG1	Lt Heinz-Martin Hadeball	Wellington	N Cloppenburg			4	W

30 November- 1 December 1941

5./NJG2	Oblt Rudolf Schoenert	Whitley	10km W Aurich	20.52	3600	10	W
5./NJG1	Lt Ludwig Meister	Halifax	5km S Bramstedt	22.21		1	W
5./NJG1	Lt Ludwig Meister	Whitley	30km NW Stade	22.25		2	W
5./NJG2	Ofw Paul Gildner	Wellington	20km ENE Leeuwarden	23.05		20	W
5./NJG1	Lt Ludwig Meister	Wellington	Hamburg	23.25		3	W

12 December 1941

1./NJG2	Lt Herbert Haas	Beaufighter	S Malta	10.05		1	M

13 December 1941

2./NJG2	Ofw Hermann Sommer	Beaufighter	S Crete	13.30		10	M

19 December 1941

1./NJG2	Lt Peter Laufs	Hurricane	Malta	10.45		6	M

27-28 December 1941

5./NJG2	Ofw Paul Gildner	Whitley	15km NW Leeuwarden	19.57		21	W
StI/NJG1	Hptm Werner Streib	Wellington	Zuiderzee	20.15		21	W
2./NJG1	Oblt Reinhold Knacke	Wellington	5km E Heeze	20.24		14	W
StI/NJG1	Hptm Werner Streib	Whitley	Zuiderzee	20.45		22	W
2./NJG1	Oblt Hans-Dieter Frank	Whitley		21.25		6	W
5./NJG2	Oblt E. Lippe-Weissenfeld	Whitley	1.5km S-W Petten	21.46	2300	15	W

28-29 December 1941

6./NJG2	Ofw Siegfried Ney	Whitley	10km W Texel	21.04		9	W
7./NJG1	Oblt Helmut Woltersdorf	Hampden	2km W Winterswijk	22.42		7	W
2./NJG1	Lt Kurt Loos	Wellington	Doetinchem	23.10		4	W

4 January 1942

StI/NJG2	Hptm Rolf Jung	Blenheim	S Malta	15.05		2*	M

10-11 January 1942

5./NJG2	Oblt Rudolf Schoenert	Wellington	17km NW Langeoog	23.15	5000	11	W

14-15 January 1942

5./NJG2	Oblt Rudolf Schoenert	Manchester	5km SW Jever	20.50	3900	12	W

15-16 January 1942

4./NJG1	Uffz Zipperlein	Whitley	Achtkarspellen	22.15		1*	W

16-17 January 1942

StII/NJG2	Hptm Helmut Lent	Whitley	Terschelling	21.40		22	W

20-21 January 1942

5./NJG2	Ofw Paul Gildner	Hampden	7km SE Gröningen	20.58		22	W
6./NJG2	Oblt Ludwig Becker	Wellington	Nr. Terschelling	21.00		10	W
6./NJG2	Oblt Ludwig Becker	Wellington	N Terschelling	21.37		11	W
6./NJG2	Oblt Ludwig Becker	Wellington	NW Terschelling	22.07		12	W

21-22 January 1942

1./NJG3	Uffz Rudolf Frank	Hampden	NE Lingen	21.03		3	W
6./NJG2	Ofw Siegfried Ney	Hampden	15km E Gröningen	21.20		10	W
5./NJG2	Ofw Hans Rasper	Wellington	6km NNE Harlingen	22.10	4000	7	W
StII/NJG2	Hptm Helmut Lent	Whitley	40km W Terschelling	22.38		23	W

23 January 1942

2./NJG2	Lt Georg Wiedow	Blenheim	SE Malta	17.45		1	M
2./NJG2	Ofw Hermann Sommer	Swordfish	SE Malta	18.45		11	M

26-27 January 1942

1./NJG3	Uffz Rudolf Frank	Whitley	W Quackenbrück	20.36		4	W
5./NJG2	Oblt Rudolf Schoenert	Whitley		20.56	3200	13	W
E./NJG2	Ofw Heinz Strüning	Whitley		22.35		10	W

6 February 1942

StII/NJG2	Hptm Helmut Lent	Hampden	70km W Terschelling	15.14		24	W

15-16 February 1942

4./NJG2	Uffz Wolter	Wellington		08.05		1	W

17-18 February 1942

6./KG53	Unknown crew	E/A		00.05			E

21-22 February 1942

9./NJG3	Ofw Heinz Kornacker	Hudson		18.10		1	W
E./NJG2	Lt Hermann Müller	Wellington	30km W Katwijk-a-S	02.30		1	W

24-25 February 1942

5./NJG2	Oblt E. Lippe-Weissenfeld	Hampden	N Terschelling	21.45	300	16	W
5./NJG2	Oblt E. Lippe-Weissenfeld	Hampden	N Terschelling	22.02	200	17	W

Rudolf Schoenert of 5./NJG 2, a pre-war civil pilot, achieved his first night victories with II./NJG 1 in 1941 and had been credited with ten by 1942 in addition to becoming Staffelkapitan of 5./NJG 2. He later commanded II./NJG 5, I./NJG 100 and NJGr. 10. He had been awarded the Ritterkreuz and Eichenlaub before returning to NJG 5 as Geschwader Kommodore on 6th March 1945. He survived the war as a Major, with 65 credited victories. He retired to Canada and died in 1985.

On 10th January he brought down a Wellington, probably from 304 Squadron, on the 14th he despatched Pilot Officer Dawkins's 207 Squadron Manchester near Jever and on 26th he shot down one of three Whitleys to be lost.

26-27 February 1942

6./NJG3	Lt Paul Szameitat	Wellington	Tondern	22.15		1	W
StII/NJG3	Hptm Gunther Radusch	Wellington	West Westerland	00.55		1	W

27-28 February 1942

5./NJG2	Lt Leopold Fellerer	Whitley	25km NNE Aurich	21.38	4100	4	W
6./NJG2	Ofw Siegfried Ney	Whitley	10km N Langeoog	22.37		11	W
5./NJG2	Uffz Heinz Vinke	Whitley	22km NE Leeuwarden	22.58		1	W

3-4 March 1942

5./NJG2	Ofw Paul Gildner	Wellington	20km N-W Terschelling	21.20		25	W

4-5 March 1942

3./NJG2	Ofw Alfons Köster	Wellington	Mediterranean	01.19		12	M

8-9 March 1942

StI ./NJG4	Unknown crew	E/A		00.30		-	W
6./NJG2	Oblt Ludwig Becker	Manchester	E Enkhuizen	03.24		13	W
5./NJG2	Ofw Paul Gildner	Manchester	3km SE Smilde	03.49		24	W
6./NJG2	Oblt Ludwig Becker	Wellington	8km N Petten	04.11		14	W

9-10 March 1942

7./NJG1	Oblt Helmut Woltersdorf	Wellington	30km E Münster	22.28		8	W
E./NJG2	Oblt Alois Lechner	Wellington		22.44		1	W

10-11 March 1942

2./NJG1	Oblt Reinkold Knacke	Stirling		23.42		15*	W

12-13 March 1942

6./NJG2	Oblt Ludwig Becker	Whitley	SE Ameland	23.04		15	W
5./NJG2	Lt Lothar Linke	Whitley	N Terschelling	23.44		3	W
5./NJG2	Ofw Paul Gildner	Wellington	NW Vlieland	02.05		25	W

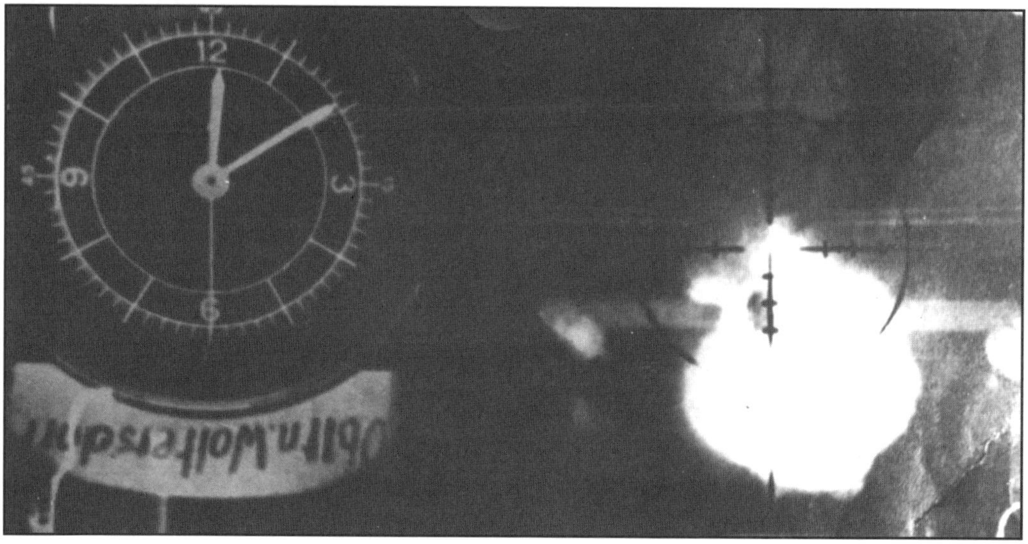

A remarkable combat photo of a Manchester being shot down near Bocholt by Oberleutnant Helmut Woltersdorf of 7./NJG 1 on 26th March 1942. The aircraft was almost certainly L7465 of 207 Squadron. None of Sergeant Markides's crew survived the attack.

13-14 March 1942

2./NJG1	Oblt Reinhold Knacke	Wellington		22.50		16*	W

25-26 March 1942

6./NJG2	Oblt Ludwig Becker	Manchester	W Makkum	22.28		16	W
2./NJG1	Lt Kurt Loos	Manchester	Turnhout	23.44		5	W
7./NJG1	Oblt Helmut Woltersdorf	Manchester	SW Bocholt	00.09		9	W
StII/NJG2	Hptm Helmut Lent	Manchester	10km N Alkmaar	00.32		25	W
6./NJG2	Oblt Ludwig Becker	Hampden	N Vlieland	00.33		17	W

26-27 March 1942

5./NJG2	Oblt E. Lippe-Weissenfeld	Wellington	Nr. De Kooi	22.27	4800	18	W
5./NJG2	Ofw Paul Gildner	Hampden	N Vlieland	22.30		26	W
5./NJG2	Oblt E. Lippe-Weissenfeld	Manchester		22.40	3800	19	W
5./NJG2	Oblt E. Lippe-Weissenfeld	Wellington	N Ijmuiden	22.55	4600	20	W
5./NJG2	Ofw Paul Gildner	Hampden	NW Terschelling	22.57		27	W
5./NJG2	Oblt E. Lippe-Weissenfeld	Wellington	Nr. Edam	23.16	5200	21	W
3./NJG1	Lt Wilhelm Johnen	Wellington	NW Wesel	23.30		1	W
StIII/NJG1	Oblt Herbert Lütje	Wellington	2km N Vriezenveen	23.50		4	W
StI/NJG1	Hptm Werner Streib	Wellington	Enkhuizen	23.51		23	W
5./NJG2	Ofw Paul Gildner	Blenheim	Waddenzee.N Wieringen	23.57		28	W
StI/NJG1	Hptm Werner Streib	Wellington	SE Zutphen	23.58		24	W
2./NJG1	Fw Richard Launer	Wellington	Xanten	23.59		2	W
3./NJG1	Oblt Siegfried Wandam	Stirling	S Doetinchem	23.59		2	W
9./NJG1	Lt Manfred Meurer	Wellington	2km E Wichmond	00.10		1	W

27-28 March 1942

StII/NJG2	Hptm Helmut Lent	Hampden	N Terschelling	22.10		26	W

On 29th March, following combat over north Germany Bf110 D5+LN of 5./NJG 3 was safely crash-landed at Schleswig. (S. Hall)

StII/NJG2	Hptm Helmut Lent	Hampden	N Terschelling	22.42		27	W
4./NJG2	Oblt Rudolf Schoenert	Halifax		23.21		14	W
6./NJG2	Ofw Siegfried Ney	Hampden	N Petten	00.08		12	W

28-29 March 1942

6./NJG2	Oblt Ludwig Becker	Stirling	N Vlieland	22.05		18	W
II./NJG2	Uffz E. Merk	Hampden	N Norderney	23.10		1	W
6./NJG3	Lt Wolfgang Jank	Wellington	Dithmarscher Bight	23.30		1	W
5./NJG2	Uffz Heinz Grimm	Wellington				4	W
5./NJG2	Lt Leopold Fellerer	Stirling		01.02	3400	5	W
StII/NJG2	Hptm Helmut Lent	Manchester	N Terschelling	22.00		28	W
2./NJG1	Oblt Reinhold Knacke	Manchester	NW Vlieland	23.31		17	W

1-2 April 1942

1./NJG3	Fw Hans-Joachim Schmidt	Hampden		00.17		1	W
4./NJG3	Unknown crew	E/A		00.20		-	W
4./NJG3	Unknown crew	E/A		00.39		-	W
4./NJG3	Unknown crew	E/A		00.50		-	W
4./NJG3	Unknown crew	E/A		01.28		-	W
6./NJG1	Oblt Ludwig Bietmann	Wellington		04.17		1	W

5-6 April 1942

4./NJG1	Oblt Wilhelm Herget	Hereford	2 km W Roly/Neuville	02.10	2600	1	W
6./NJG1	Oblt Heinrich Petersen	Wellington		02.28		1	W

6-7 April 1942

6./NJG1	ObLt E-W von Bonin	Manchester		02.38		4	W
7./NJG1	Oblt Helmut Woltersdorf	Wellington	SW Zwolle	04.13		10	W

8-9 April 1942

I./NJG3	Fw Goerke	Manchester		00.49		1	W
5./NJG2	Ofw Paul Gildner	Wellington	20km NW Vlieland	02.07		29	W

10-11 April 1942

2./NJG1	Oblt Reinhold Knacke	Hampden	10km E Enkhuizen	22.58		18	W

StII/NJG2	Hptm Helmut Lent	Wellington	20km SE Den Helder	00.23		39	W
StI/NJG1	Hptm Werner Streib	Wellington	15km S Nijmegen	00.41		25	W
StI/NJG1	Hptm Werner Streib	Wellington	12km SW Venlo	00.49		26	W

12-13 April 1942

E./NJG2	Lt Hermann Müller	Boston	2km SW Klundert	00.20		2	W
StII/NJG2	Hptm Helmut Lent	Hampden	N Terschelling	00.32		30	W
2./NJG1	Oblt Hans-Dieter Frank	Wellington	N Maarheeze	01.53		7	W
E./NJG2	Oblt Horst Patuschka	Wellington		02.38		1	W
7./NJG1	Oblt Helmut Woltersdorf	Wellington		03.15		11	W

13-14 April 1942

4./NJG2	Hptm Hans-Georg Schütze	Stirling		01.09		3	W

14-15 April 1942

2./NJG2	Lt Hans-Werner Linnemann	Wellington	Egypt			1	M
III./NJG1	Egfr Georg Worch	Wellington	1km W Kilder	02.18		1	W
E./NJG2	Oblt Alois Lechner	Wellington		03.30		2	W
3./NJG1	Oblt Siegfried Wandam	Wellington	NW Weert	03.53		3	W
I./NJG1	Uffz Walter Schienbein	Stirling	SW Deurne	03.30		1	W

17-18 April 1942

StII/NJG2	Hptm Helmut Lent	Wellington	25km N Terschelling	01.55		31	W
II./NJG3	Oblt Ernst Schmidt	Wellington	N Rendsburg	03.00		2	W
7./NJG3	Oblt Reinhold Eckardt	Wellington	SW Stade	03.18		12	W

21 April 1942

7./NJG1	Oblt Helmut Woltersdorf	Ventura		14.02		12	W

22 April 1942

5./NJG2	Lt Lothar Linke	Hudson	WNW Texel	11.00		4	W

22-23 April 1942

5./NJG2	Ofw Paul Gildner	Hampden	NW Ameland	00.07		30	W
E./NJG2	Ofw Wilhelm Beier	Boston		01.15		15	W
4./NJG2	Hptm Hans-Georg Schütze	Wellington		02.30		4	W

23-24 April 1942

6./NJG3	Unknown crew	E/A		03.00		-	W
4./NJG2	Hptm Hans-Georg Schütze	Wellington		03.35		5	W

24-25 April 1942

7./NJG1	Oblt Helmut Woltersdorf	Manchester	S Ameland	23.32	2100	13	W

25-26 April 1942

4./NJG2	Hptm Hans-Georg Schütze	Wellington		23.34		6	W
5./NJG2	Lt Lothar Linke	Hampden	NW Ameland	00.04		5	W
4./NJG3	Unknown crew	Hampden	NW Ameland	00.06		-	W
4./NJG2	Oblt Rudolf Schoenert	Wellington	25km SW Helgoland	00.24		15	W
StII/NJG3	Unknown crew	E/A		01.05		-	W
StII/NJG3	Hptm Gunther Radusch	Stirling	S Schelde Estuary	02.08		2	W

Formerly an infantry soldier, Egmont, Prinz zur Lippe-Weissenfeld was remustered into the Luftwaffe and flew with I and II Gruppen of NJG 1 and with II./NJG 2, achieving fifteen victories by 1942. On 26/27th March 1942 he destroyed three Wellingtons and a Hampden to bring his score to 21 and was awarded the Ritterkreuz on 16th April. In October he was sent to command I./NJG 3 and moved to lead III./NJG 1 in June 1943. He finally commanded NJG 5 from 20th February 1944, having been awarded the Eichenlaub in August 1943. On 12th March 1944, while flying to visit his old unit, III./NJG 1, he crashed in bad weather in Belgium and was killed. His score at the time of his death was 51 night victories.

5./NJG2	Uffz Heinz Grimm	Wellington	NW Vlieland	03.37		5	W
II./NJG2	Lt Karl-Heinz Vollkopf	Wellington	NW Texel	04.28		1	W
8./NJG 4	Lt Heinz-Martin Hadeball	Wellington	Siefersheim			5*	W

27-28 April 1942

4./NJG1	Oblt Hubert Rauh	Wellington	Montplaisir	00.37	4800	1	W
StII/NJG3	Hptm Günther Radusch	Stirling	Romo	01.05		3	W
E./NJG2	Hptm Herbert Bönsch	Boston		01.16		7	W
5./NJG1	Fw Fritz Schellwat	Wellington	24km NW Namur	02.14		1	W
E./NJG2	Hptm Herbert Bönsch	Wellington		02.15		8	W

28-29 April 1942

7./NJG3	Oblt Reinhold Eckardt	Stirling	NW Ahrensburg	23.37	13	W
II./NJG3	Oblt Ernst Schmidt	Wellington	Amrum	01.04	3	W
E./NJG2	Ofw Heinz Strüning	Boston	5km SE Noord-Schans	01.15	11	W
StII/NJG3	Hptm Günther Radusch	Wellington		02.18	4	W
StII/NJG3	Oblt Willi Elstermann	Wellington	Sylt	02.15	1	W
6./NJG3	Lt Paul Szameitat	Wellington	N Flensburg	02.30	2	W
6./NJG3	Lt Gustav Tham	Wellington	BockHorst	03.09	1	W
St.II./NJG3	Oblt Günther Köberich	Manchester	Off Rantum/Sylt		1*	W

29-30 April 1942

6./NJG3	Lt Paul Szameitat	Manchester	E Rantum	02.31	3	W
5./NJG3	Unknown crew	E/A		02.33	-	W

3-4 May 1942

2./NJG2	Hptm Heinz Harmstorf	B-24	Derna	02.09	2	M
4./NJG2	Fw Hans Berschwinger	Halifax		02.31	3	W

4-5 May 1942

7./NJG3	Lt Knosch	Stirling	Montcornet	03.26	1	W

5-6 May 1942

9./NJG4	Oblt H-M. Hadeball	Stirling	Maubeuge	01.36		6	W

6-7 May 1942

9./NJG4	Oblt Hubert Rauh	Halifax		00.25		2	W
E./NJG2	Oblt H. Sayn-Wittgenstein	Blenheim	40km W Walcheren	01.20		1	W
8./NJG4	Unknown crew	E/A		01.58		-	W
7./NJG3	Lt Reinhold Eckardt	Wellington	8km W Trelau	02.53		14	W
StI./NJG4	Hptm Wilhelm Herget	Halifax	10km SW St Hubert			2*	W

7-8 May 1942

5./NJG2	Oblt Leopold Fellerer	Hudson	1km SW Den Helder	04.51	300	6	W

8-9 May 1942

I./NJG3	Fw Goerke	Manchester	5km E Schleswig	01.00		2	W
E./NJG2	Ofw Heinz Strüning	Stirling		01.02		12	W

15-16 May 1942

StII/NJG2	Hptm Helmut Lent	Hudson	N Terschelling	22.45		32	W

16-17 May 1942

6./KG53	Unknown crew	E/A		22.30		-	E

17-18 May 1942

4./NJG2	Hptm Hans-Georg Schütze	Wellington	NE Pellworm	00.49		7	W
4./NJG2	Oblt Rudolf Schoenert	Stirling	15km NE Leeuwarden	00.54		16	W

18-19 May 1942

6./KG53	Unknown crew	E/A		00.45		-	E

19-20 May 1942

9./NJG4	Oblt Hubert Rauh	Wellington		23.51		3	W
9./NJG4	Unknown crew	E/A		00.55		-	W
E./NJG2	Ofw Heinz Strüning	Stirling		00.56		13	W
9./NJG4	Hptm Wilhelm Herget	Halifax	N-W St.Hubert	01.09		3	W
8./NJG4	Ofw Wilhelm Engel	Stirling	Munscheweiler	01.13		2	W
6./NJG1	Oblt Fritz Gutezeit	Stirling	1km E St.Truiden	02.24		2	W
E./NJG2	Hptm Erich Simon	Stirling		03.25		1	W

20-21 May 1942

6./KG53	Unknown crew	E/A		00.10		-	E

29-30 May 1942

6./KG53	Unknown crew	E/A		22.55		-	E
4./NJG2	Fw Hans Berschwinger	Stirling		02.03		4	W

Chapter Four

Kammhuber versus Harris

31st May 1942 - March 4th 1943

As has already been noted, Josef Hammhuber's *Himmelbett* defensive line had relied upon the principal of, 'one box, one fighter, one bomber' and within these parameters the system worked remarkably well. But this was only while the British obliged them by sending out bombers with no cohesive formations. However, the relatively new commander of Bomber Command, Air Marshal Arthur Harris, was a forceful, single-minded officer who had other ideas. With pressure on him to release squadrons to Coastal Command following the disastrous Butt Report of August 1941, he had to achieve a major success quickly. His plan was to raise a force of 1,000 aircraft to attack a single target, using tactics that would swamp the defences. The industrial targets in the Ruhr valley were of course the most strategically effective, but they were almost always covered in smoke and haze from the factories. Therefore for the first "Thousand Plan" attack, code-named *Millennium I*, he chose Köln (Cologne). He raised the force to 1,047 aircraft, each briefed to fly the same course and speed, bringing into existence the 'Bomber Stream'. Such was the planned cohesion, that all aircraft were to attack in a 90 minute time-band.

The attack struck the German defences with shattering force. Great damage was done the old city and 41 bombers were lost, a sustainable 3.9%. However *Millennium II*, against Essen on 1/2nd June was not a success, with towns being attacked all over the Ruhr due to poor visibility. Even so, it was clear that the Germans would have to re-think their tactics if such attacks continued.

The Luftwaffe nightfighter strength in the west was now as follows:

Stab./NJG 1	*Oberst* Falck	Deelen	Bf110
I./NJG 1	*Maj* Streib	Venlo	Bf110
II./NJG 1	*Maj* Ehle	St Trond	Bf110
III./NJG 1	*Maj* v.Graeve	Twente	Bf110
Stab./NJG 2	*Obstlt* Hulshoff	Gilze Rijn	Ju88C
II./NJG 2	*Maj* Lent	Gilze Rijn	Ju88C
III./NJG 2	*Hptm* Bönsch	Gilze Rijn	Ju88C
Stab./NJG 3	*Maj* Schalck	Stade	Bf110
I./NJG 3	*Maj* Knoetzsch	Vechta	Bf110
II./NJG 3	*Obstlt* Nacke	Stade	Bf110
Stab./NJG 4	*Maj* Stoltenhoff	Metz	Bf110/Do217
II./NJG 4	Unknown	Leipheim	Bf110
III./NJG 4	*Maj* Holler	Mainz-Finthen	Bf110/Do217.

Between May 1942 and April 1943 a considerable nightfighter expansion was to take place. In September 1942 *I./NJG 2* returned from Sicily and took its Ju88Cs to Melsbroek, commanded by *Major* Erich Jung. During September also, *Stab./NJG 5* was formed at Doberitz under *Major* Schaffer and *I./NJG 5* at Stendal under *Hauptmann* Siegfried Wandam, both with Bf110s. On 1st October 1942 *II./NJG 2* was renumbered as

IV./NJG 1 and moved its Bf110s to Leeuwarden under *Major* Helmut Lent. *III./NJG 2* became *II./NJG 2*, with *Hauptmann* Herbert Bönsch and *III./NJG 2* disappeared until August 1943. A month later, On 1st November 1942, *IV./NJG 3* was formed at Grove with Bf110s under *Major* Erich Simon. However, the Allied invasion of North Africa, *Operation Torch* on 8th November, saw *II./NJG 2* leave Gilze Rijn for Sicily. In December 1942 *II./NJG 5* was established at Parchim, equipped with Bf110s and Do217s under command of *Hauptmann* Rudolf Schoenert, and *IV./NJG 3* at Leiphaim and Lechfeld, commanded by the charismatic and controversial *Major* Heinrich, Prinz zu Sayn Wittgenstein and equipped with Bf110s. On 1st January 1943 *IV./NJG 4* was established at Mainz-Finthen under *Major* Heinrich Wohlers with Bf110s.

30-31 May 1942

E./NJG2	Hptm Horst Patuschka	Wellington		00.17		2	W
E./NJG2	Ofw Heinz Strüning	Boston		00.27		14	W
E./NJG2	Hptm Horst Patuschka	Wellington		00.28		3	W
I./NJG1	StFw Gerhard Herzog	Wellington	SE Zaltbommel	00.45		9	W
4./NJG1	Lt Franz Brinkhaus	Wellington	35km NW Köln	00.50	4000	1	W
StI/NJG1	Hptm Werner Streib	Whitley	15km SE Tilburg	00.56		27	W
StI/NJG1	Oblt Reinhold Knacke	Wellington	10km E Weert	01.00	5400	19	W
I./NJG1	StFw Gerhard Herzog	Wellington	W Dortrecht	01.14		10	W
E./NJG2	Hptm Erich Simon	Wellington		01.14		2	W
E./NJG2	Oblt H. Sayn-Wittgenstein	Wellington		01.16		2	W
2./NJG 1	Oblt Hans-Dieter Frank	Wellington	W Maarheeze	01.30		8*	W
E./NJG2	Oblt H. Sayn-Wittgenstein	Manchester		01.36		3	W
5./NJG1	Oblt Rolf Bokemeyer	Wellington	20km W Bonn	01.43		2	W
4./NJG1	Oblt Walter Barte	Wellington	NNE Maastricht	01.46		3*	W
4./NJG1	Lt Franz Brinkhaus	Halifax	30km WNW Köln	01.48	4000	2	W
2./NJG1	Oblt Hermann Reese	Wellington	Suchteln	01.56	3500	4	W
4./NJG1	Oblt Walter Barte	Wellington	30km N Hasselt	01.57	3400	4	W
E./NJG2	Oblt Alois Lechner	Wellington		01.57		3	W
StI/NJG1	Oblt Reinhold Knacke	Halifax	3km ESE Weert	02.04	4500	20	W
StIII/NJG1	Oblt Manfred Meurer	Hampden	3km N Deventer	02.05		2	W
StI/NJG1	Hptm Werner Streib	Whitley	Hertogenbosch	02.05		28	W
2./NJG1	Hptm Wolfgang Thimmig	Halifax	3km NW Maarheeze	02.06		8	W
E./NJG2	Ofw Wilhelm Beier	Wellington		02.07		16	W
4./NJG1	Oblt Walter Barte	Manchester	Kinroy	02.14	3500	5	W
E./NJG2	Oblt Alois Lechner	Wellington		02.23		4	W
E./NJG2	Ofw Wilhelm Beier	Wellington		02.25		17	W
7./NJG1	Oblt Helmut Woltersdorf	Wellington	8km SE Apeldoorn	02.25		14	W
StI/NJG1	Oblt Reinhold Knacke	Wellington	3km SW Middelbeerer	02.30	3500	21	W
6./NJG1	Lt Helmut Niklas	Wellington	5km SW Charleroi	02.46		2	W
7./NJG1	Oblt Helmut Woltersdorf	Wellington	10km SE Zutphen	02.55		15	W
6./NJG1	Lt Helmut Niklas	Wellington	4km S Landen	03.11		2	W
E./NJG2	Oblt Alois Lechner	B-24		03.56		5	W

31 May-1 June 1942

7./ZG 26	Oblt Alfred Wehmeyer	Wellington	Africa	23.10		1	M

1-2 June 1942

StI/NJG1	Oblt Kurt Loos	Wellington	2km S Horst	00.25	4500	6	W
E./NJG2	Oblt H. Sayn-Wittgenstein	Halifax		00.43		4	W
StI/NJG1	Hptm Werner Streib	Wellington	5km SSE Venlo	00.46		29	W
I./NJG1	Lt Jakob Bender	Wellington	Kerkdriel	01.32		2	W

A Luftwaffe pilot - believed to be Oberfeldwebel Paul Gildner of 4./NJG 1 - prepares for take-off. Note the Englandblitz emblem of the nightfighter force beneath the cockpit; a diving eagles with its claws on a map of England. This emblem is still carried by Bundes Luftwaffe fighters - but the map of England has, happily, been removed.

via H.Scholl

III./NJG1	Hptm Alfred Haesler	Halifax	Bredenbruch	01.48		1	W
II./NJG1	Lt H-W. Schnaufer	Halifax	15km S Louvain	01.55		1	W
E./NJG2	Hptm Erich Simon	Halifax		02.04		3	W
2./NJG1	Fw Heinz Pähler	Manchester	NW Asselt	02.08		3	W
2./NJG1	Oblt Hermann Reese	Wellington	SW Ess	02.08		5	W
7./NJG1	Oblt Helmut Woltersdorf	Halifax	SE Winterswijk	02.08		16	W
7./NJG3	Oblt Reinhold Eckardt	Wellington		02.09	2600	15	W
E./NJG2	Hptm Erich Simon	Wellington		02.25		4	W
7./NJG3	Oblt Reinhold Eckardt	Wellington		02.31	3700	16	W
StII/NJG1	Hptm Walter Ehle	Blenheim	NNE Brussels	02.34		6	W
I./NJG1	Egfr Herbert Hubatsch	Wellington	Turnhout	02.54		1	W
E./NJG2	Ofw Heinz Strüning	B-24		03.10		15	W
E./NJG2	Ofw Heinz Strüning	Wellington		03.25		16	W

2-3 June 1942

StII/NJG2	Hptm Helmut Lent	Hampden	E Medemblik	01.06		33	W
StI/NJG1	Hptm Werner Streib	Wellington	5km W Viersen	01.35		30	W
3./NJG1	Oblt Siegfried Wandam	Wellington	Molenberssel	01.35		4	W
StIII/NJG1	Oblt Manfred Meurer	Halifax	Nr. Emmerich	01.54		3	W
8./NJG1	Oblt Werner Rowlin	Wellington	6km N Gildehaus	01.58		2	W
E./NJG2	Ofw Rolf Bussmann	Halifax		02.25		7	W
5./NJG1	Fw Fritz Schellwat	Lancaster	4km E Waterloo	02.30		2	W
6./NJG1	Oblt E-W von Bonin	Wellington	10km NNW Brussels	02.41		5	W
E./NJG2	Hptm Horst Patuschka	Halifax		02.50		4	W
E./NJG2	Oblt Alois Lechner	Halifax		03.00		6	W
5./NJG2	Ofw Paul Gildner	Stirling	70km W Petten	03.10		31	W

3-4 June 1942

6./NJG2	Oblt Ludwig Becker	Stirling	2km N De Kooi	00.27	2000	19	W
5./NJG2	Oblt E. Lippe-Weissenfeld	Wellington	SE Vlieland	00.50		22	W
StII/NJG2	Hptm Helmut Lent	Halifax	W St.Maartensvlotburg	00.55		34	W
1./NJG3	Ofw Georg Klever	Wellington		01.06		2	W
StIII/NJG3	Lt König	Lancaster	Bad Zwischenahn	01.46		1	W
5./NJG2	Uffz Heinz Vinke	Stirling	Zuiderzee	02.16		2	W
StIII/NJG1	Oblt Herbert Lütje	Wellington		02.25		5	W
III./NJG1	Oblt Martin Bauer	Manchester	S Apeldoorn	02.35		1	W

5-6 June 1942

StII/NJG2	Hptm Helmut Lent	Wellington	S Hoorn	00.34		35	W
StII/NJG2	Hptm Helmut Lent	Wellington	E Amsterdam	01.16		36	W
E./NJG2	Oblt H. Sayn-Wittgenstein	Wellington		01.16		5	W
6./NJG2	Oblt Ludwig Becker	Wellington	Slootdorp	01.21	5200	20	W
E./NJG2	Oblt H. Sayn-Wittgenstein	Wellington		01.50		6	W
4./NJG1	Oblt Walter Barte	Stirling	SW Tirlemont	02.27		6	W
6./NJG1	Oblt Heinrich Petersen	Stirling	NW Calais	03.01		2	W
E./NJG2	Ofw Heinz Strüning	Stirling		03.35		17	W

6-7 June 1942

6./NJG2	Oblt Ludwig Becker	Manchester	Sea.30km N Ameland	00.44	3500	21	W
4./NJG2	Fw Hans Berschwinger	Wellington	NW Aurich	01.05	5300	5	W
5./KG40	Fw Kling	Beaufighter	Ramsgate	01.09		1	W
6./NJG2	Oblt Ludwig Becker	Stirling	SW Holwerd	01.10	3600	22	W
6./NJG2	Oblt Ludwig Becker	Wellington	E Ameland	01.47	4600	23	W
5./NJG2	Oblt E. Lippe-Weissenfeld	Stirling	W Terschelling	01.47		23	W
4./NJG2	Oblt Rudolf Schoenert	Wellington	Borkum	02.27		17	W

7-8 June 1942

8./JG54	Oblt Günther Fink	R-5		23.35	1200	1	E
8./JG54	Oblt Günther Fink	R-5		23.50		2	E
8./JG54	Oblt Günther Fink	R-5		00.05	500	3	E
8./JG54	Oblt Günther Fink	R-5		00.20		4	E
8./JG54	Lt Hans-Joachim Heyer	R-5		00.38		1	E
8./JG54	Lt Hans-Joachim Heyer	R-5		00.55	1500	2	E
III./NJG1	Fw Richard Philipp	Halifax	2km S Vardingholt	00.55		2	W
I./NJG1	Fw Walter Spille	Wellington		01.59		1	W

8-9 June 1942

5./NJG2	Oblt Leopold Fellerer	Wellington	3km NE Petten	02.15		7	W
6./NJG2	Oblt Ludwig Becker	Wellington	W Texel	02.16	5100	24	W
5./NJG2	Oblt Paul Gildner	Halifax		02.18		32	W
5./NJG2	Oblt Leopold Fellerer	Wellington	NW Callantsoog	02.19		8	W
E./NJG2	Hptm Erich Simon	B-24		02.24		5	W
6./NJG1	ObLt E-W von Bonin	Halifax	4km N-E Brussels	02.38		6	W
E./NJG2	Oblt H. Sayn-Wittgenstein	Wellington		03.05		7	W
E./NJG2	Hptm Erich Simon	Halifax	SW Hook-of-Holland	03.19		6	W
8./NJG1	Oblt Werner Rowlin	Halifax	2km S Zaltbommel	03.44		3	W

9-10 June 1942

9./NJG4	Lt Fritz Graef	Wellington		01.35		1	W

10-11 June 1942

8./JG54	Oblt Günther Fink	PS-84		23.30		5	E
8./JG54	Oblt Günther Fink	R-5		23.35		6	E
8./JG54	Lt Hans-Joachim Heyer	PS-84		23.36		3	E
8./JG54	Oblt Günther Fink	R-5		23.52	1300	7	E
8./JG54	Oblt Günther Fink	R-5		23.53		8	E

There were few resources to spare for night fighting in the east. In June 1942 some pilots of III./JG 54 found great success in their Bf109Fs however. Günther Fink (left) scored nine, Erwin Leykauf (centre) scored eight, including six in one night, and Reinhard Seiler (right) claimed no less then fifteen. Fink would go on to win the Ritterkreuz, but would die in action against US heavies in 1943. Seiler, already a Ritterkreuzträger, would win the Eichenlaub in 1944 and survive the war as the last Kommodore of JG 104, with 100 victories in World War II, plus nine more in Spain.

11-12 June 1942

8./JG54	Oblt Günther Fink	R-5		23.10		9	E
StIII/JG54	Hptm Reinhard Seiler	R-5		23.35		1	E
8./JG54	Lt Hans-Joachim Heyer	SB-2		00.01		4	E
StIII/JG54	Oblt Werner Feise	SB-2		00.10		3	E
StIII/JG54	Hptm Reinhard Seiler	R-5		00.30		2	E
5./NJG2	Oblt E. Lippe-Weissenfeld	Lancaster	N Ameland	03.08		24	W

14 June 1942

I./NJG2	Lt Georg Wiedow	Beaufighter	100km NE Derna	13.30		2	M

14-15 June 1942

StIII/JG54	Hptm Reinhard Seiler	R-5		22.45	800	3	E
StIII/JG54	Hptm Reinhard Seiler	R-5		23.15	1000	4	E
8./JG54	Oblt Günther Fink	PS-84		23.28	1500	10	E
9./JG54	Oblt Waldemar Wubke	R-5		23.30	800	1	E
8./JG54	Lt Erwin Leykauf	R-5	Mostki	23.45		1	E

15 June 1942

1./NJG2	Oblt Albert Schulz	Maryland		07.45		7	W

15-16 June 1942

2/NJG2	Ofw Hermann Sommer	Twin-eng	Mediterramean			12	M
StIII/JG54	Hptm Reinhard Seiler	R-5		23.20		5	E
StIII/JG54	Hptm Reinhard Seiler	Yak-1		23.30		6	E
StIII/JG54	Hptm Reinhard Seiler	R-5		24.00		7	E
StIII/JG54	Hptm Reinhard Seiler	R-5		00.20		8	E

16-17 June 1942

2./NJG1	Lt Wilhelm Johnen	Wellington		01.00		2	W
9./NJG4	Hptm Wilhelm Herget	Stirling	W Limont-Fontaine	02.14		4	W
E./NJG2	Hptm Herbert Bönsch	Wellington	60km SW Hardenberg	02.25		9	W
StI/NJG1	Oblt Reinhold Knacke	Halifax	W Veldhoven	02.28	4200	22	W
8./NJG1	Oblt Wilhelm Dormann	Halifax	8km SW Hardenberg	03.00		1	W
StIII/NJG1	Oblt Gerhard Friedrich	Wellington	5km W Ijmuiden	03.08		1	W
E./NJG2	Oblt H. Sayn-Wittgenstein	B-24	20km NW Leiden	03.09		8	W
E./NJG2	Oblt H. Sayn-Wittgenstein	Wellington	15km NNW Leiden	03.23		9	W
StI/NJG1	Oblt Reinhold Knacke	Wellington	2km NE Ouddorp	04.20	5300	23	W

17-18 June 1942

8./JG54	Lt Hans-Joachim Heyer	R-5		00.02		5	E
8./JG54	Lt Hans-Joachim Heyer	R-5		00.25		6	E

19-20 June 1942

4./NJG2	Oblt Rudolf Schoenert	Wellington	W Emden	01.49		18	W
8./NJG1	Lt August Geiger	Stirling	4km SE Neuenhaus	02.22		2	W
8./NJG1	Oblt Herbert Lütje	Wellington	5km SW Raalte	02.32		6	W
3./NJG3	Uffz Wolf Gütter	Stirling		02.45		6	W
StIII/NJG1	Oblt Martin Bauer	Wellington	S Emmen	02.48		2	W
3./NJG3	Lt Armin Hubmann	Whitley		02.50		1	W
E./NJG2	Lt Hermann Müller	B-24		02.59		3	W
6./NJG2	Oblt Ludwig Becker	Wellington	30km NW Vlieland	03.00		25	W
StIII/NJG1	Oblt Gerhard Friedrich	Wellington	10km N Zandvoort	03.13		2	W
E./NJG2	Ofw Heinz Strüning	Wellington		03.27		18	W

20 June 1942

1./NJG2	Lt Heinz Rökker	Beaufort	S Crete	17.20		1	W

20-21 June 1942

StII/NJG2	Hptm Helmut Lent	Hampden	N Ameland	00.59		37	W
5./NJG2	Oblt E. Lippe-Weissenfeld	Halifax	25km NW Gröningen	01.43		25	W
5./NJG2	Oblt E. Lippe-Weissenfeld	Wellington	20km NW Gröningen	01.45		26	W
5./NJG2	Oblt E. Lippe-Weissenfeld	Wellington	N Ameland	01.56		27	W
./NJG2	Lt Johann Werth	Stirling	4km NW Wognum	02.00		1	W
StII/NJG2	Oblt Rudolf Sigmund	Stirling	NW Zwolle	02.05		2	W
./NJG2	Unknown crew	E/A		02.38			W

22-23 June 1942

StIII/JG54	Hptm Reinhard Seiler	PS-84		23.05		9	E
StIII/JG54	Hptm Reinhard Seiler	PS-84		23.08		10	E
8./JG54	Lt Erwin Leykauf	R-5		23.10		2	E
8./JG54	Lt Erwin Leykauf	R-5		23.15		3	E
8./JG54	Lt Erwin Leykauf	R-5		23.45		4	E
8./JG54	Lt Erwin Leykauf	R-5		23.50		5	E
8./JG54	Lt Erwin Leykauf	R-5		23.55		6	E
8./JG54	Lt Erwin Leykauf	R-5		24.00		7	E
4./NJG2	Oblt Rudolf Schoenert	Wellington	NW Baltrum	01.17	2800	19	W
4./NJG2	Oblt Rudolf Schoenert	Stirling	N Aurich	02.07		20	W
E./NJG2	Ofw Wilhelm Beier	Stirling		03.01		18	W

A sight that was becoming very familiar in Holland; a wrecked British bomber lies forlornly in a Dutch polder. The photo caption identifies it as a Stirling, but the fuselage framework shows that it is a Wellington. (G.Zwanenberg MBE)

23-24 June 1942

StII/NJG2	Hptm Helmut Lent	Wellington	10km N Terschelling	01.12		38	W
StII/NJG2	Hptm Helmut Lent	Wellington	NW Vlieland	01.46		39	W

24-25 June 1942

8./JG54	Oblt Günther Fink	R-5		00.35		11	E

25-26 June 1942

1./NJG2	Lt Heinz Rökker	Wellington	50km SW Marsa Mat'k	22.45	400	2	A
StIII/JG54	Hptm Reinhard Seiler	R-5		23.42		11	E
StIII/JG54	Hptm Reinhard Seiler	R-5		23.50		12	E
9./NJG3	Oblt Hans-Joachim Jabs	Stirling				1	W
8./JG54	Lt Erwin Leykauf	R-5		00.01		8	E
1./NJG2	Lt Heinz Rökker	Wellington	SW Marsa Matruk	00.09		3	A
5./JG54	Lt Wolfgang Kretschmer	E/A		00.26		1	E
StIII/JG54	Hptm Reinhard Seiler	R-5		00.37		13	E
6./NJG2	Oblt Ludwig Becker	Stirling	Wieringemeer	00.39		26	W
5./NJG2	Uffz Heinz Vinke	Halifax	Zuiderzee	00.42		3	W
StIII/JG54	Oblt Werner Feise	MBR-2		00.44		4	E
II./NJG2	Lt Kaiser	Stirling	40km N Juist	00.52		1	W
8./NJG1	Oblt Herbert Lütje	Wellington	100km NE Nordhorn	00.58		7	W
5./NJG2	Oblt E. Lippe-Weissenfeld	Wellington	Terschelling	01.05		28	W
E./NJG2	Ofw Heinz Strüning	B-24		01.13		19	W
8./NJG1	Lt August Geiger	Wellington	20km NE Rheine	01.20		3	W
1./NJG3	Uffz Wolfgang Heymann	Wellington		01.20		1	W
8./NJG1	Lt August Geiger	Whitley	10km NW Lingen	01.30		4	W
StIII/NJG2	Hptm Herbert Bönsch	Boston	1km S Eudorp	01.32		10	W
5./NJG2	Lt Lothar Linke	Stirling	NE Wieringen	01.32		6	W
6./NJG1	ObLt E-W von Bonin	Blenheim	16km N Tirlemont	01.38		7	W
StII/NJG2	Oblt Rudolf Sigmund	Halifax	Assen	01.50		3	W
5./NJG2	Oblt E. Lippe-Weissenfeld	Wellington	10km N Vlieland	01.52		29	W
8./NJG1	Lt August Geiger	Stirling	6km NW Nordhorn	01.58		5	W
III./NJG1	Hptm Alfred Haesler	Halifax	8km NW Eldenzaal	02.05		2	W

7./NJG1	Oblt Fritz Carstens	Halifax		02.09	2	W
6./NJG3	Lt Paul Szameitat	Halifax	Cuxhafen	02.11	4	W
StIII/NJG3	Lt König	Wellington	S Fehmann	02.14	2	W
II./NJG2	Lt Robert Denzel	Wellington	E De Kooi	02.15	1	W
5./NJG3	Oblt Werner Hoffmann	Hudson	6km SE Heide	02.26	1	W
StII/NJG2	Hptm Helmut Lent	Wellington	6km NW Enkhuizen	02.37	40	W
3./NJG3	Lt Armin Hubmann	Whitley	Meppen	02.50	2	W
StII./NJG2	Maj Kurt Holler	Whitley	N Vlieland	02.54	1*	W
StII/NJG2	Hptm Helmut Lent	Whitley	Noorswijk	02.56	41	W
8./NJG1	Oblt Werner Rowlin	Wellington	5km NW Hengelo	02.56	4	W
E./NJG2	Oblt H. Sayn-Wittgenstein	Wellington		03.07	10	W
5./NJG3	Unknown crew	E/A		03.08	-	W
E./NJG2	Hptm Erich Simon	Blenheim		03.18	7	W
5./NJG3	Oblt Werner Hoffmann	Whitley	6km N Busum	03.24	2	W
StII./NJG2	Maj Kurt Holler	Wellington	NW Terschelling	03.27	2*	W
E./NJG2	Lt Hermann Müller	Wellington		03.58	4	W
II./NJG2	Oblt Franz Buschmann	Wellington		04.02	1	W
II./NJG2	Lt Günther Löwa	Whitley	10km SW Den Helder	04.26	1	W
II./NJG2	Lt H.G Bötel	Wellington	10km NW Tersch'ling	04.29	1	W
II./NJG2	Lt H.G Botel	Wellington	15km NW Vlieland	04.38	2	W
4./NJG2	Oblt Hermann Greiner	Wellington	S Harlingen	06.00	1	W

26-27 June 1942

3./NJG2	Ofw Alfons Köster	Beaufighter	Mersa Matruh	23.44	13	M
1./NJG2	Fw Werner Heyne	Wellington	Mediterranean	01.34	1	M

27-28 June 1942

StIII/JG54	Hptm Reinhard Seiler	R-5		23.48	14	E
StIII/JG54	Hptm Reinhard Seiler	R-5		24.00	15	E
3./NJG2	Ofw Alfons Köster	Blenheim	Mediterranean	00.07	14	M
5./NJG2	Uffz Heinz Vinke	Wellington	E Enkhuizen	00.54	4	W
8./NJG1	Lt August Geiger	Wellington	NNE Lingen	01.29	6	W
6./NJG2	Lt Gollasch	Wellington	24km E Leeuwarden	02.13	1	W
5./NJG2	Fw Karl-Heinz Scherfling	Wellington	25km W Den Helder	02.55	3	W
StII/NJG2	Oblt Rudolf Sigmund	Halifax	20km SSW Den Helder	03.00	4	W

28-29 June 1942

1./NJG2	Lt Heinz Rökker	Wellington	60km SE Marsa Matruh	23.58	4	M
StII/NJG2	Unknown crew	E/A		00.31	-	W
StII/NJG2	Unknown crew	E/A		00.45	-	W
StII/NJG2	Unknown crew	E/A		00.45	-	W
StII/NJG2	Unknown crew	E/A		01.00	-	W

29-30 June 1942

3./NJG2	Hptm Paul Semrau	Halifax	Mediterranean	00.15	10	M
3./NJG2	Hptm Paul Semrau	Wellington	Mediterranean	00.35	11	M
StII/NJG2	Oblt Rudolf Sigmund	Halifax	40km S Leeuwarden	01.08	5	W
StIII/NJG3	Lt Rüssmann	Wellington	S Papenburg	01.45	1	W
1./NJG3	Uffz Alfred Brackmann	Halifax		01.46	1	W
5./NJG3	Unknown crew	E/A		02.33	-	W
5./NJG2	Oblt Leopold Fellerer	Halifax	S Hoorn	03.02	9	W

Left: The only known picture of Leutnant Günther Löwa from 5./NJG 2. He had a short career; On 29/30th June he claimed a Whitley, but his Bf110F-4 did not return. No trace was ever found. (S.Hall). Georg Wiedow of I./NJG 2 (right) claimed four victories in the Mediterranean theatre; the last, a Wellington near Tobruk, on 5th July 1942. He failed to return from a mission on the next night.

5./NJG2	Oblt E. Lippe-Weissenfeld	Wellington	S Ameland	03.08		30	W
E./NJG2	Oblt H. Sayn-Wittgenstein	Wellington		03.26		11	W

1-2 July 1942

4./KG40	Uffz Saveus	E/A		03.14	1500	1	W

2-3 July 1942

5./NJG2	Oblt E. Lippe-Weissenfeld	Hampden	S Koudum	00.54		31	W
5./NJG2	Oblt E. Lippe-Weissenfeld	Wellington	N Urk	01.09		32	W
1./NJG2	Fw Werner Heyne	Wellington	Mediterranean	01.17		2	M
1./NJG3	Uffz Rudolf Frank	Wellington	5km N Cloppenburg	01.26		5	W
5./NJG2	Fw Heinz Grimm	Hampden	15km WNW Zwolle	01.40		6	W
8./NJG1	Oblt Herbert Lütje	Wellington	3km N Nordhorn	01.40		8	W
5./NJG2	Fw Karl-Heinz Scherfling	Stirling	NNW Gröningen	01.50		4	W
3./NJG2	Hptm Paul Semrau	Wellington	Mediterranean	02.02		12	M
1./NJG3	Uffz Rudolf Frank	Wellington	20km SW Vechta	02.31		6	W
8./NJG1	Lt August Geiger	Wellington	2km NW Lochem	02.49		7	W
6./NJG3	Lt Paul Szameitat	Wellington		02.53		5	W
2./NJG3	Oblt Rudolf Altendorf	Wellington		02.55		7	W
5./NJG2	Oblt E. Lippe-Weissenfeld	Stirling		03.05		33	W
9./NJG2	Oblt H. Sayn-Wittgenstein	Stirling		03.05		12	W
St.II./NJG2	Hptm Helmut Lent	Wellington				42*	W

3-4 July 1942

4./NJG3	Ofw Karl Haisch	Lancaster	Blavand	02.46		1	W

5 July 1942

6./JG54	Hptm Karl Sattig	U-2		00.30	1000	1	E
6./JG54	Hptm Karl Sattig	U-2		00.57	1000	2	E
StI/NJG2	Lt Georg Wiedow	Wellington	Nr. Tobruk	01.28		3	M
3./NJG2	Hptm Paul Semrau	Wellington	Medierranean	02.10		13	M
StI/NJG2	Lt Georg Wiedow	Wellington	Nr. Tobruk	03.20		4	M

5-6 July 1942

3./NJG2	Hptm Paul Semrau	Wellington	Mediterranean	01.55		14	M
2./NJG2	Hptm Heinz Harmstorf	Wellington	Mediterranen	02.46		3	M

6-7 July 1942

5./JG54	Hptm Joachim Wandel	U-2	Demjansk	23.50	1500	1	E
5./JG54	Hptm Joachim Wandel	TB-3	Kraseja	00.05	1400	2	E

5./JG54	Hptm Joachim Wandel	U-2	Kraseja	00.32	900	3	E
5./JG54	Hptm Joachim Wandel	U-2	Ramuschevo	00.56	1200	4	E
5./JG54	Hptm Joachim Wandel	SB-2		22.50	1600	5	E
5./JG54	Hptm Joachim Wandel	U-2	Ramuschevo	23.09	1200	5	E
5./JG54	Hptm Joachim Wandel	U-2	Ramuschevo	23.18	1100	7	E

7-8 July 1942

5./JG54	Hptm Joachim Wandel	U-2	Ramuschevo	00.33	1100	8	E

8-9 July 1942

StII/NJG2	Oblt Rudolf Sigmund	Halifax	21km NW Gröningen	01.52		6	W
5./NJG2	Oblt Paul Gildner	Lancaster	30km NW Ameland	02.44		33	W
St.II./NJG2	Hptm Helmut Lent	Wellington				43*	W

11-12 July 1942

II./NJG3	Hptm Hans Baer	Lancaster		21.24		1	W

13-14 July 1942

8./NJG2	Hptm Horst Patuschka	Wellington	Mediterranean	03.20		5	M

19-20 July 1942

5./JG54	Hptm Joachim Wandel	U-2		00.25	500	9	E
5./JG54	Hptm Joachim Wandel	U-2		01.15	1400	10	E
5./NJG2	Oblt Rudolf Schoenert	Halifax	Nr. Borkum	02.51		21	W
5./NJG2	Oblt E. Lippe-Weissenfeld	Halifax	N Terschelling	02.52		34	W

20-21 July 1942

3./NJG2	Ofw Alfons Köster	Wellington	Mediterranean	23.03		15	M
3./NJG2	Ofw Alfons Köster	Wellington	Mediterranean	23.15		16	M

21-22 July 1942

8./NJG2	Ofw Heinz Strüning	Halifax	14km S Utrecht	01.45		20	W
StI/NJG1	Oblt Reinhold Knacke	Wellington	8km SE Roermond	01.49	3600	24	W
StI/NJG1	Oblt Reinhold Knacke	Wellington	11km E Eindhoven	02.06	3500	25	W

22-23 July 1942

2./NJG2	Hptm Heinz Harmstorf	Wellington	Mediterranean	22.50		4	M

23-24 July 1942

I./NJG3	Lt Witzleb	Blenheim	NE Vechta	01.46		1	W
StIII/NJG2	Hptm Herbert Bönsch	Stirling	5km. SW Oss	02.21	4500	11	W
9./NJG2	Oblt H. Sayn-Wittgenstein	Lancaster		03.10		13	W
3./NJG1	Uffz Heinz Oloff	Stirling	s'Hertogenbosch	03.25		1*	W

25-26 July 1942

I./NJG2	Oblt Heinz-Horst Hissbach	Wellington	SW Crete			1	M
StI/NJG1	Oblt Reinhold Knacke	Blenheim	Schandeloo	01.19	1000	26	W
StI/NJG1	Oblt Reinhold Knacke	Stirling	3km N Horst	01.31	4000	27	W
StI/NJG1	Oblt Reinhold Knacke	Wellington	Repelen	01.50	4000	28	W
8./NJG1	Oblt Herbert Lütje	Wellington	Wielen	02.01		9	W
StIII/NJG1	Oblt Gerhard Friedrich	Wellington	6km W Zieuwent	02.12		3	W
2./NJG1	Oblt Hans-Dieter Frank	Halifax	N Venlo	02.14		9	W

Wilhelm Beier was one of the pioneers of night intruding with I./NJG 2, achieving his first kill in December 1940. After fourteen victories he was awarded the Ritterkreuz and was transferred to 3./NJG 1 in early 1942, where his successes continued. Late in the year he flew with Stab IV./NJG 1, returning to 3./NJG 1 in 1943, where he claimed his last of 38 night kills. He was then moved to various training posts and, at the armistice, was undergoing nightfighting training on the Me262. He died in July 1977.

1./NJG1	Oblt Kurt Loos	Halifax	1km N s'Hertogenbosch	02.26	3000	7	W
8./NJG2	Ofw Heinz Strüning	Wellington	NE Amersfoort	02.30		21	W
2./NJG1	Fw Heinz Pähler	Wellington		02.52		4	W
9./NJG2	Ofw Rolf Bussmann	Wellington	W Bergen-on-Sea	03.26		8	W
9./NJG2	Ofw Rolf Bussmann	Stirling	N Goeree	04.10		9	W

26-27 July 1942

5./NJG2	Oblt Rudolf Schoenert	Halifax				22	W
5./NJG2	Fw Karl-Heinz Scherfling	Wellington				5	W
9./NJG3	Oblt Hans-Joachim Jabs	Wellington				2	W
St.II./NJG2	Hptm Helmut Lent	Halifax	NW Vlieland			44*	W
St.II./NJG2	Hptm Helmut Lent	Wellington				45*	W
2./NJG2	Fw Alfred Siewert	Wellington	Mediterranean	00.47		1	M
5./NJG3	Lt Gerhard Raht	Halifax	SW Pellworm	01.09		1	W
4./NJG2	Lt Kaiser	Halifax		01.10		2	W
5./NJG3	Unknown crew	E/A		01.12		-	W
6./NJG3	Lt Gustav Tham	Wellington	S Neumünster	01.20		2	W
4./NJG2	Ofw Hans Berschwinger	Halifax		01.25		6	W
9./NJG2	Ofw Wilhelm Beier	Havoc	24km NE Eindhoven	01.32		19	W
5./NJG3	Unknown crew	E/A		01.56		-	W
5./NJG2	Lt Lothar Linke	Lancaster	20km ESE Leeuwaarden	02.05		7	W
5./NJG3	Unknown crew	E/A		02.37		-	W
2./NJG2	Ofw Hermann Sommer	Wellington	50km SE Tobruk	23.35		13	M

28-29 July 1942

1./NJG2	Lt Heinz Rökker	Wellington	E Tobruk	23.10		5	M
5./NJG2	Oblt Rudolf Schoenert	Wellington		03.28		23	W
StI/NJG1	Oblt Reinhold Knacke	Mosquito	Haps	01.10	1800	29	W
9./NJG2	Oblt H. Sayn-Wittgenstein	Fulmar		01.11		14	W
6./NJG3	Unknown crew	E/A		01.21		-	W
5./NJG2	Oblt Paul Gildner	Stirling	20km NW Terschelling	01.27		34	W
StII/NJG3	Lt Günther Köberich	Wellington	10km W Rendsburg	01.32		2	W

Hauptmann Karl Helmer (left) of 8./NJG 3 had no known victories credited. He engaged a bomber near Hirson on 29/30th July and the Bf110 received the worst of the encounter. Both engines sustained severe damage and his Bordfunker and Bordschutze baled out, but Helmer was killed when D5+FE crashed.
(S.Hall)

4./NJG3	Unknown crew	E/A		01.58		-	W
8./NJG2	Hptm Horst Patuschka	Havoc		02.00		6	W
6./NJG3	Uffz Kicke (Ritter?)	Wellington	nr Halle	02.19	3000	1	W
8./NJG1	Lt August Geiger	Wellington	5km NNW Neuenhaus	02.35		8	W
IV./NJG1	Lt Wolfgang Küthe	Wellington	Zuiderzee	03.04		1	W
III./NJG1	Oblt Hagenow.von	Stirling	Heelsum	03.10		1	W
4./NJG3	Unknown crew	E/A		03.14		-	W
9./NJG2	Ofw Rolf Bussmann	Wellington	14km NW Stade	03.15		10	W
9./NJG2	Ofw Wilhelm Beier	Wellington	20km N Enschede	03.20		20	W
6./NJG2	Lt Gollasch	Wellington	Zuiderzee	03.21		2	W

29-30 July 1942

8./NJG1	Lt August Geiger	Whitley	8km WNW Rijssen	00.58		9	W
7./NJG3	Oblt Reinhold Eckardt	Stirling	7km WNW. Rocroi	01.15	2500	17	W
5./NJG1	Hptm Ludwig Bietmann	Wellington	30km ESE Namur	01.30		2	W
7./NJG3	Oblt Reinhold Eckardt	Halifax	200m S Brabant	02.51	3900	18	W
7./NJG3	Oblt Reinhold Eckardt	Lancaster	Braine-le-Compte	03.06	3800	19	W
5./NJG1	ObLt E-W von Bonin	Wellington	50km E Antwerp	03.17	1300	8	W
8./NJG2	Ofw Heinz Strüning	Wellington		03.55		22	W

31 July-1 August 1942

StI/NJG1	Hptm Werner Streib	Blenheim	15km SSE Kalelaar	00.35	1300	31	W
9./NJG2	Hptm Walter Fenske	Stirling		01.37		8	W
StI/NJG1	Hptm Werner Streib	Wellington	N Stokkem	01.44	4300	32	W
3./NJG1	Hptm Siegfried Wandam	Wellington		01.45	4000	5	W

Reinhold Eckardt was originally a Bf110 pilot with 2./ZG 76 and flew in the Norwegian campaign and the Battle of Britain, achieving three day victories before being transferred to 6./NJG 1. His first night kill came on the night of 9th January 1941. After twelve victories he was awarded the Ritterkreuz in August 1941. On the night of 29th July 1942 he claimed a Stirling and a Halifax, then attacked a Lancaster. This bomber, his 22nd opponent, also went down over Belgium, but one of the British turret gunners had hit his Bf110 and he ordered his crew to bale out near Melsbroek. They got out safely, but his opening parachute caught on the tail unit of the Messerschmitt and he crashed to his death.

II./NJG1	Lt Hans Autenrieth	Wellington	20km S M-Gladbach	02.03	3100	2	W
II./NJG1	Lt Hans Autenrieth	Wellington	2km S Berg	02.20	1800	3	W
I./NJG1	Hptm Walter Barte	Wellington		02.28		7*	W
8./NJG2	Hptm Horst Patuschka	Hampden		02.25		7	W
StI/NJG1	Hptm Werner Streib	Wellington	12km S-W Venlo	02.28	3300	33	W
II./NJG1	Lt H-W. Schnaufer	Wellington	1km SW Loon-op-Zand	02.47		2	W
3./NJG1	Fw Walter Spille	Four-eng	500m S Maesyck	02.53	3800	2	W
9./NJG2	Hptm H. Sayn-Wittgenstein	Hampden		03.17		15	W
II./NJG1	Lt H-W. Schnaufer	Wellington	Huldenberg	03.17	3800	3	W
8./NJG2	Hptm Horst Patuschka	Wellington		03.27		8	W
9./NJG2	Hptm H. Sayn-Wittgenstein	Halifax		03.30		16	W
6./NJG1	Lt Ludwig Wolf	Wellington	18km S Weert	03.30		1	W
5./NJG1	Fw Fritz Schellwat	Hampden	2km E Antwerp	03.34	2300	3	W
II./NJG1	Lt H-W. Schnaufer	Whitley	Gilly	03.45	2800	4	W
9./NJG2	Hptm H. Sayn-Wittgenstein	Wellington		04.42		17	W

Leutnant Wolfgang Jank's Do217J of 5./NJG 3, crashed at Schleswig on 4th August. He had claimed a Wellington in March and it is believed that he later transferred to JG 54 in the east, where he claimed 5-6 more night victories. *(S.Hall)*

2-3 August 1942

5./JG54	Hptm Joachim Wandel	U-2		01.35	400	11	E
5./JG54	Hptm Joachim Wandel	U-2		01.45	300	12	E
5./JG54	Hptm Joachim Wandel	U-2	NE Ramuschevo	02.00	400	13	E

4-5 August 1942

6./KG54	Unknown crew	Beaufighter		01.30	4500	-	W
9./NJG2	Hptm Walter Fenske	B-24		03.50		9	W

5-6 August 1942

2./NJG1	Oblt Hans-Dieter Frank	Halifax	NW Posterholt	00.17	4300	10	W
1./NJG1	Oblt Reinhold Knacke	Wellington	Hau	01.06	5000	30	W
StI/NJG1	Hptm Werner Streib	Halifax	15km N Eindhoven	01.18	4700	34	W
8./NJG1	Lt August Geiger	Halifax	15km S Zwolle	01.35	4000	10	W
1./NJG1	Oblt Kurt Loos	Halifax	20km W Nijmegen	02.28	5500	8	W

6-7 August 1942

StII/NJG1	Hptm Walter Ehle	Halifax	2km SW Gruitrode	02.40	4500	7	W
1./NJG1	Oblt Reinhold Knacke	Stirling	500m S Hötheim	02.42	4000	31	W
5./NJG1	Hptm Ludwig Bietmann	Stirling	12km NE Roermond	02.57	4200	3	W
1./NJG1	Oblt Reinhold Knacke	Wellington	5km NE s'Hertogen'sch	02.57	4500	32	W
9./NJG2	Hptm H. Sayn-Wittgenstein	Halifax		04.15		18	W

9-10 August 1942

1./NJG1	Oblt Reinhold Knacke	Wellington	NE Doetinchem	03.07	3700	33	W
1./NJG1	Oblt Reinhold Knacke	Halifax	8km NW Helmond	03.52	3600	34	W
8./NJG1	Oblt Herbert Lütje	Wellington	8km NW Wierden	04.18	4800	10	W
8./NJG1	Lt August Geiger	Wellington	8km SSE Deventer	04.34	1200	11	W

11-12 August 1942

5./NJG1	Hptm Ludwig Bietmann	Halifax	1½km NNE Namur	00.54	3500	4	W
II./NJG1	Lt Hans Autenrieth	Wellington	700m SW Vaals	01.33	1300	4	W
5./NJG2	Oblt Paul Gildner	Halifax	Nr. Ameland	01.55	4500	35	W
7./NJG4	Ofw Reinhard Kollak	Stirling	20km E Namur	01.57	5400	9	W
1./NJG1	Oblt Kurt Loos	Halifax	Schelde Estuary	03.05	600	9	W
StII/NJG1	Hptm Walter Ehle	Wellington	NW Leuven	03.12	3000	8	W
3./NJG1	Fw Walter Spille	Wellington	400m S Lierop	03.29	2800	3	W
I./NJG1	StFw Gerhard Herzog	Wellington	E Eindhoven	03.30		11	W

12-13 August 1942

9./NJG4	Oblt Kurt Martinek	Stirling	2km . Rotterdam	01.11	4100	1	W
StII/NJG1	Lt Hans Autenrieth	Wellington	2km NW Eisden	01.12	2500	5	W
8./NJG2	Hptm Horst Patuschka	Halifax		02.40		9	W

13-14 August 1942

7./NJG2	Uffz Fritz Hobusch	Halifax		02.41		1	W

15-16 August 1942

1./NJG1	Oblt Reinhold Knacke	Halifax	2½km E Kevelaer	03.26	5100	35	W
StIII/NJG1	Oblt Manfred Meurer	Wellington	Rozendaal	04.17	3400	4	W
8./NJG2	Hptm Horst Patuschka	Manchester		05.05		10	W

Leutnant Rudi Rohr (left) and Unteroffizier Erwin Gerhard Cobi of 5/NJG 1 were both killed in combat north of Essen on 24/25th August 1942. No credited victories have been found. (S.Hall)

16-17 August 1942

StII/NJG3	Maj Günther Radusch	Halifax	Sonderborg	02.42	3000	5	W
StII/NJG3	Maj Günther Radusch	Halifax		02.52	3000	6	W

17-18 August 1942

8./NJG1	Oblt Herbert Lütje	Wellington	3km W Emlichheim	01.38	4700	11	W
6./NJG2	Oblt Ludwig Becker	Stirling	30km W Terschelling	01.46	1800	27	W
StII/NJG1	Maj Helm	Wellington	Box 6F	02.03	4200	1	W
6./NJG2	Oblt Ludwig Becker	Wellington	5km S Harlingen	02.35	3500	28	W

18-19 August 1942

9./NJG2	Oblt Alois Lechner	DB-3	Schaustenhof	22.50		7	E
9./NJG2	Oblt Alois Lechner	DB-3	Labiau	23.28		8	E
5./NJG3	Fw Herbert Altner	Halifax	1km ESE Sonderborg	00.08	4500	1	W
5./NJG3	Lt Hans Krause	Stirling	W Esbjerg	00.48		1	W

24-25 August 1942

St ./NJG4	Maj Kurt Holler	Stirling	nr Thynes, SSE Namur	23.41	4400	3	W
St ./NJG4	Maj Kurt Holler	Stirling	nr Givet-Pondrôme	23.55	4100	4	W
7./NJG4	Oblt Hans-Karl Kamp	Stirling		00.10	3300	4	W
StII/NJG1	Lt Hans Autenrieth	Wellington	5km .SSW. Liège	00.18	4000	6	W
St ./NJG4	Maj Kurt Holler	Wellington	4km SE Rheine	00.25	3100	5	W
5./NJG1	Uffz Heinz Kleinert	Lancaster	SW Nivelles	01.39	4500	1	W
4./NJG1	Oblt Walter Barte	Lancaster	S Herentals	01.46	3500	8	W
7./NJG4	Ofw Reinhard Kollak	Stirling	Leuze 27km NW Mons	01.46	3600	10	W
7./NJG4	Fw Heinz Macke	Stirling	Hélenne-en-Temps	02.00	4300	1	W

9./NJG4	Hptm Wilhelm Herget	Wellington	2km SW Marly	02.15	6400	5	W
1./NJG1	Oblt Kurt Loos	Lancaster	N Aulnoye	02.15	4100	10	W
9./NJG4	Oblt Kurt Martinek	Stirling	Nr. Thy-le-Chateau	02.17	100	2	W
StII/NJG1	Lt H-W. Schnaufer	Wellington	1km E Loonbeek	02.54	2700	5	W
5./NJG1	Fw Fritz Schellwat	Wellington	20km E Leuven	02.55	3500	4	W

27-28 August 1942

StIII/NJG2	Hptm Erich Simon	Stirling		23.18	3500	8	W
III./NJG2	Oblt Johann Werth	Wellington		23,18		2*	W
2./NJG1	Oblt Hans-Dieter Frank	Wellington	10km N Helmond	23.25	5800	11	W
StIII/NJG1	Hptm Wolfgang Thimmig	Wellington	Epe railway station	23.25	4200	9	W
1./NJG1	Oblt Kurt Loos	Stirling	6km E Culembourg	23.35		11	W
8./NJG1	Oblt Herbert Lütje	Wellington	3km SE Rechteren	23.45	1400	12	W
StIII/NJG1	Hptm Wolfgang Thimmig	Halifax	4km S Haaksbergen	23.54	3500	10	W
StI/NJG1	Hptm Werner Streib	Wellington	Eindhoven	23.57	3200	36	W
7./NJG1	Oblt Martin Bauer	Stirling	3km S Delden	00.08	2800	3	W
9./NJG2	Ofw Wilhelm Beier	Boston	40km W Ijmuiden	00.25		21	W
7./NJG1	Oblt Martin Bauer	Halifax	2km NW Leer	00.26	4600	4	W
3./NJG1	Hptm Siegfried Wandam	Wellington	30km W Ijmuiden	00.30	5000	6	W
6./NJG3	Oblt Rudolf Altendorf	Wellington	N-W Meppen	00.42	1900	8	W
1./NJG1	Oblt Kurt Loos	Stirling	2km W Elst	01.04		12	W
StI/NJG1	Hptm Werner Streib	Stirling	Eichem	01.04	3000	36	W
I./NJG1	StFw Gerhard Herzog	Lancaster	S-W Roermond	01.10		12	W
6./NJG2	Oblt Ludwig Becker	Wellington	6km SE Medemblik	01.22	3200	29	W
6./NJG2	Oblt Ludwig Becker	Stirling	W Bergen-on-Sea	01.43	3600	30	W
StIII/NJG2	Lt Hermann Müller	Halifax		01.43	1200	5	W
5./NJG2	Oblt E. Lippe-Weissenfeld	Wellington		01.50	1200	35	W
./NJG2	Maj Karl Hülshoff	Wellington		02.10	2800	5	W
StIII/NJG2	Lt Hermann Müller	Lancaster		02.16	1200	6	W
III./NJG2	Oblt Johann Werth	Wellington		02.30		3*	W
5./NJG2	Uffz Heinz Vinke	Wellington		02.32	3800	5	W

28-29 August 1942

StII/NJG1	Hptm Walter Ehle	Wellington	E Wihogne	23.02	4700	9	W
5./NJG1	Hptm Ludwig Bietmann	Wellington	SW Hasselt	23.03	5200	5	W
9./NJG4	Hptm Wilhelm Herget	Halifax	Dinant	23.04	4100	6	W
5./NJG1	Hptm Ludwig Bietmann	Wellington	1km S-E Grand Hallet	23.12	3000	6	W
9./NJG4	Hptm Wilhelm Herget	Hampden	NW Philippeville	23.22	2400	7	W
8./NJG2	Hptm Horst Patuschka	Wellington		23.26		11	W
9./NJG4	Oblt Kurt Martinek	Wellington	Petit-Doische, Namur	23.46	5500	3	W
StII/NJG1	Hptm Walter Ehle	Wellington	NE Liege	23.51	3000	10	W
7./NJG4	Lt Gunther Bellwied	Hampden	6km. N Kleve	23.54	2000	1	W
8./NJG4	Lt Ludwig Meister	Wellington	NNE Worms	00.12		4	W
5./NJG1	Fw Fritz Schellwat	Halifax	Kloster Maria Laach	00.10	3000	5	W
8./NJG4	Lt Ludwig Meister	Wellington	Bad Kreuznach	00.15	3300	5	W
3./NJG1	Hptm Siegfried Wandam	Lancaster		00.15	4400	7	W
5./NJG2	Oblt Leopold Fellerer	Stirling	Aichelbach	00.30	3000	10	W
8./NJG4	Lt Ludwig Meister	Stirling	Arlenbach	00.43	5300	6	W
StII/NJG1	Lt H-W. Schnaufer	Halifax	1km E Tombeek	01.16	3600	6	W
9./NJG4	Oblt Kurt Fladrich	Hampden	1km S Lierre	01.52	2100	1	W
9./NJG4	Lt Fritz Graef	Halifax	2km SE Soize	02.46	3400	2	W

Ludwig Meister (right) flew with II./NJG 1 and I./NJG 4 and finally, after recovering from wounds, assumed command of III./NJG 4 in December 1944, leading it until the surrender. He was credited with 39 victories, including a B-17 in daylight, and was awarded the Ritterkreuz on 9th June 1944. On the left is a photo of Meister taken by his Bordfunker, probably on a test flight. Meister claimed his 4th - 6th victories on the night of 28/29th August.

9./NJG4	Ofw Adolf Brandstetter	Halifax	Maubeuge	02.50	4200	1	W
StII/NJG1	Hptm Walter Ehle	Wellington	Grez Doiceau	02.52	3000	11	W
St ./NJG4	Maj Kurt Holler	Wellington	St. Lambert	03.00	2200	6	W
St ./NJG4	Maj Kurt Holler	Stirling	Meruland	03.15	4500	7	W
9./NJG2	Ofw Wilhelm Beier	Manchester	NE Moerbeke	03.16		22	W

1-2 September 1942

9./NJG4	Ofw Erich Rahner	Wellington	2.5km S Feize/Charleville	02.58	3500	1	W
9./NJG4	Hptm Wilhelm Herget	Halifax	5km S Dixmuiden	03.30	3700	8	W
StII/NJG1	Hptm Walter Ehle	Stirling	Mosseghem	04.13	4000	12	W

2-3 September 1942

9./NJG4	Oblt Kurt Martinek	Halifax	12km SW Namur	01.47	4200	4	W
3./NJG1	Uffz Heinz Oloff	Wellington		01.47	2800	2	W
StII/NJG1	Hptm Walter Ehle	Lancaster	22km SW Liege	01.59	3200	13	W
9./NJG4	Oblt Kurt Martinek	Wellington	2km S Warmont	02.11	3400	5	W
9./NJG2	Ofw Wilhelm Beier	Stirling	40km W Den Haag	03.08		23	W
9./NJG4	Ofw Bruno Eikmeier	Wellington	8km E St.Just	05.00		1	W

3-4 September 1942

5./NJG2	Oblt Paul Gildner	Wellington	SW Norden West'warsch	02.20	1500	36	W
6./NJG2	Lt Wolfgang Küthe	Wellington	S Gröningen	02.49	1500	2	W

4-5 September 1942

StII/NJG2	Hptm Helmut Lent	Halifax	15km SE Leeuwarden	02.50	5800	46	W
5./NJG2	Uffz Heinz Vinke	Halifax	9km E Enkhuizen	02.51	4700	6	W
8./NJG1	Oblt Herbert Lütje	Wellington	NE Altenkeine	02.53	4700	13	W
5./NJG2	Oblt E. Lippe-Weissenfeld	Halifax	5km SW Leeuwarden	03.39	1000	36	W

8./NJG1	Oblt Herbert Lütje	Stirling	8km NW Hasselt	03.59	5000	14	W
5./NJG2	Lt Lothar Linke	Wellington	N Vlieland	04.03	2200	8	W
6./NJG2	Oblt Ludwig Becker	Wellington	N Ameland	04.15	4200	31	W
6./NJG2	Oblt Ludwig Becker	Wellington	NW Ameland	04.45	3200	32	W
6./NJG2	Oblt Ludwig Becker	Wellington	W Schiermonnikoog	05.09	1000	33	W

6-7 September 1942

3./NJG1	Hptm Siegfried Wandam	Halifax		04.00	3000	8	W
8./NJG2	Hptm Horst Patuschka	Wellington		04.41		12	W
9./NJG2	Ofw Wilhelm Beier	Manchester		04.49		24	W
StII/NJG1	Hptm Walter Ehle	Wellington	7km S Tilburg	04.55	3000	14	W
9./NJG2	Ofw Wilhelm Beier	Halifax		05.03		25	W
9./NJG2	Ofw Wilhelm Beier	Wellington		05.11		26	W

8-9 September 1942

5./NJG1	Hptm Ludwig Bietmann	Halifax	5km SW Ardennes	23.30	5000	7	W
St ./NJG4	Maj Kurt Holler	Wellington	23km S Sedan	23.55	4900	8	W
5./NJG2	Oblt Leopold Fellerer	Wellington	Nakenheim	00.34	5500	11	W
5./NJG1	Uffz Heinz Kleinert	Wellington	8km SSW Namur	03.07	3000	2	W

9-10 September 1942

7./NJG2	Uffz Fritz Hobusch	Stirling	W Thyboron	03.52		2	W

10-11 September 1942

8./NJG2	Ofw Heinz Strüning	Wellington	12km NE Helmond	22.50		23	W
9./NJG2	Hptm H. Sayn-Wittgenstein	Stirling		23.15		19	W
1./NJG1	Ofw Karl Wieland	Halifax		23.20	5200	3	W
3./NJG1	Hptm Siegfried Wandam	Stirling		23.23	3000	9	W
8./NJG2	Hptm Horst Patuschka	Stirling		23.36		13	W
1./NJG2	Oblt Albert Schulz	Wellington	NW Bevliet	23.45	4500	8	W
9./NJG2	Hptm H. Sayn-Wittgenstein	Halifax		23.50		20	W
9./NJG2	Hptm H. Sayn-Wittgenstein	B-24		23.54		21	W
8./NJG1	Lt August Geiger	Wellington	2km E Osterwick	00.02	4000	12	W
StIII/NJG1	Oblt Manfred Meurer	Wellington	1km W Bienen	00.06	2600	5	W
1./NJG1	Ofw Karl Wieland	Wellington		00.15	4000	1	W
5./NJG1	Ofw Fritz Schellwat	Wellington	12km W Roermond	00.34	3500	6	W
StI/NJG1	Hptm Werner Streib	Wellington	20km E Hertogenbosch	00.34	3200	37	W
9./NJG2	Oblt Karl Becker	Boston		00.39		1	W
9./NJG2	Ofw Wilhelm Beier	Wellington	50km W Katwijk-a-S	01.19		27	W
8./NJG2	Hptm Horst Patuschka	Wellington		01.37		14	W

13-14 September 1942

6./NJG2	Oblt Ludwig Becker	Halifax	10km S Leeuwarden	02.21	5700	34	W
StII/NJG2	Oblt Rudolf Sigmund	Wellington	W Texel	02.30	3500	7	W
5./NJG2	Lt Lothar Linke	Wellington	Zuiderzee	02.54	3900	9	W
1./NJG3	Uffz Rudolf Frank	Wellington	10km NE Hardenberg	03.10	3600	7	W
StIII/NJG1	Oblt Manfred Meurer	Lancaster	10km N Appeldoorn	03.26	3500	6	W
6./NJG2	Lt Eberhard Gardiewski	Wellington	NW Terschelling	03.43	3900	1	W
3./NJG3	Uffz Wolf Gütter	Wellington	25km ENE Lingen	03.45	5000	1	W
StII/NJG2	Hptm Helmut Lent	Wellington	NW Terschelling	05.02	3000	47	W
9./NJG1	Hptm Wilhelm Dormann	Wellington	2km W Elburg	05.48	4000	2	W

Hauptmann Dr. Horst Patuschka, a former bomber pilot, joined the nightfighters in summer 1941 and quickly rose to lead 7./NJG 2. In October 1942 he was promoted to lead the reformed II./NJG 2 (formerly III./NJG 2) and took the unit to the Middle East after the Allied invasion. He crashed near Bizerta on the night of 6th March 1943 and was killed. The Ritterkreuz was awarded posthumously. At the time of his death he had been credited with 23 night victories.

14-15 September 1942

5./NJG2	Uffz Heinz Vinke	Wellington	N Texel	23.48	3700	7	W

16-17 September 1942

5./NJG2	Uffz Heinz Vinke	Stirling	6km NW Bergen-a-S	22.45	3700	8	W
9./NJG2	Hptm H. Sayn-Wittgenstein	Stirling		22.52		22	W
6./NJG2	Lt Eberhard Gardiewski	Wellington		23.07	4000	2	W
7./NJG1	Oblt Martin Bauer	Lancaster	4km SW Lippramsdorf	23.11	5000	5	W
8./NJG1	Oblt Herbert Lütje	Wellington	1km N Ahaus	23.22	4700	15	W
8./NJG2	Lt Heinz Strüning	Wellington		23.34		24	W
5./NJG2	Oblt Albert Schulz	Wellington	2km W Egmond	00.10		9*	W
8./NJG2	Hptm Horst Patuschka	Wellington		00.12		15	W
7./NJG4	Ofw Reinhard Kollak	Stirling	8km SW Gembloux	00.34	2600	11	W
9./NJG4	Oblt Hubert Rauh	Halifax	E edge of Maubeuge	00.37	4700	4	W
8./NJG2	Hptm Horst Patuschka	Stirling		00.45		16	W
4./NJG1	Oblt Walter Barte	Wellington	SW Diest	00.55	2700	9	W
StII/NJG1	Hptm Walter Ehle	Stirling	S Tirlemont	01.08	4000	15	W
9./NJG2	Lt Wilhelm Beier	Boston		01.12		28	W
StI/NJG1	Hptm Werner Streib	Wellington	7km SE Eindhoven	01.12	3100	38	W
9./NJG2	Lt Wilhelm Beier	Boston		01.13		29	W
7./NJG4	Ofw Reinhard Kollak	Lancaster		01.20	2600	12	W
4./NJG1	Oblt Walter Barte	Wellington	W Hasselt	01.23	2600	10	W
9./NJG2	Lt Wilhelm Beier	Stirling		01.37		30	W
8./NJG2	Oblt Gerhard Stammler	Whitley		02.12		1	W

19-20 September 1942

7./NJG4	Ofw Reinhard Kollak	Stirling	6km NE Revigny	22.59	2300	13	W
7./NJG4	Lt Helmut Bergmann	Whitley	10km S Verdun	00.01	3000	1	W
4./NJG4	Lt Ferdinand Christiner	Halifax	Blesmes	01.07	2600	1	W
5./NJG4	Hptm H. Niebelschütz.von	Four-eng	SE Chalons	01.14	2500	1	W
9./NJG4	Hptm Wilhelm Herget	Wellington	20km N Reims	01.15	3300	9	W
6./NJG2	Lt Gollasch	Lancaster	2½km W Krehlen	02.36	2800	3	W
9./NJG4	Hptm Wilhelm Herget	Lancaster	W Warlimont	03.28	4600	10	W

21-22 September 1942

4./NJG3	Ofw Karl Maisch	Wellington	2km SW Amrum	23.28	1400	2	W
4./NJG3	Ofw Karl Maisch	Wellington	50km NW Westerland	00.18	2300	3	W
StII/NJG3	Maj Gunther Radusch	Wellington	18km W Blidsel	01.01	1800	7	W

23-24 September 1942

5./NJG2	Oblt E. Lippe-Weissenfeld	Wellington	60km NW Vlieland	23.36	200	37	W
6./NJG2	Uffz Pfeiffer	Stirling		03.55	700	1	W
5./NJG2	Fw Karl-Heinz Scherfling	Halifax		05.03	1600	6	W

24-25 September 1942

7./NJG2	Lt Karl Heinz Brandes	Lancaster		01.22		1	W
7./NJG2	Lt Karl Heinz Brandes	Lancaster		02.13	1900	2	W

29-30 September 1942

8./NJG2	Lt Wilhelm Beier	Beaufighter		00.19		31	W

1-2 October 1942

StII/NJG3	Maj Gunther Radusch	Stirling		01.04	1700	8	W
StII/NJG3	Maj Gunther Radusch	Stirling		01.22	800	9	W
2./NJG1	Oblt Hans-Dieter Frank	Stirling	10km NW Venlo	21.34	3400	12	W
1./NJG1	Oblt Reinhold Knacke	Halifax		21.40	5000	36	W
StIII/NJG1	Oblt Manfred Meurer	Wellington	2km NNE Haldern	21.47	3300	7	W
2./NJG1	Oblt Hans-Dieter Frank	Wellington	flugplatz Volkel	22.30	2600	13	W

5-6 October 1942

2./NJG4	Ofw Gerhard Jecke	Halifax	Marquillies	22.30	4200	1	W
6./NJG1	ObLt E-W von Bonin	Wellington	NE Maastricht rw st	22.38	5100	9	W
II./NJG1	Lt Hans Autenrieth	Halifax	1km N Boxmeer	22.47	1200	7	W

6-7 October 1942

1./NJG3	Lt Hans Krause	Wellington	Ulzen	23.04	4000	2	W
1./NJG3	Lt Hans Krause	Wellington	6km N Hagenau	23.22	3100	3	W
11./NJG1	Oblt Hermann Greiner	Halifax	25km W Harlingen	23.24	2800	2	W
1./NJG1	Ofw Karl Wieland	Halifax		23.55	3800	2	W

8-9 October 1942

12./NJG1	Fw Timm	Mosquito		08.05	20	1	W

13-14 October 1942

II./NJG3	Ofw Karl Fleischmann	Stirling	W Kiel	23.09	3000	1	W
11./NJG1	Oblt Paul Gildner	Halifax	15km N Schierm'ikoog	23.19	2300	37	W
12./NJG1	Oblt Ludwig Becker	Stirling	35km NW Texel	23.49	3600	35	W
12./NJG1	Oblt Ludwig Becker	Wellington	30km NW Vlieland	00.20	2700	36	W

15-16 October 1942

1./NJG1	Oblt Reinhold Knacke	Lancaster		21.50	4300	37	W
II./NJG1	Lt Hans Autenrieth	Halifax	2km E Roermond	21.54	3700	8	W
2./NJG1	Oblt Hans-Dieter Frank	Wellington	10km SW Arnheim	22.05	3500	14	W
10./NJG1	Lt Wilhelm Beier	B-24	50km W Den Haag	22.13	3000	33	W
10./NJG1	Lt Wilhelm Beier	Stirling	70km W Den Haag	22.16	3000	34	W

Reinhold Knacke flew as a Bf110 pilot with 3./ZG 1 in the French campaign of May - June 1940 and was one of the original pilots to be selected for night fighting duties, joining 2./NJG 1. He first scored in May 1941 and soon became one of the top Experten, rising to command 1./NJG 1 on 4th April 1942 and being awarded the Ritterkreuz in July. By the end of 1942 his score stood at forty victories. On the night of 3rd February 1943, after shooting down a Stirling and a Halifax - bring his score to 44 - he ordered his crew to bale out following damage sustained in his last fight. His parachute failed to open and he was found dead near Achterveld. He was posthumously awarded the Eichenlaub.

10./NJG1	Lt Wilhelm Beier	Stirling	50km W Den Haag	22.19	3000	35	W
2./NJG1	Oblt Hans-Dieter Frank	Stirling	15km NE Hertoge'sch	22.31	3400	15	W
2./NJG4	Oblt Rudolf Altendorf	Halifax	Waterloo	22.48	3700	9	W
10./NJG1	Lt Wilhelm Beier	Manchester	50km W Leiden	22.52	2700	36	W
StIII/NJG1	Oblt Manfred Meurer	Halifax	1km N Hellendorn	22.53	3000	8	W
2./NJG1	Oblt Hans-Dieter Frank	Stirling	10km W Breda	22.56	3800	16	W
E./NJG2	Oblt Machat	Wellington	Hazeldonk	23.03	3700	1	W
12./NJG1	Lt Gollasch	Wellington		23.21	800	4	W
4./NJG1	Oblt Wilhelm Telge	Wellington	15km W Maastricht	23.24	2500	1	W
4./NJG2	Hptm Horst Patuschka	Wellington		23.29	3100	17	W
2./NJG1	Ofw Bruno Eikmeier	Halifax	5km W Rhenen	23.33	3000	2	W
4./NJG2	Hptm Horst Patuschka	Wellington		00.25	4800	18	W

20-21 October 1942

2./NJG1	Lt Erich Richter	Lancaster		20.20	1000	1	W

23-24 October 1942

E./NJG2	Hptm Godbersen	Stirling		02.30		1	W

24-25 October 1942

StII/NJG4	Oblt Gerhard Friedrich	Halifax	Bar-le-Duc	21.50	5000	4	W
St ./NJG4	Maj Kurt Holler	Wellington	Rethel	00.33	2700	9	W
4./NJG2	Lt Rolf Bussmann	Stirling	10km NW Schouwen Is	02.48	2700	11	W
1./NJG4	Lt Ludwig Meister	Wellington	9km SE Valenciennes	02.54	1700	7	W
4./NJG4	Fw Heinz Wilberg	Halifax	Campiegne	03.00	2900	1	W
4./NJG2	Hptm Horst Patuschka	Wellington	10km W Goeree Island	03.10	30	19	W

29-30 October 1842

3./NAufkSt	Unknown crew	E/A	Makarova	22.30		-	E
3./NAufkSt	Unknown crew	E/A	Makarova	22.30		-	E
3./NAufkSt	Unknown crew	E/A	Makarova	22.30		-	E

9-10 November 1942

StIV/NJG1	Maj Helmut Lent	Halifax	Ameland	20.37	3000	48	W
4./NJG2	Hptm Horst Patuschka	Stirling		22.42	2800	18	W
3./NJG2	Ofw Alfons Köster	Halifax		22.42		17*	W
12./NJG1	Oblt Ludwig Becker	Wellington	5km W Dokkum	22.55	2600	37	W

17-18 November 1942

StII/NJG4	Hptm Heinrich Wohlers	Halifax	2km NE Consigny	21.05	4000	3	W

22-23 November 1942

StI ./NJG4	Hptm Wilhelm Herget	Wellington	N Laon	20.46	1800	11	W
./NJG1	Lt Lothar Linke	Ventura	25km N Terschelling	21.54	300	10	W
StII/NJG4	Hptm Heinrich Wohlers	Wellington	10km S Brienne	22.55	2800	4	W
7./NJG4	Oblt Hans-Karl Kamp	Wellington	26km NE Laon	23.48	3000	5	W
1./NJG4	Lt Ludwig Meister	Lancaster	Le Lambert	23.48	3700	8	W

28-29 November 1942

7./NJG4	Lt Helmut Bergmann	Stirling	2km E Laon	00.49	3000	2	W

2-3 December 1942

2./NJG4	Uffz Ericher	Halifax	Resteigne	03.55	4500	1	W
8./NJG4	Lt Heinz-Martin Hadeball	Halifax	Laumersheim	04.25	4400	7	W
8./NJG4	Fw Paul Faden	Hampden	Monzernheim	04.39	2300	1	W
8./NJG4	Uffz Peter Deppermann	Stirling	5km E Kirn	05.13	2500	1	W
StIII/NJG4	Oblt Fritz Carstens	Halifax	Couvron-et-Aumenc't	06.17	1600	3	W

6-7 December 1942

8./NJG4	Oblt Gerhard Friedrich	Halifax	Courouvre	19.29	3000	5	W
8./NJG4	Ofw Wilhelm Engel	Wellington	15km N Worms	20.12	3400	3	W
5./NJG1	Ofw Fritz Schellwat	Stirling	SW Tirlemont	21.43	2700	7	W
7./NJG4	Lt Helmut Bergmann	Wellington	36km NE Laon	21.54	2000	3	W

8-9 December 1942

2./NJG3	Fw Helmut Schuppan	Stirling		21.48	2600	1	W
4./NJG3	Oblt Hans Gref	Stirling		22.03	1400	1	W
5./NJG5	Hptm Leopold Fellerer	E/A				12	W

16-17 December 1942

1./NJG1	Lt Werner Rapp	Stirling	Gortel	20.09	500	1	W

17-18 December 1942

1./NJG3	Hptm Walter Milius	Wellington	Salzbergen	19.20	500	3	W
3./NJG3	Fw Nagel	Halifax	S Grafeld	19.33	900	1	W
StII/NJG3	Maj Günther Radusch	Lancaster	SE Ribe	19.53	1300	10	W
1./NJG3	Hptm Walter Milius	Stirling	flugplatz Vechta	19.56	500	4	W
1./NJG3	Hptm Walter Milius	Stirling	S Münster	20.16	600	5	W
StIV/NJG1	Maj Helmut Lent	Lancaster	N Sloten	20.22	600	49	W
StIV/NJG1	Maj Helmut Lent	Halifax	E Urk	20.38	500	50	W
5./NJG2	Oblt Albert Schulz	Wellington		20.52	2000	10	W
./NJG1	Oblt Werner Husemann	Stirling		20.53	2200	1	W
8./NJG1	Oblt Herbert Lütje	Lancaster	6km NE Zwolle	21.10	2500	16	W

Ground crew check a Bf110 of NJG 3. Note the spindly undercarriage and the extended step. (H.Scholl)

12./NJG1	Lt Lothar Linke	Halifax	N Petten	21.18	1150	11	W
5./NJG2	Oblt Albert Schulz	Wellington		21.25	2500	11	W
6./NJG3	Unknown crew	E/A		21.48		-	W
11./NJG1	Oblt Hans-Joachim Jabs	Halifax	W Egmont	22.40	1100	3	W
11./NJG1	Oblt Hans-Joachim Jabs	Stirling	20km SW Amsterdam	22.54	50	4	W
1./NJG3	Lt Martin Fünfstück	Stirling	W Cloppenburg	22.55		1	W

20-21 December 1942

7./NJG1	Oblt Martin Bauer	Halifax	6km W Bocholt	19.45	4700	6	W
1./NJG1	Oblt Reinhold Knacke	Halifax		19.45	3000	38	W
StI/NJG1	Hptm Werner Streib	Wellington	25km SE Herto'bosch	20.13	3600	39	W
1./NJG1	Oblt Reinhold Knacke	Wellington		20.19	300	39	W
2./NJG1	Lt Erich Richter	Lancaster	Den Hout	20.20	1000	2	W

21-22 December 1942

7./NJG4	Ofw Kurt Karsten	Halifax	25km SE Sedan	19.32	4100	1	W
StII/NJG4	Hptm Heinrich Wohlers	Four-eng	5 km S Void-Vacon	19.41	2500	5	W
5./NJG4	Lt Ferdinand Christiner	Lancaster	4km W Rumont	20.04	4000	2	W
StII/NJG4	Hptm Heinrich Wohlers	Stirling	Minecourt	22.58	2400	6	W
1./NJG4	Lt Ludwig Meister	Stirling		23.00		9*	W
StII/NJG4	Hptm Heinrich Wohlers	Wellington	2½km W Matignicourt	23.13	1500	7	W
1./NJG4	Lt Rehdaus	Lancaster	15km SE Maubeuge	23.16	2600	1	W
7./NJG4	Oblt Hans-Karl Kamp	Stirling	24km S Flôret	23.20	2700	8	W
StI ./NJG4	Hptm Wilhelm Herget	Lancaster	1km W Steenbuyne	23.30	3000	12	W
StI/NJG1	Lt H-W. Schnaufer	Lancaster	3km NNE Poelkapelle	23.53	2700	7	W
1./NJG1	Oblt Reinhold Knacke	Lancaster		24.00	500	40	W

31 December 1942-1 January 1943

1./NJG1	Hptm Reinhold Knacke	Lancaster	10km E Ess	20.15	6800	41	W
3./NJG3	Fw Nagel	Wellington		02.02		2	W

Oblt Ludwig Wolf of 6/NJG 1. He claimed a Wellington in the early morning of 1st August 1942, but this was to be his only combat success. He crashed while turning onto final approach to St. Trond in the early hours of 4th January 1943. Both he and his Bordfunker Feldwebel Ernst Kurz died in the crash of their Bf110F-4 G9+DP. (S.Hall)

1-2 January 1943

StIV/NJG1	Maj Helmut Lent	Bomber		02.49		51	W

3-4 January 1943

7./NJG1	Uffz Christian Koltringer	Lancaster	5km E Arnheim	20.15	5600	1	W
3./NJG1	Oblt Manfred Meurer	Lancaster	20km NNE Roermond	20.23	4000	9	W
1./NJG1	Hptm Reinhold Knacke	Lancaster	5km S Maastricht	20.19	5200	42	W

4-5 January 1943

10./NJG1	Oblt Rudolf Sigmund	Halifax	NNW Scheveningen	20.30	4300	8	W

6-7 January 1943

II./NJG2	Fw Werner Heyne	Wellington	Tunis			3	M

8-9 January 1943

12./NJG1	Oblt Ludwig Becker	Lancaster	7km SE Zutphen	19.23	800	38	W
StII/NJG3	Maj Günther Radusch	Lancaster	W Sylt	20.00	2500	11	W
StIV/NJG1	Maj Helmut Lent	Halifax	10km W Texel	20.24	4000	52	W
4./NJG3	Fw Ernst Baader	Lancaster	over the Sea	21.10	2700	1	W
11./NJG1	Oblt Hans-Joachim Jabs	Halifax	10km NW Vlieland	22.15	2700	5	W

9-10 January 1943

12./NJG1	Lt Lothar Linke	Halifax	Schiermonnikoog	18.22	1100	12	W
StI/NJG1	Maj Werner Streib	Lancaster	8km W Venlo	19.15	6500	40	W
2./NJG1	Lt Erich Richter	Lancaster	12km NW Utrecht	20.02	3600	3	W

13-14 January 1943

StI/NJG1	Maj Werner Streib	Lancaster	SW Apeldoorn	19.16	6200	41	W
StI/NJG1	Maj Werner Streib	Lancaster	15km SW Hoog-Soeren	19.44	5000	42	W
1./NJG1	Oblt Horst Pause	Lancaster	W Schouwen	20.56	5000	1	W

The grim reality: German troops recover the bodies of a shot-down Halifax crew, winter 1942-43.

17-18 January 1943

6./NJG3	Lt Paul Szameitat	Stirling	70km NW Westerland	19.33	1500	6	W
4./NJG3	Ofw Karl Haisch	Lancaster	Nr. Kolding	19.50	1800	4	W
11./NJG1	Fw Heinz Vinke	Lancaster	35km N Terschelling	20.25	3900	9	W
10./NJG3	Fw Hermann Holz	Halifax	15km W Amrun	21.34		1	W
4./NJG3	Ofw Karl Haisch	Lancaster	10km SW Schleswig	21.45	3900	5	W
6./NJG3	Lt Paul Szameitat	Lancaster	90km NW Westerland	21.49	1600	7	W
6./NJG3	Uffz Hans Meissner	Lancaster	10km SW Schleswig	21.50		1	W
2./NJG3	Oblt Paul Zorner	Halifax	45km NW Juist	21.55	2100	1	W
4./NJG3	Lt Gerhard Raht	Lancaster	Hademarschen	21.59	3000	2	W
StI/NJG3	Maj E. Lippe-Weissenfeld	Halifax	5km N Leer	22.13	5200	38	W
StII/NJG3	Maj Günther Radusch	Lancaster	5km W Eiderstedt	22.14	5500	12	W
11./NJG1	Oblt Hans-Joachim Jabs	Stirling	10km N Ameland	22.17	3200	6	W
6./NJG3	Lt Paul Szameitat	Lancaster	25km W Fano	22.18	1000	8	W
2./NJG1	Hptm Hans-Dieter Frank	Lancaster	10km W Ijmuiden	22.24	150	17	W
7./NJG3	Oblt Martin Drewes	Stirling	Linten	22.32	5600	1	W
4./NJG3	Fw Ernst Baader	Lancaster	Lubeck	22.43	4300	2	W
12./NJG1	Oblt Ludwig Becker	Stirling	20km N Ameland	23.03	3800	39	W
12./NJG1	Oblt Ludwig Becker	Stirling	40km N Terschelling	23.28	3800	40	W
10./NJG1	Uffz Karl Pfeiffer	Lancaster	10km W Vlieland	23.29		2	W

18-19 January 1943

6./NJG3	Ofw Anton Offner	Stirling		20.01	5000	1	W
5./NJG2	Uffz Tolksdorf	Wellington	Mediterranen	20.25	200	1	M

21-22 January 1943

8./NJG1	Fw Kleinhaus	Stirling	Enschelde	19.45	4000	1	W
12./NJG1	Uffz Georg Kraft	Lancaster	30km W Den Helder	19.52	5600	1	W
StIV/NJG1	Maj Helmut Lent	Wellington	N Schiermonnikoog	20.11	100	53	W

23-24 January 1943

11./NJG1	Lt Wolfgang Küthe	Lancaster	E Stavoren	20.56		3	W

26-27 January 1943

11./NJG1	Lt Wolfgang Küthe	Wellington	Harlingen	00.28	700	4	W
12./NJG1	Uffz Georg Kraft	Wellington	Terschelling	03.03	1400	2	W

27-28 January 1943

1./NJG1	Hptm Reinhold Knacke	Lancaster	12km SSSW Nijmegen	20.10	6300	43	W
3./NJG1	Oblt Manfred Meurer	Halifax	Handel	20.53	5300	10	W
11./NJG1	Fw Heinz Vinke	Wellington		00.28		10	W

29-30 January 1943

II./NJG2	Hptm Horst Patuschka	Wellington	Mediterranen	04.06	3700	21	M

30 January 1943

11./NJG1	Lt Wolfgang Küthe	Wellington	10km N Terschelling	13.31		5	W

30-31 January 1943

12./NJG1	Oblt Ludwig Becker	Lancaster	15km E Texel	03.10	6500	41	W
1./NJG3	Lt Weiss	Lancaster		03.24		1	W

2-3 February 1943

StI/NJG1	Hptm Werner Streib	Lancaster	20km S Eindhoven	21.12	5800	43	W
StI/NJG1	Hptm Werner Streib	Four-eng	15km SW Venlo	21.30	5000	44	W
2./NJG1	Hptm Hans-Dieter Frank	Stirling	12km ESE Rotterdam	22.01	4400	18	W

3-4 February 1943

1./NJG1	Hptm Reinhold Knacke	Stirling	5km S Amerongen	19.46	3500	44	W
1./NJG1	Hptm Reinhold Knacke	Halifax	N-N-W Thiel	20.05	3300	45	W
StIII/NJG1	Hptm Wolfgang Thimmig	Stirling	E Hengelo	20.10	4200	11	W
10./NJG1	Ofw Karl-Heinz Scherfling	Halifax	20km E Assen	20.30	5300	7	W
3./NJG1	Oblt Manfred Meurer	Stirling	5km S Amersfoort	20.41	4200	11	W
7./NJG1	Lt Werner Rapp	Halifax	5km NW Quakenbruck	20.48	5000	2	W
9./NJG1	Hptm Wilhelm Dormann	Stirling	6km NNW Nijmegen	20.54	3000	3	W
1./NJG1	Oblt Horst Pause	Stirling	W Middelburg	21.05	5000	2	W
12./NJG1	Uffz Alois Ott	Wellington	3km E Estermeer	21.37	5000	1	W
12./NJG1	Lt Lothar Linke	Wellington	25km W Vlieland	22.19	4400	13	W
12./NJG1	Lt Lothar Linke	Halifax	W Vlieland	22.34	4800	14	W
1./NJG1	Uffz Johannes Vogel	Stirling	16km SW Utrecht	23.00	3200	1	W
7./NJG1	Uffz Christian Költringer	Stirling	4km SW Utrecht	23.35	3600	2	W

4 February 1943

11./NJG1	Uffz Naumann	B-17	E Terschelling	12.25	3500	1	W
12./NJG1	Ofw Heinz Grimm	B-17	50km NW Texel	12.35	6700	7	W
11./NJG1	Oblt Hans-Joachim Jabs	B-17	50km NW Texel	12.51	10000	7	W

8-9 February 1943

II./NJG2	Hptm Horst Patuschka	Halifax	Mediterranen	21.30	3800	22	M
II./NJG2	Hptm Horst Patuschka	Wellington	Mediterranen	22.13	1500	23	M

11-12 February 1943

2./NJG3	Oblt Paul Zorner	Lancaster	40km WNW Borkum	20.44	5300	2	W

14-15 February 1943

3./NJG1	Oblt Manfred Meurer	Halifax	50km E Eindhoven	20.05	5200	12	W
3./NJG1	Oblt Manfred Meurer	Halifax	NNE Maastricht	20.20	4700	13	W
2./NJG1	Ofw Hermann Schönthier	Wellington	40km SE Eindhoven	20.20	4800	1	W

On 24/25th February 1943 Oberleutnant Paul Gildner of 1./NJG 1 (right) crashed near Gilze-Rijn with one engine of his Bf110 on fire and was killed. He was awarded the Eichenlaub posthumously. Next day Hauptmann Ludwig Becker of 12./NJG 1 (left) took off to intercept American heavy bombers over the North Sea, having just been awarded the Eichenlaub. He did not return. Both men had been credited with 44 victories, two of Gildner's having been by day. Attrition was taking its toll of the 'old guard'.

6./NJG1	Lt Johannes Hager	Halifax	N Maastricht	20.42	3300	1	W
5./NJG1	Ofw Fritz Schellwat	Stirling	3km NNW Hechteren	20.56	4500	8	W
3./NJG1	Oblt Martin Bauer	Stirling	2km SW Boxmeer	21.05	3400	7	W
3./NJG1	Oblt Manfred Meurer	Wellington	5km W Roermond	21.06	4400	14	W
IV./NJG1	Oblt Paul Gildner	Wellington	20km W Schouwen	21.47	3000	38	W
IV./NJG1	Oblt Paul Gildner	B-17	45km W Vlissingen	22.07		39	W

18-19 February 1943

2./NJG3	Fw Helmut Schüppan	Lancaster	N Borkum	20.10	700	2	W
12./NJG1	Uffz Georg Kraft	Lancaster	35km NW Vlieland	21.21	1200	3	W

19 February 1943

4./NJG3	Oblt Hans Gref	B-24	N Helgoland	20.32	1600	2	W
2./NJG3	Oblt Paul Zorner	Boston	N Norderney	20.34	2700	3	W
5./NJG3	Fw Herbert Altner	Wellington	N Suderlugem	20.55	1800	2	W
11./NJG1	Oblt Hans-Joachim Jabs	Stirling	10km N Ameland	21.00	1600	8	W
IV./NJG1	Oblt Paul Gildner	Halifax	20km N Vlieland	21.05	1100	40	W
IV./NJG1	Oblt Paul Gildner	Halifax	15km N Vlieland	21.16	1000	41	W
11./NJG1	Oblt Hans-Joachim Jabs	Stirling	10km NW Schierm'koog	21.18	1900	9	W
11./NJG1	Oblt Hans-Joachim Jabs	Stirling	50km N Terschelling	21.44	2500	10	W
11./NJG1	Lt Wolfgang Küthe	Stirling	40km N Texel	22.55	1100	6	W

20-21 February 1943

10./NJG1	Oblt Rudolf Sigmund	Wellington	55km N Texel	19.45		9	W

25 February 1943

10(N)./ZG 1	Ofw Josef Kociok	E/A		09.35	80	2	E

25-26 February 1943

StII./NJG4	Hptm Heinrich Wohlers	Lancaster				8*	W
10./NJG5	Uffz Heintze	Wellington	Nr. Bussard-Bretagne	00.10	250	1	W

26 February 1943

10./NJG1	Oblt Rudolf Sigmund	B-24	60km N Ameland	12.24	4500	10	W
12./NJG1	Uffz Georg Kraft	B-24	70km NW Terschelling	12.27	3500	4	W

26-27 February 1943

2./NJG3	Oblt Paul Zorner	Stirling	30km WNW Schi'koog	20.46	1400	4	W
3./NJG1	Oblt Manfred Meurer	Lancaster	10km NE Roermond	21.18	5900	15	W
3./NJG1	Oblt Manfred Meurer	Wellington	20km NE Eindhoven	21.35	3600	16	W
11./NJG1	Fw Heinz Vinke	Stirling	50km W Kateijklaan Z	22.28	1400	11	W
3./NJG1	Ofw Fritz Kruse	Wellington	15km NE Eindhoven	22.30	4900	1	W
1./NJG1	Oblt Horst Pause	Wellington	40km W Walcheren	22.35	3100	3	W

27-28 February 1943

12./NJG1	Uffz Georg Kraft	Lancaster	30km WNW Schi'koog	20.38	200	5	W
4./NJG2	Oblt Machat	Wellington		04.35	1200	2	W

1-2 March 1943

StIV/NJG1	Maj Helmut Lent	Halifax	8km N Ameland	21.39	1400	54	W
11./NJG1	Lt Rolf Bussmann	Lancaster	40km NW Vlieland	22.23	800	12	W
7./NJG1	Lt August Geiger	Halifax	2km NW Zutphen	23.52	3600	13	W
3./NJG1	Oblt Manfred Meurer	Lancaster	10km SE Rotterdam	23.56	6000	17	W
8./NJG1	Hptm Herbert Lütje	Halifax	3km N Zuidlo	00.08	5800	17	W
3./NJG1	Ofw Fritz Kruse	Halifax	20km E Deventer	00.13	3700	2	W
10./NJG1	Lt Oskar Köstler	Lancaster	20km NW Vlieland	00.36	4300	1	W
7./NJG1	Lt August Geiger	Halifax	1km W Markelo	00.37	4400	14	W
1./NJG1	Oblt Horst Pause	Lancaster	8km E Tiel	00.43	3000	4	W
11./NJG1	Lt Wolfgang Küthe	Stirling	12km S Leeuwarden	00.57	2300	7	W
1./NJG1	Fw Heinrich Neppel-Schmid	Stirling	St.Maartensdijk	01.04	3700	1	W
12./NJG1	Lt Robert Denzel	Halifax	40km W Texel	01.05	2600	2	W

2-3 March 1943

12./NJG1	Lt Heinz Grimm	Wellington	50km NW Texel	20.42	200	8	W

3-4 March 1943

8./NJG3	Fw Wilhelm Kürreck	Stirling	W Stade	21.29		1	W
4./NJG3	Ofw Karl Haisch	Lancaster	W Heligoland	21.45	3200	6	W
1./NJG3	Uffz Rudolf Frank	Stirling	S DelmenHorst	22.10	800	8	W
12./NJG1	Lt Lothar Linke	Halifax	15km NW Ameland	22.21	2300	15	W
4./NJG3	Ofw Karl Haisch	Stirling	W Heligoland	22.26	3000	7	W
8./NJG3	Oblt Walter Borchers	Wellington	E Emden	22.30	3500	1	W
10./NJG1	Oblt Rudolf Sigmund	Stirling	20km W Texel	22.32	2000	11	W

4 March 1943

8./NJG1	Hptm Herbert Lütje	B-17	W Den Helder	11.31	4200	18	W
7./NJG1	Lt August Geiger	B-17		11.33		15	W
8./NJG1	Oblt Karl Goetze (shared)	B-17		11.36	5000	5	W
10./NJG1	Lt Oskar Köstler (shared)	B-17	60km W Texel	11.36	510	2	W
8./NJG1	Lt Hans Pancritius	B-17		11.40	6000	1	W

Chapter Five

The Battle of the Ruhr

5th March - 24th July 1943

By the spring of 1943, Air Marshal Harris of Bomber Command felt that he had amassed sufficient resources to attempt something hitherto considered practically impossible; the destruction of the great Ruhr industrial complex of cities and towns containing the factories producing steel, armaments and other necessities vital to Germany's war effort. He had at his disposal 53 squadrons, of which 16 were twin-engined units but, more importantly, his pathfinder units were now fitted with the new H^2S terrain mapping radar, which could pierce the great layer of smoke haze that covered the area and had, previously, effectively screened the various targets from effective attack. For the RAF, this would be something new; a concentrated bombing campaign against a specific objective.

For the *Luftwaffe* too, things had changed drastically since the initial shock of a cohesive bomber stream introduced in mid-1942. The *Kammhuber Line*, while still effective, had now been supplemented by *Dunkel Nachtjagd* (Dark night fighting), where fighters were scrambled, assembled at various fighter beacons, and directed towards an incoming bomber stream. The sheer size of the RAF formations made them difficult to miss and the scene was now set for a night campaign of air battles unprecedented in aerial warfare.

At the beginning of the campaign, the strength and disposition of the *Luftwaffe* nightfighter force was as follows:

Stab./NJG 1	*Oberst* Falck	Deelen	Bf110
I./NJG 1	*Maj* Streib	Venlo	Bf110
II./NJG 1	*Maj* Ehle	St.Trond	Bf110/Do217
III./NJG 1	*Hptm* Thimmig	Twente	Bf110
IV./NJG 1	*Maj* Lent	Leeuwarden	Bf110
Stab./NJG 2	*Obstlt* Hulshoff	Comiso, Sicily*	Ju88C
I./NJG 2	*Maj* Jung	Catania/Benghasi*	Ju88C
II./NJG 2	*Hptm* Dr. Patuschka	Comiso*	Ju88C
** Middle East*			
Stab NJG 3	*Maj* Schalk	Stade	Bf110
I./NJG 3	*Maj* Prinz zur Lippe-Weissenfeld	Vechta	Bf110/Do217
II./NJG 3	*Hptm* Radusch	Schleswig	Bf110/Do217
III./NJG 3	*Hptm* Nacke	Stade	Bf110
IV./NJG 3	*Maj* Simon	Grove	Bf110/Do217
Stab NJG 4	*Maj* Stoltenhoff	Metz	Bf110/Do217
I./NJG 4	*Maj* Herget	Florennes	Bf110/Do217/Ju88
IV./NJG 4	*Maj* Wohlers	Mainz-Finthen	Bf110
Stab NJG 5	*Maj* Schaffer	Doberitz	Bf110
I./NJG 5	*Hptm* Wandam	Stendal	Bf110
II./NJG 5	*Hptm* Schoenert	Parchim	Bf110/Do217
IV./NJG 5	*Hptm* Prinz zu Sayn Wittgenstein	Intserburg	Bf110

During the course of the battle, several new units would be formed. In March, *NJG 101* was established, *Stab* and *I Gruppe* at Scheissheim, under *Oberstleutnant* Stoltenhoff and *Hauptmann* Theodor Rossiwall respectively, *II Grupp*e at Munich-Riem under *Major* Hans-Dietrich Knoetzsch and *III Gruppe* at Echterdingen under Hauptmann Helmut Peters. All these units were initially equipped with Bf110s and Do217s. In April, *II./NJG 4* was established at Leipheim, equipped with Bf110s. The *Gruppenkommandeur* is not known, but was possibly *Major* Otto Materne. Also, *III./NJG 5* was formed at Neuruppin under *Hauptmann* Walter Borchers, equipped with Bf110s and Do217s. In May, *III./NJG 4* came into being at Mainz-Finthen, with Bf110s and Do217s, commanded by *Major* Kurt Holler, and *IV./NJG 6* was formed at Otopeni and Zilistea with Bf110s. The initial *Gruppenkommandeur* is again unknown, but the unit was taken over within a month by *Hauptmann* Herbert Lütje. Finally, on 15th July, the formation of *NJG 101* was completed by the addition of *IV Gruppe* at Kitzingen, under *Major* Ludwig Fehling and equipped with Bf110s and Do217s. From small beginnings in 1940, the *Luftwaffe* had developed a formidable force.

5-6 March 1943

3./NJG1	Oblt Manfred Meurer	Halifax	W Düsseldorf	20.57	6400	18	W
3./NJG1	Oblt Manfred Meurer	Wellington	4km NW Weeze	21.20	5400	19	W
8./NJG1	Hptm Herbert Lütje	Halifax	5km SE Staphorst	21.35	4200	19	W
3./NJG1	Lt Walter Schön	Halifax	N Nijmegen	21.37	3300	1	W
12./NJG1	Lt Robert Denzel	Lancaster	W Texel	22.02	3500	3	W
StIV/NJG1	Maj Helmut Lent	Halifax	10km W Harlingen	22.20	3500	55	W
StIV/NJG1	Maj Helmut Lent	Halifax	10km E Wieringer	22.34	4000	56	W
11./NJG1	Fw Heinz Vinke	Wellington	10km E Amsterdam	22.35	3700	12	W
11./NJG1	Lt Rolf Bussmann	Lancaster	15km W Texel	22.48	500	13	W
12./NJG1	Lt Lothar Linke	Wellington	W Texel	22.52	2400	16	W

7-8 March 1943

10./NJG1	Ofw Karl-Heinz Scherfling	Lancaster	In sea	20.15	1800	8	W
2./NJG3	Oblt Paul Zorner	Wellington	NE Norden	20.50	1700	5	W

8-9 March 1943

6./NJG4	Oblt Hans Autenrieth	Stirling	15km NE Verdun	21.45	2800	9	W
StIII/NJG4	Lt Helmut Bergmann	Stirling	10km NW Charleville	22.13	3700	4	W
7./NJG4	Oblt Hans-Karl Kamp	Halifax	W Verdun	22.29	5100	7	W
10./NJG4	Ofw Wilhelm Engel	Stirling	Philippsburg	22.35	3300	4	W
1./NJG4	Oblt Ludwig Meister	Halifax	9km E Le Cateau	00.30	3700	10	W

9-10 March 1943

StIII/NJG4	Lt Helmut Bergmann	Stirling	Nr. Charleville	22.13	2600	5	W
4./NJG3	Ofw Karl Haisch	Wellington	50km NW List	22.30	2100	8	W
10./NJG4	Oblt Gerhard Friedrich	Lancaster	NW Rinuthall	23.35	4500	6	W
3./NJG4	Hptm Hubert Rauh	Halifax	Wallingen	01.56	3000	5	W
7./NJG4	Ofw Reinhard Kollak	Lancaster	N Vitry-les-Reims	01.59	4800	14	W
4./NJG4	Uffz Barten(?)	Halifax	Bausen/Strasbourg	02.15	4400	-	W
7./NJG4	Ofw Reinhard Kollak	Lancaster	Taguebec	02.30	4600	15	W

10-11 March 1943

5./NJG3	Oblt Gerhard Raht	Lancaster	S Middelfahrt	22.12	4500	3	W

11-12 March 1943

7./NJG4	Oblt Hans-Karl Kamp	Stirling	20km NE Montmirail	21.40	3700	8	W
7./NJG4	Ofw Reinhard Kollak	Halifax	1km W Malmaison	22.00	4500	16	W

The Battle of the Ruhr became a battle of wills between Arthur Harris (right) and Josef Kammhuber. Both men were energetic, decisive and totally dedicated to their own ideas. Harris - 'Bert' to his friends 'Butch' to his crews and 'Bomber' to the press - implacably pursued his goal of victory by bombing to a totally ruthless degree. Kammhuber however, was to be replaced by Josef Schmid late in 1943 due to major differences of opinion with his superiors over the night defence of Germany.

4./NJG4	Lt Jakob Schaus	Halifax	22km SE Aigny	22.32	5000	1	W
10./NJG4	Lt Wilhelm Engel	Halifax	10km NE Heidelberg	22.46	5600	5	W
4./NJG4	Hptm Materne	Halifax	N Rosières	23.43	3000	1	W
4./NJG4	Lt Jakob Schaus	Halifax	18km SE Charleville	00.06	2500	2	W
6./NJG4	Oblt Hans Autenrieth	Stirling	3km S-W Chalons	00.18	1900	10	W
StIII/NJG4	Oblt Kornacher	Halifax	Maufresnes	00.45	4500	1	W
6./NJG4	Fw Gerhard Rase	Halifax	Sogny en L'Aigle	00.47	3000	1	W

12-13 March 1943

3./NJG1	Oblt Manfred Meurer	Halifax	NW Venlo	21.16	5000	20	W
8./NJG1	Hptm Herbert Lütje	Halifax	3km ESE Alsatte	21.18	5500	20	W
9./NJG1	Hptm Wilhelm Dormann	Stirling	8km SSE Zevenaa	21.19	4500	4	W
3./NJG1	Oblt Manfred Meurer	Lancaster	3-5km E Bergen	21.25	5300	21	W
10./NJG1	Lt Oskar Köstler	Wellington	Zuiderzee	21.32	2800	3	W
6./NJG1	Oblt E-W von Bonin	Halifax	4-5km NNW Amerika	21.40	4800	10	W
10(N)./ZG1	Ofw Josef Kociok	MBR-2	Russia	21.43	3200	3	E
10(N)./ZG1	Ofw Josef Kociok	DB-3	Russia	21.45	3000	4	E
3./NJG1	Oblt Manfred Meurer	Halifax	Grafwegen	21.48	4600	22	W
10./NJG1	Lt Oskar Köstler	Wellington	20km NE Alkmaar	21.58	2500	4	W
11./NJG1	Lt Wolfgang Küthe	Wellington		22.22	3000	8	W
3./NJG1	Oblt Manfred Meurer	Wellington	25km N-E Herto'bosch	22.25	4400	23	W
10./NJG1	Lt Oskar Köstler	Lancaster	W Den Helder	22.27	2500	5	W
11./NJG1	Fw Heinz Vinke	Wellington	50km W Ijmuiden	22.55	2700	13	W
10./NJG1	Oblt Rudolf Sigmund	Stirling	60km W Den Helder	23.16	3200	12	W

13-14 March 1943

10./NJG5	Uffz Günther Franz	Wellington		22.52	2700	1	W
2./NJG3	Oblt Paul Zorner	Wellington	3km N Norderney	03.51	900	6	W

14-15 March 1943

7./NJG3	Oblt Martin Drewes	Halifax	Store Heddinge	21.42	200	2	W

15-16 March 1943

NJS LF6	Oblt Günter Bertram	R-5	E Brjansk	23.50		1*	E

16-17 March 1943

10./NJG4	Oblt Gerhard Friedrich	Halifax	Börnloss	00.16	3400	7	W

18 March 1943

StIII/NJG3	Oblt Walter Borchers	B-24	Jade estuary	15.45	6500	2	W

19-20 March 1943

10./NJG3	Oblt Werner Spiedel	Lancaster	NW Skagen	23.28		6	W
NJS LF6	Oblt Günter Bertram	R-5		23.50		2*	E

24-25 March 1943

8./NJG1	Oblt Dietrich Schmidt	Halifax	NE Enkhuizen	01.36	150	1	W

27-28 March 1943

1./NJG5	Oblt Ulrich Wulff	Halifax	SE Bergerdorf	23.50	3600	1	W

29-30 March 1943

7./NJG1	Lt August Geiger	Wellington	5km SW Ahaus	22.52	5500	16	W
3./NJG1	Oblt Martin Bauer	Wellington	W Boxmeer	23.15	3800	8	W
7./NJG1	Lt August Geiger	Wellington	WNW Borculo	23.15	5400	17	W
StI/NJG1	Maj Werner Streib	Wellington	N Arnheim	23.15	3500	45	W
StIV/NJG1	Maj Helmut Lent	Wellington	NW Lemmer	23.46	5000	57	W
11./NJG1	Fw Heinz Vinke	Wellington	15km W Zandyvoort	00.20	2600	14	W
5./NJG3	Oblt Paul Szameitat	Halifax	Sudschmedeby	00.36	4000	9	W
5./NJG3	Oblt Gerhard Raht	Halifax	Welmbuttel	00.50	5300	4	W
StII/NJG5	Hptm Rudolf Schoenert	Lancaster	NW Rerik	01.30	6000	24	W
11./NJG5	Fw Georg Bargel	Halifax	Buschau	02.35	4000	1	W
1./NJG3	Oblt Hans Krause	Lancaster	N Hanover	03.12	5000	4	W
7./NJG1	Lt August Geiger	Halifax	10km SE Zutphen	03.47	5500	18	W
7./NJG1	Lt Werner Rapp	Lancaster	15km S Amsterdam	04.11	4800	3	W
7./NJG1	Uffz Christian Koltringer	Lancaster	1km SW Klosterhaar	04.17	6700	3	W
7./NJG1	Lt August Geiger	Lancaster	7km NE Deventer	04.27	5300	19	W
8./NJG1	Hptm Herbert Lütje	Lancaster	7km S Raalte	04.28	6200	21	W
7./NJG1	Lt August Geiger	Lancaster	3km NW Lieveide	04.46	5500	20	W
7./NJG1	Lt Werner Rapp	Stirling	3km SSE Marken	04.49	4300	4	W

3-4 April 1943

12./NJG1	Uffz Baldischweiler	Halifax	W Harderwijk	22.45	5300	1	W
2./NJG1	Hptm Hans-Dieter Frank	Halifax	5km SE Rhenen, Arnhem	22.45	4500	19	W
8./NJG1	Hptm Herbert Lütje	Lancaster	3km NW Winterswijk	22.45	6200	22	W
StI/NJG1	Maj Werner Streib	Halifax	20km W Arnheim	22.52	5000	46	W
7./NJG1	Lt August Geiger	Lancaster	5km N Gemen	23.00	6000	21	W
StI/NJG1	Maj Werner Streib	Halifax	NW Kleve	23.00	5200	47	W
3./NJG1	Oblt Martin Bauer	Lancaster	3km SE Dingrew	23.01	6000	9	W
2./NJG1	Hptm Hans-Dieter Frank	Lancaster	4km N Klave	23.02	6200	20	W
6./NJG1	ObLt E-W von Bonin	Halifax	N Nederweert	23.21	5000	11	W
2./NJG1	Hptm Hans-Dieter Frank	Stirling	3km SW Den Haag	23.25	3100	21	W
StI/NJG1	Maj Werner Streib	Halifax	43km N Roermond	23.30	4500	48	W
8./NJG1	Hptm Herbert Lütje	Halifax	22km S Zwolle	23.37	5500	23	W
3./NJG1	Lt Walter Schön	Halifax	9km S Zwolle	23.44	4800	2	W
8./NJG1	Uffz Emil Heinzelmann	Halifax	10km S Zwolle	23.50	4600	1	W

Ofw Karl Haisch of 4./NJG 3 was credited with at least eight victories. He was killed in an accident on 20th March while flying a Bf108. (S.Hall)

6./NJG1	ObLt E-W von Bonin	Lancaster	20km NE Helmond	23.54	4600	12	W
StIV/NJG1	Maj Helmut Lent	Lancaster	15km E Texel	00.24	5900	58	W
12./NJG1	Oblt Eberhard Gardiewski	Halifax	70km W Terschelling	00.47	4000	3	W
10./NJG3	Oblt Werner Spiedel	Manchester	50km NW Skagen	04.32		7	W

4-5 April 1943

12./NJG3	Hptm Günter Vorwinkel	Stirling		23.22		1	W
9./NJG3	Oblt Siegfried Hahn	Halifax	SW Neumunster	23.35	5100	1	W
8./NJG3	Fw Wilhelm Kurreck	Lancaster	10km S Neumunster	23.50	5600	2	W
2./NJG3	Fw Alfred Brackmann	Stirling		23.57		2	W
5./NJG3	Oblt Paul Szameitat	Lancaster	Hindenburgdamm	00.22	3500	10	W
IV./NJG1	Oblt Walter Borchers	Wellington	70km W Terschelling	00.46	2900	3	W

8-9 April 1943

7./NJG4	Ofw Karstenauer	Halifax	9km NW Compiegne	00.10	5000	1	W

9-10 April 1943

11./NJG1	Fw Heinz Vinke	Lancaster	16km NW Amsterdam	22.42	7000	15	W
10./NJG1	Lt Oskar Köstler	Lancaster	20km NE Harderwijk	22.45	5800	6	W
3./NJG1	Ofw Fritz Kruse	Lancaster		23.23	5400	3	W
StI/NJG1	Maj Werner Streib	Lancaster	17km NE s'Herto'bosch	23.25	5800	49	W
2./NJG1	Hptm Hans-Dieter Frank	Lancaster		23.46	5700	22	W
11./NJG1	Fw Heinz Vinke	Lancaster	15km W Utrecht	23.49	2800	16	W
11./NJG1	Hptm Hans-Joachim Jabs	Lancaster	40km W Alkmaar	23.56	1500	11	W
7./NJG4	Ofw Kurt Karsten	Halifax	9km NW Schleptrufer	00.10		2*	W
11./NJG1	Fw Heinz Vinke	Stirling		00.14	8400	17	W

10-11 April 1943

NJS LF6	Oblt Günter Bertram	Li-2		20.56		3*	E
StIII/NJG4	Lt Helmut Bergmann	Halifax	6.5km N Sedan	01.40	3600	6	W
3./NJG4	Ofw Walter Piwarz	Halifax	Hirson	02.00	45000	1	W
9./NJG4	Oblt Kurt Fladrich	Lancaster	1km N La Thour	02.28	5300	2	W
9./NJG4	Uffz Thiell	Halifax	8km NNW Trier	02.30	3000	1	W
8./NJG4	Fw Gross	Halifax	Saarburg	02.33	3200	1	W
12./NJG4	Lt Heinz-Martin Hadeball	Stirling	8km SW Bacharach	02.35	4800	8	W

10./NJG4	Oblt Gerhard Friedrich	Wellington	Luinbach-Weuir	03.23	3900	8	W
StI ./NJG4	Hptm Wilhelm Herget	Wellington	Nivelles	03.23	3400	13	W
StII/NJG1	Maj Walter Ehle	Stirling	1km S Tongerloo	03.45	4000	16	W
StI ./NJG4	Hptm Wilhelm Herget	Wellington	1km S Le Havre	03.55	4300	14	W
7./NJG4	Ofw Reinhard Kollak	Stirling	1km NW Le Guevière	04.16	4200	17	W

11-12 April 1943

NJS LF6	Oblt Günter Bertram	An-6	Russia	21.25		4*	E
NJS LF6	Oblt Günter Bertram	An-6	Russia	21.35		5*	E
12./NJG1	Uffz Georg Kraft	Mustang	20km W Den Helder	22.26	2000	6	E

13-14 April 1943

10(N)./ZG1	Oblt Egon Mangold	SB-2	Russia	20.52		1	E
5./NJG2	Uffz Tolksdorf	Whitley	Korba	01.10		2	M
10./NJG5	Lt Josef Pützkuhl	Lancaster	Russia	06.21	1800	1	W

14-15 April 1943

NJS LF6	Oblt Gunter Bertram	Li-2	Russia	21.00		6*	E
StIV./NJG4	Hptm Heinrich Wohlers	Stirling	Biblis			9*	W
12./NJG4	Lt Heinz-Martin Hadeball	Wellington	Mutterscheid	00.20	5400	9	W
11./NJG4	Hptm Karl Floitgraf	Stirling	Nr. Worms	00.21	4000	1	W
StIV/NJG4	Hptm Heinrich Wohlers	Halifax	Niederleistadt	00.23	5000	10	W
StIV./NJG4	Hptm Heinrich Wohlers	Halifax		01.15		11*	W
5./NJG2	Fw Lubenka	Wellington	Tunisia	01.24		1	M
5./NJG2	Fw Lubenka	Wellington	10km SW Korba	01.43		2	M
4./NJG4	Lt Franz Draude	Wellington	Nr. Yainville	01.46	2900	1	W
4./NJG4	Hptm Otto Materne	Stirling	St.Souplet	01.48	2000	2	W
2./NJG4	Oblt Rudolf Altendorf	Stirling	1km W Charleville	01.50	2900	10	W
9./NJG4	Oblt Kurt Fladrich	Stirling	12km E Reims	01.59	3300	3	W
7./NJG4	Ofw Reinhard Kollak	Lancaster	27km E Reims	02.11	3000	18	W
2./NJG4	Oblt Rudolf Altendorf	Halifax	Chimay	02.18	5100	11	W
StI/NJG4	Lt Fritz Gräef	Stirling	Regiessart	02.23	3400	3	W
St ./NJG4	Maj Kurt Holler	Stirling	30km NE St.Quentin	02.30	3700	10	W
7./NJG4	Hptm Hans-Karl Kamp	Wellington	5km E St.Quentin	02.50	3800	9	W
StIII/NJG4	Lt Helmut Bergmann	Wellington	5km SW St.Quentin	02.55	3800	7	W
7./NJG4	Hptm Hans-Karl Kamp	Halifax	1km N Reims	02.57	3800	10	W
St ./NJG4	Maj Kurt Holler	Wellington	2km N Perpignan	02.58	3900	11	W
4./NJG4	Hptm Otto Materne	Stirling	20km SSW Chalons-s-M	03.04	1400	3	W
StIII/NJG4	Lt Helmut Bergmann	Halifax	15km SW St.Quentin	03.22	3100	8	W

16-17 April 1943

StIII/NJG4	Lt Helmut Bergmann	Halifax	Sevigry	23.03	5800	9	W
StIV/NJG5	Hptm H. Sayn-Wittgenstein	DB-3	Russia	23.13		23	E
StIV/NJG5	Hptm H. Sayn-Wittgenstein	DB-3	Russia	23.15		24	E
StIII/NJG4	Lt Helmut Bergmann	Stirling	8km E Brunnchemmer	23.31	2700	10	W
7./NJG4	Ofw Reinhard Kollak	Lancaster	1km N Northeim	23.53	5300	19	W
StIII/NJG4	Lt Helmut Bergmann	Wellington	Nr. Rethel	23.56	4000	11	W
II./NJG4	Lt Friedrich Tober	Halifax	5km NE Laimont	00.03	2500	1	W
St ./NJG4	Maj Kurt Holler	Stirling	15km S Sedan	00.14	4000	12	W
2./NJG4	Lt Norbert Pietrek	Stirling		00.15	35000	1	W
12./NJG4	Lt Heinz-Martin Hadeball	Stirling	1km N Rapperalle	00.20	3300	10	W

Werner Streib (centre) chats to an unidentified Ritterkreuz winner in the of summer 1943.
(via H.Scholl)

3./NJG4	Ofw Erich Rahner	Wellington	Froidliers	00.20	4500	2	W
7./NJG101	Oblt Jakob Bender	Halifax	NE Boblingen	00.30	3800	3	W
10./NJG4	Ofw Richard Launer	Four-eng	Worms	00.34	4100	3	W
4./NJG4	Hptm Otto Materne	Halifax	1km SSE Vaurouleuse	00.35	4500	4	W
12./NJG4	Lt Heinz-Martin Hadeball	Wellington	1km N Lotzburen	00.49	1600	11	W
1./NJG4	Hptm Alfred Haesler	Halifax	14km NE Laon-Missy	01.10	300	3	W
StIV/NJG4	Hptm Heinrich Wohlers	Halifax	W Kastenbach	01.15	4200	12	W
7./NJG101	Lt Heinz Bock	Halifax	Holzweilerhof	01.44	3300	1	W
11./NJG4	Fw Paul Faden	Wellington	S Bitburg	01.51	2600	2	W
7./NJG4	Hptm Hans-Karl Kamp	Wellington	4km NE Villiers	02.09	3700	11	W
9./NJG4	Ofw Heinz Macke	Stirling	Commenchon	02.12	4600	2	W
9./NJG4	Oblt Kurt Fladrich	Wellington	8km SE Soissons	02.13	4000	4	W
StIII/NJG4	Lt Helmut Bergmann	Halifax	W Hamm	02.19	4100	12	W
12./NJG4	Lt Herbert Jung	Halifax	Beitelsheim	02.32	4000	1	W
12./NJG4	Uffz Peter Deppermann	Halifax	2km N Neuenweiler	03.01	3000	2	W
11./NJG4	Fw Paul Faden	Lancaster	S Bitburg	03.07	4000	3	W
10./NJG4	Lt Otto Blohm	Halifax	Hadamar, Limburg	03.12	3600	1	W
III./NJG101	Lt Josef Nabrich	Lancaster	3km S Stuttgart	03.17		1*	W
9./NJG4	Uffz Thiell	Halifax	3km S Marle	03.38	3300	2	W
4./NJG4	Hptm Otto Materne	Stirling	S Quigny-les-Roses	03.48	3500	5	W
6./NJG4	Oblt Hans Autenrieth	Halifax	15km WSW Chalons	03.50	3600	11	W
2./NJG4	Oblt Rudolf Altendorf	Lancaster	Givry	04.02	2400	12	W
StI ./NJG4	Hptm Wilhelm Herget	Halifax	Rance	04.06	3000	15	W
1./NJG4	Oblt Ludwig Meister	Halifax	NW St.Hubert	04.08	3000	11	W
StI ./NJG4	Hptm Wilhelm Herget	Lancaster	Villers-Deux-Église	04.23	2500	16	W
7./NJG4	Hptm Hans-Karl Kamp	Halifax	Roye	04.36	3100	12	W
3./NJG4	Ofw Adolf Brandstetter	Halifax	3km N Guise	04.37	2200	2	W
7./NJG4	Ofw Kurt Karsten	Lancaster	E Paulavert	04.45	2300	3	W

17 April 1943

1./NJG3	Ofw Erich Heitmann (shared)	B-17		13.30	50	1	W
10./NJG1	Oblt Rudolf Sigmund	B-17		13.40	7000	13	W
IV./NJG1	Oblt Walter Borchers	B-17		13.45	20	15	W

17-18 April 1943

10(N)./ZG1	Fw Melschert	Boston	Russia	21.50		1	E
10(N)./ZG1	Fw Melschert	Boston	Russia	21.54		1	E
10(N)./ZG1	Fw Melschert	DB-£F	Russia	21.54		1	E
II./NJG2	Fw Liebscher	Wellington	Tunis			1	M

18-19 April 1943

1./NJG2	Lt Heinz Rökker	Wellington	S Marettimo	01.15	800	6	M
3./NJG2	Uffz Ternieden	Wellington	Tunis	03.12	1000	1	M
4./NJG2	Fw Fritz Hobusch	Whitley	N Cape Bone	05.20	1200	3	M
4./NJG2	Fw Fritz Hobusch	Whitley	N Cape Bone	05.33	1300	4	M

19-20 April 1943

10(N)./ZG1	Oblt Egon Mangold	PS-84	Russia	21.02		2	E
4./NJG3	Hptm Hans Baer	Lancaster	1km S-W Vester V'ted	00.00	1300	2	W
StIV/NJG1	Maj Helmut Lent	Mosquito	W Staverer	03.38	100	59	W

20-21 April 1943

4./NJG5	Ofw Heinz Säwert	Stirling	W Spitz Hiddenau	00.33	3000	6	W
12./NJG3	Ofw Rudolf Mangelsdorf	Wellington	50km S Copenhagen	00.44		1	W
4./NJG5	Hptm Werner Hoffmann	Halifax	1.5 km E Eggesin	00.45	4000	3	W
10./NJG5	Uffz Heintze	Wellington	500 m. S Güt-Borkum	00.50	20	2	W
4./NJG5	Hptm Werner Hoffmann	Halifax	500m S Güt-Borkum	00.50	3800	4	W
11./NJG3	Fw Herbert Koch	Halifax	Klagshamm	00.50	500	1	W
7./NJG3	Uffz Berg	Boeing	Hillinge	01.58	2400	1	W
5./NJG5	Lt Peter Ehrhardt	Wellington	Schwerin-Herneberg	02.00	1700	1	W
12./NJG1	Oblt Lothar Linke	Mosquito	E Middenmeer	02.07	6000	17	W
StII/NJG5	Hptm Rudolf Schoenert	Lancaster	Häfen Gjetser	02.23	47000	25	W
5./NJG5	Lt Peter Ehrhardt	Lancaster	2km E Rerik	02.28	3000	2	W
4./NJG3	Hptm Hans Baer	Lancaster	5km N-W Romo	02.42	200	3	W
StII/NJG3	Maj Günther Radusch	Stirling	In sea	03.07	1000	13	W
12./NJG3	Hptm Alfred Heldt	Lancaster		03.12		1	W
4./NJG3	Hptm Hans Baer	Lancaster	25km W Romo	03.21	200	4	W
4./NJG5	Lt Viktor Sorko	Stirling	12km S Karlshult	03.25	2500	1	W
StII/NJG3	Maj Günther Radusch	Stirling	In sea	03.32	200	14	W
IV./NJG3	Oblt Hans-Hermann Müller	Stirling	NE Vejle	03.40	8	1	W
11./NJG3	Oblt Günther Köberich	Stirling	20km W Ringkobing	03.54	600	3	W

22-23 April 1943

StIV/NJG5	Hptm H. Sayn-Wittgenstein	DB-3	Russia	23.25	4000	25	E
StIV/NJG5	Hptm H. Sayn-Wittgenstein	Mitchell	Russia	00.30		26	E

23-24 April 1943

4./NJG2	Fw Zaularts	Wellington	Mediterranen	05.10	500	1	W

26-27 April 1943

10(N)./ZG1	Ofw Josef Kociok	SB-2	Russia	22.25		5	E
10(N)./ZG1	Ofw Josef Kociok	Yak-4	Russia	22.30		6	E
11./NJG1	Fw Heinz Vinke	Stirling	Zuiderzee	02.08		17	W
12./NJG1	Lt Heinz Grimm	Wellington	W Terschelling	02.21	3800	9	W
11./NJG1	Fw Heinz Vinke	Halifax	N Texel	02.34	4500	19	W

Major Hans-Dieter Frank joined the Luftwaffe in 1937 and came to nightfighting in the spring of 1941 after serving first with I./ZG 1 over Scandinavia and France in 1940. His unit was 8./NJG 1 and, by mid-June 1942, was a Hauptmann and leader of 2 Staffel, flying He219s. He was awarded the Ritterkreuz on 20th June and, on the very next night, claimed six bombers destroyed. On 1st July he became Kommandeur I./NJG 1, but was to hold this position for less than three months. On the night of 29th September he was killed with his crew in a mid-air collision. It is very likely that the collision was caused by an attack by Wing Commander J.R.D.Braham of 141 Squadron. With 55 victories credited, he was awarded the Eichenlaub posthumously. He claimed his 24th, 25th and 26th victories - a Stirling and two Wellingtons - on 12th May 1943.

3./NJG1	Ofw Fritz Kruse	Lancaster	1km SE Ezaart	03.09	4800	4	W
7./NJG1	Oblt August Geiger	Wellington	Bornebroek	03.11	4000	22	W
2./NJG1	Hptm Hans-Dieter Frank	Wellington	3km W Brakel	03.15	3800	23	W
12./NJG1	Oblt Lothar Linke	Wellington	Bergeneerzee	03.36	4700	18	W
2./NJG1	Hptm Hans-Dieter Frank	Wellington	5km W Dougen	03.37	4600	24	W
./NJG1	Oblt Werner Husemann	Stirling	17km W Hilversum	03.37	2800	2	W
12./NJG1	Uffz Georg Kraft	Halifax	Zuiderzee	04.29	3900	7	W

27 April 1943

1./NJG6	Ofw Richard Launer	E/A	Russia(?)	10.34		4	

27-28 April 1943

10(N)./ZG1	Oblt Wübbenhorst	DB-3	Russia	20.27		1	W
10./NJG3	Fw Walter Schlette	Manchester	30km NE Skagen	23.24		1	W

28-29 April 1943

6./NJG3	Lt Weiss	Wellington	Logumkloster	00.16	1500	1	W
11./NJG3	Uffz Günter Holtfreter	Stirling	S Rye	00.17	1400	1	W
7./NJG3	Uffz Berg	Stirling	Kalundborg	00.56	2000	2	W
6./NJG3	Lt Weiss	Wellington	SW Haderslev	00.57	2100	2	W
4./NJG3	Hptm Hans Baer	Stirling	25km W Romo	01.17	1600	5	W
10./NJG3	Oblt Werner Spiedel	Lancaster	N Hjorring	01.29		8	W
12./NJG3	Ofw Rudolf Mangelsdorf	Stirling	S-W Ringkobing	01.45	2000	2	W
12./NJG3	Lt Günter Rogge	Stirling	60km W Esbjerg	01.48	2200	1	W
5./NJG3	Oblt Paul Szameitat	Lancaster	NW Kolding	03.51	3500	11	W

30 April-1 May 1943

11./NJG1	Fw Heinz Vinke	Halifax	Muiderberg	02.26	6300	20	W
9./NJG1	Hptm Wilhelm Dormann	Lancaster	S Winterswijk	02.36	6000	5	W
StIII/NJG1	Hptm Wolfgang Thimmig	Halifax	E Stadtloan	02.43	6200	12	W
11./NJG1	Lt Hermann Greiner	Stirling	3.5km SW Dokkum	02.52	4000	3	W
12./NJG1	Lt Heinz Grimm	Stirling	NNW Leeuwarden	02.52		10	W
StIII/NJG1	Hptm Wolfgang Thimmig	Lancaster	12km SE Zwolle	03.20	6200	13	W

Halifax JB873 of 78 Squadron did not return from Bochum on the night of 13th May 1943. It was shot down at Haasrode by Leutnant Heinz-Wolfgang Schnaufer of Stab II./NJG 1 as his ninth combat success. Two members of Pilot Officer Dane's crew were killed. Like so many of the top Experten, Schnaufer was a relatively 'slow starter' but would end the war with 121 credited combat victories as the top scoring nightfighter pilot of all time and one of the most highly decorated officers of the German armed forces.

7./NJG1	Oblt August Geiger	Halifax	NW Nijmegen	03.21	5000	23	W

2-3 May 1943

StIV/NJG5	Hptm H. Sayn-Wittgenstein	DB-3	SW Eydtkau	22.37	3800	27	E

3-4 May 1943

NJS LF6	Oblt Günter Bertram	Mitchell	Russia	20.50		7*	E
NJS LF6	Oblt Günter Bertram	Li-2	Russia	21.50		8*	E

4-5 May 1943

StIV/NJG1	Maj Helmut Lent	Stirling	8km S Enkhuizer	00.08	4500	60	W
StIV/NJG1	Maj Helmut Lent	Stirling	1km SE Hommaerts	00.18	3600	61	W
StIII/NJG1	Hptm Wolfgang Thimmig	Halifax	S Gronau	00.34	5700	14	W
12./NJG1	Lt Robert Denzel	Stirling	23km ENE Leeuwarden	00.36	4800	4	W
StIII/NJG1	Hptm Wolfgang Thimmig	Halifax	N-W Schuttorf	00.50	6000	15	W
12./NJG1	Lt Robert Denzel	Halifax	20km SSE Leeuwarden	00.52	4800	5	W
12./NJG1	Oblt Lothar Linke	Wellington	7km ESE Meppel	00.58	6400	19	W
10./NJG1	Uffz Karl Pfeiffer	Stirling	3km E Gröningen	01.00	3800	9	W
12./NJG1	Lt Robert Denzel	Halifax	20km NE Leeuwarden	01.07	5800	6	W
3./NJG1	Ofw Fritz Kruse	Halifax	N Spakenburg	01.10	5900	5	W
12./NJG1	Oblt Lothar Linke	Halifax	1km S Workum	01.14	5200	20	W
8./NJG1	Oblt Dietrich Schmidt	Lancaster	Nordhorn	01.15	6000	2	W
9./NJG1	Hptm Wilhelm Dormann	Halifax	4km SW Doetinchem	01.30	6000	6	W
12./NJG1	Oblt Lothar Linke	Lancaster	SE Utrecht	01.40	6000	21	W
10./NJG1	Ofw Karl-Heinz Scherfling	Wellington	42km SE Groninger	01.44	4500	9	W
2./NJG1	Hptm Hans-Dieter Frank	Stirling	W Zandvoort	01.47	3500	25	W
7./NJG1	Oblt August Geiger	Halifax	W Zwolle	02.09	5300	24	W
12./NJG1	Oblt Lothar Linke	Halifax	SE Vlieland	02.10	4700	22	W
2./NJG1	Hptm Hans-Dieter Frank	Wellington	20km S Amsterdam	02.16	4100	26	W

Halifax JB924 of 78 Sqn, downed at Wynaldum by Major Helmut Lent, the Gruppen Kommandeur of IV./ NJG 1.
(G.Zwanenberg MBE)

8./NJG1	Uffz Emil Heinzelmann	Wellington	3km SE Zwolle	02.16	4000	2	W
8./NJG1	)blt Dietrich Schmidt	Wellington				3	W
2./NJG1	Hptm Hans-Dieter Frank	Wellington	Werkendam	02.38	2800	26	W
12./NJG1	Lt Heinz Grimm	Halifax	Terschelling	02.44		11	W
11./NJG1	Lt Hermann Greiner	Halifax		02.47	3800	4	W

5-6 May 1943

12./NJG1	Lt Heinz Grimm	Halifax	Terschelling	23.50	400	12	W

9-10 May 1943

NJS LF6	Oblt Günter Bertram	An-6	Russia	23.55		9	E

10 May 1943

10(N)./ZG1	Oblt Marotke	DB-3	Russia			1	E

12-13 May 1943

2./NJG1	Ofw Bruno Eikmeier	Halifax	2km W Zuidland	01.23	4700	3	W
12./NJG1	Lt Robert Denzel	Lancaster	Zuiderzee	01.38	5600	7	W
12./NJG1	Uffz Georg Kraft	Wellington	30km NW Harderwijk	02.02	4500	8	W
12./NJG1	Oblt Lothar Linke	Halifax	4km SE Harlingen	02.02	6000	23	W
11./NJG1	Fw Heinz Vinke	Lancaster	Eesterga	02.02	6200	21	W
7./NJG1	Oblt August Geiger	Halifax	Vreden	02.06	5600	25	W
10./NJG1	Oblt Rudolf Sigmund	Wellington	40km W Petten	02.09	4500	14	W
3./NJG1	Oblt Manfred Meurer	Wellington	18km ESE Eindhoven	02.10	6000	24	W
9./NJG1	Hptm Wilhelm Dormann	Lancaster	5km NE Emmerich	02.15	6000	7	W
7./NJG1	Oblt August Geiger	Wellington	S-W Eldenzaal	02.17		26	W
11./NJG1	Fw Heinz Vinke	Lancaster	18km NNE Utrecht	02.17	6300	22	W
7./NJG1	Oblt August Geiger	Halifax	12km S Hengelo	02.20	5200	27	W
2./NJG1	Hptm Hans-Dieter Frank	Wellington	15km NW Ereda	02.21	3700	28	W
3./NJG1	Lt Wilhelm Beier	Halifax	16km NNW Amsterdam	02.30	5100	37	W
1./NJG1	Hptm Liedke	Halifax	30km ESE Arnheim	02.43	4000	1	W
StII/NJG1	Maj Walter Ehle	Lancaster	3km N Roermond	02.44	4800	17	W
3./NJG1	Ofw Fritz Krüse	Wellington	Zeist	02.58	2700	6	W
3./NJG1	Lt Walter Schön	Halifax	2km E Roermond	03.04	5300	3	W
12./NJG1	Lt Robert Denzel	Wellington	30km W Petten	03.07	4800	8	W
12./NJG1	Lt Heinz Grimm	Wellington	45km W Den Helder	03.25	4800	13	W

1./NJG1	Fw Karl Leopold	Wellington		03.26	3900	1	W
12./NJG1	Lt Robert Denzel	Stirling	45km W Petten	03.28	3700	9	W
8./NJG1	Uffz Emil Heinzelmann	Stirling	9km SW Harderwijk	03.28	2200	3	W
2./NJG1	Hptm Hans-Dieter Frank	Stirling	Rotterdam	03.39	4000	29	W
2./NJG1	Ofw Hermann Schönthier	Stirling	35km W Voorne	03.52	3200	2	W
13-14 May 1943							
12./NJG1	Oblt Lothar Linke	Lancaster	3km N Lemmer	23.22	6000	24	W
8./NJG1	Hptm Herbert Lütje	Lancaster	5km SE Almelo	23.42	5200	24	W
12./NJG1	Oblt Lothar Linke	Lancaster	13km W Sneek	23.51	4700	25	W
8./NJG1	Hptm Herbert Lütje	Lancaster	5km N Eldenzaal	23.52	5200	25	W
8./NJG1	Hptm Herbert Lütje	Lancaster	3km S Horstel	23.54	5200	26	W
7./NJG1	Lt Hans-Heinz Augenstein	Lancaster	6km E Lingen	23.57	6000	1	W
StI/NJG3	Hptm E. Lippe-Weissenfeld	Halifax	10km NNW Hengelo	01.11	3000	39	W
2./NJG1	Lt Heinz Strüning	Halifax	10km N Breskens	01.49	5200	25	W
2./NJG1	Hptm Hans-Dieter Frank	Wellington	Hilversum	02.02		30*	W
5./NJG1	Ofw Fritz Schellwat	Stirling	SE Erkelenz	02.02	4200	9	W
StII/NJG1	Lt H-W. Schnaufer	Stirling	Heerlen	02.14	3700	8	W
8./NJG1	Hptm Herbert Lütje	Halifax	5km W Dedemswaart	02.25	200	27	W
2./NJG1	Lt Heinz Strüning	Halifax	60km W Walcherer	02.26	3300	26	W
3./NJG1	Oblt Manfred Meurer	Halifax		02.31	6200	25	W
5./NJG1	Ofw Fritz Schellwat	Wellington	4km E Mönchengladbach	02.38	4000	10	W
8./NJG1	Hptm Herbert Lütje	Halifax	Dalen	02.45	5200	28	W
9./NJG1	Lt Karl-Heinz Vollkopf	Stirling	5km NW Bocholt	02.50	3000	2	W
9./NJG1	Hptm Wilhelm Dormann	Halifax	10km W Hengelo	02.53	4000	8	W
2./NJG1	Hptm Hans-Dieter Frank	Wellington	10km NW Hilversum	02.53	4000	31	W
StIV/NJG1	Maj Helmut Lent	Halifax	4km N Harlingen	02.54	5000	62	W
8./NJG1	Hptm Herbert Lütje	Halifax	19km E Hoogeveen	02.55	5000	29	W
StII/NJG1	Lt H-W. Schnaufer	Halifax	Nr. Blanden	03.07	4700	9	W
3./NJG1	Lt Walter Schön	Stirling	NW Aachen	03.07	4300	4	W
11./NJG1	Fw Heinz Vinke	Wellington	35km W Ijmuiden	03.07	2600	23	W
8./NJG1	Oblt Dietrich Schmidt	Lancaster	6km SW Ruhlertwist	03.44	4200	4	W
10./NJG1	Uffz Karl Pfeiffer	Lancaster	20km W Den Helder	03.59	4600	3	W
10./NJG1	Uffz Scherer	Wellington	16km E Wieringen	04.06	200	1	W

Sometimes they were just unlucky: On 25/26th May 1943 two Stirlings were blown out of the sky by a massive explosion. Major Walter Ehle, Kommandeur II./NJG 1, had attacked a 77 Squadron Halifax near Julich and Sergeant Lewis's aircraft blew up. The explosion caught both Flight Lieutenant Berthiaume's EF301 of 7 Squadron (opposite page) and Sergeant Collins's EH887 of 218 Squadron and both were torn apart. Ehle was credited with five victories this night, but he refused to count the ones above in his personal score as he had not even seen them and certainly could not have fired at either one.

14 May 1943

6./NJG3	Uffz Hans Meissner	B-17	50km NW Sylt	12.45	3000	2	W

15 May 1943

8./NJG3	Oblt Heinrich Fröhlich	B-17		10.50		1	W
10(N)./ZG1	Ofw Josef Kociok	PS-84	Russia	11.34	1000	7	E

15-16 May 1943

10(N)./ZG1	Ofw Josef Kociok	PS-84	Russia	21.41	1200	8	E
10(N)./ZG1	Ofw Josef Kociok	SB-2	Russia	22.05	3500	9	E
10(N)./ZG1	Ofw Josef Kociok	PS-84	Russia	22.54	3800	10	E

18-19 May 1943

1./NJG1	Uffz Hans Kurz	Wellington		00.25	150	1	W

21-22 May 1943

1./NJG2	Lt E. Gross	Wellington	Mediterranean	03.13	2200	1	M

23-24 May 1943

11./NJG1	Maj Rolf Leuchs	Halifax	5km N Enkhuizen	00.55	5700	1	W
St./NJG1	Hptm Kurt Fladrich	Wellington		01.03	350	5	W
7./NJG1	Lt Werner Rapp	Halifax	3km NW Greven	01.03	5800	1	W
StIII/NJG1	Hptm Wolfgang Thimmig	Wellington	9km SW Enschelde	01.08	5000	16	W
StIII/NJG1	Hptm Wolfgang Thimmig	Wellington	5km E Haaksbergen	01.25	4800	17	W
StIII/NJG1	Hptm Wolfgang Thimmig	Halifax	12km W Almedo	01.34	5400	18	W
9./NJG1	Hptm Wilhelm Dormann	Halifax	2km N Dingen	01.35	6000	9	W
3./NJG1	Oblt Manfred Meurer	Wellington	13km NW Eindhoven	01.55	3600	26	W

11./NJG1	Fw Heinz Vinke	Lancaster	40km W Egmont-am-Zee	02.01	5100	24	W
9./NJG1	Lt Karl-Heinz Vollkopf	Halifax	3km S Ahaus	02.04	6000	3	W
7./NJG1	Oblt August Geiger	Lancaster	48km E Meppel	02.07	5600	28	W
3./NJG1	Oblt Manfred Meurer	Wellington	E Ess	02.12	3700	27	W
10./NJG1	Ofw Karl-Heinz Scherfling	Stirling	N Texel	02.14	3600	10	W
2./NJG1	Lt Heinz Strüning	Stirling	18km S Utrecht	02.14	3000	27	W
StIV/NJG1	Maj Helmut Lent	Lancaster	SW Workum	02.16	6200	63	W
St./NJG1	Hptm Kurt Fladrich	Wellington		02.18	4000	6	W
11./NJG1	Fw Heinz Vinke	Halifax	2km W Leiden	02.24	4900	25	W
7./NJG1	Oblt August Geiger	Lancaster	18km E Devemer	02.30	6000	29	W
3./NJG1	Oblt Manfred Meurer	Lancaster	32km SE Nijmegen	02.33	5800	28	W
10./NJG1	Ofw Karl-Heinz Scherfling	Halifax	30km NW Vlieland	02.36	5400	11	W
7./NJG1	Lt Hans-Heinz Augenstein	Halifax	6km SW Raalte	02.39	6000	2	W
12./NJG1	Lt Heinz Grimm	Lancaster	5km N Amsterlmeer	02.49	6700	14	W
12./NJG1	Oblt Eberhard Gardiewski	Halifax	E edge of Texel	03.41	3700	4	W

25-26 May 1943

3./NJG1	Oblt Manfred Meurer	Wellington	Oostrum	01.24	5200	29	W
6./NJG1	ObLt E-W von Bonin	Wellington	8km N Maastricht	01.36	4700	13	W
3./NJG1	Oblt Manfred Meurer	Lancaster	10km S-E Roermond	01.36	6800	30	W
StII/NJG1	Oblt Wilhelm Telge	Halifax	25km N-E Hasselt	01.40	4600	2	W
StII/NJG1	Maj Walter Ehle	Halifax	1km S Jülich	01.51	4800	18	W
1./NJG1	Hptm Liedke	Lancaster	NW Hertogenbosch	01.51	5900	2	W
StII/NJG1	Maj Walter Ehle	Stirling	2km S Jülich	01.52	4700	19	W
StII/NJG1	Maj Walter Ehle	Stirling	6km S Jülich	01.52	4700	20	W
StII/NJG1	Maj Walter Ehle	Stirling	8km S Jülich	01.55	4500	21	W
3./NJG1	Oblt Manfred Meurer	Lancaster	5km W Nijmegen	02.08	6600	31	W
1./NJG1	Ofw Hermann Sommer	Lancaster	E Horst	02.10	5600	14	W
E./NJG2	Lt Bruno Heilig	Stirling	10km NE Antwerp	02.15	6300	1	W
StII/NJG1	Oblt Wilhelm Telge	Wellington	5km S Mechelen	02.26	3900	3	W
2./NJG1	Lt Heinz Struning	Wellington	Loosduinen	02.33	4700	28	W
11./NJG1	Fw Heinz Vinke	Stirling	5km NW Ijmuiden	02.33	2900	26	W
StII/NJG1	Maj Walter Ehle	Wellington	3km N Nederweert	02.35	3500	22	W
12./NJG1	Uffz Georg Kraft	Stirling	25km NW Den Helder	02.48	2600	9	W
12./NJG1	Uffz Georg Kraft	Stirling	40km W Vlieland	03.04	800	10	W
11./NJG1	Oblt Hermann Greiner	Lancaster	35km W Den Helder	03.05	4000	5	W
11./NJG1	Oblt Hermann Greiner	Lancaster	15km N Bergen-am-Zee	03.25	1000	6	W
2./NJG1	Lt Werner Hopf	Lancaster	2km SSE Vlissingen	03.42	5500	1	W

27-28 May 1943

10(N)./ZG1	Lt Reichelt	DB-3	Russia	20.49		1	E
3./NJG1	Oblt Manfred Meurer	Lancaster	2km NNE Barlo	00.38	3000	32	W
2./NJG1	Fw Heinz Oloff	Halifax	10km NW Arnheim	00.47	5400	3	W
7./NJG1	Lt Hans-Heinz Augenstein	Wellington	3km SW Hengelo	00.54	5000	3	W
9./NJG1	Hptm Wilhelm Dormann	Halifax	4km NE Wesel	00.55	6000	10	W
./NJG1	Hptm Gunter Friedrich	Four-eng	5km NE Geargodorf	00.58	5600	1	W
10./NJG1	Uffz Karl Pfeiffer	Stirling	70km NW Leeuwarden	00.58	1300	4	W
8./NJG1	Uffz Emil Heinzelmann	Halifax	4km NW Ahaus	00.59	5400	4	W
7./NJG1	Lt Hans-Heinz Augenstein	Halifax	3km NE Nienborg	01.05	5500	4	W
StI/NJG1	Maj Werner Streib	Lancaster	18km NW Arnheim	01.15	5300	50	W
7./NJG1	Lt Hans-Heinz Augenstein	Halifax	Leer	01.21	5000	5	W

On 29/30th May 1943 Halifax JB793 of 419 Squadron was caught en route to Wuppertal by Oberleutnant Wilhelm Telge of Stab II./NJG 1. The disturbed ground on the lower left indicates that Flight Sergeant Winegarden attempted a crash-landing. (E.Nonnenmacher)

3./NJG1	Oblt Manfred Meurer	Wellington	21km SE Wanroij	01.30	5500	33	W
9./NJG1	Lt Karl-Heinz Vollkopf	Wellington	17km N Rheine	01.30	3600	4	W
7./NJG1	Lt Hans-Heinz Augenstein	Halifax	10km ENE Deventer	01.37	5000	6	W
12./NJG1	Lt Heinz Grimm	Halifax	3km W Eosterland	01.44	2400	15	W
10./NJG1	Hptm Rudolf Sigmund	Lancaster	25km SW Leeuwarden	01.45	5600	15	W
7./NJG1	Lt Werner Rapp	Halifax	6km WSW Emmen	01.47	4500	6	W
2./NJG1	Lt Heinz Struning	Mosquito	15km SE Gouda	01.54	5400	29	W
3./NJG1	Lt Walter Schon	Lancaster	3km NNE Leiden	02.09	6000	5	W
12./NJG1	Lt Heinz Grimm	Wellington	35km NW Den Helder	02.12	2800	16	W

29-30 May 1943

2./NJG1	Lt Schmidt	Halifax	Heer	00.07	5200	1	W
3./NJG1	Lt Wilhelm Beier	Halifax	20km NNW Maastricht	00.27	5000	38	W
StII/NJG1	Oblt Wilhelm Telge	Halifax	Leopoldsburg	00.28	5700	4	W
2./NJG1	Fw Heinz Oloff	Wellington	20km SSE Amsterdam	00.29	2000	4	W
3./NJG1	Oblt Manfred Meurer	Halifax	5km NNE Roermond	00.33	6200	34	W
StII/NJG1	Maj Walter Ehle	Halifax	9km SE Aachen	00.37	5600	23	W
6./NJG1	ObLt E-W von Bonin	Wellington		00.43	5400	14	W
StII/NJG1	Lt H-W. Schnaufer	Stirling	S Baelen	00.48	3500	10	W
3./NJG1	Oblt Manfred Meurer	Wellington	12km NE Roermond	00.50	5400	35	W
StII/NJG1	Maj Walter Ehle	Halifax	6km NE Jülich	01.05	4700	24	W
3./NJG1	Oblt Manfred Meurer	Stirling	SE Roermond	01.09	2500	36	W
5./NJG1	Hptm Hoffmann	Halifax		01.40	4500	1	W
StII/NJG1	Lt H-W. Schnaufer	Halifax	Budingen	01.43	6500	11	W
3./NJG1	Oblt Manfred Meurer	Lancaster	20km W Geldern	01.44	4300	37	W
1./NJG4	Oblt Ludwig Meister	Stirling	3km SW Cambrai	02.07	3100	12	W
5./NJG1	Ofw Fritz Schellwat	Lancaster	2km E Hachtel	02.18	4200	11	W
StII/NJG1	Lt H-W. Schnaufer	Stirling	500m E Schaffen AD	02.22	4500	12	W
2./NJG4	Oblt Rudolf Altendorf	Wellington	E Geraadsbergen	02.28	3400	13	W
4./NJG1	Lt Johannes Hager	Stirling	E Aachen	02.28	4400	2	W
5./NJG1	Ofw Fritz Schellwat	Lancaster	Leer	02.37	5100	12	W

1./NJG1	Lt Rolf Bussmann	Halifax	W Vlissinger	02.50	3700	14	W
2./NJG4	Oblt Rudolf Altendorf	Halifax	Rassaix	02.57	4300	14	W

30-31 May 1943

10./NJG1	Ofw Dörrer	Lancaster		20.59		1	W

3-4 June 1943

4./NJG2	Ofw Detering	Swordfish	Mediterranean	03.19	100	1	M

10-11 June 1943

11./NJG5	Fw Karl Maisch	Whitley	SW St.Meine	01.14	3000	1	W
St./NJG1	Hptm Gunter Friedrich	Four-eng		01.22		2	W

11 June 1943

10./NJG1	Uffz Karl Pfeiffer	B-17		19.35	4300	5	W

11-12 June 1943

10(N)./ZG1	Oblt Janke	TB-7	Russia	21.20		1	E
1./NJG1	Ofw Reinhard Kollak	Halifax	S Haamstede	00.41	6200	20	W
1./NJG1	Ofw Reinhard Kollak	Wellington	N Westkapelle	01.00	5400	21	W
StI/NJG1	Maj Werner Streib	Halifax	14km SE Roermond	01.05	4700	51	W
1./NJG1	Lt Werner Baake	Wellington	1km N Neeroeteren	01.10	5500	1	W
4./NJG1	Oblt Walter Barte	Lancaster		01.12	6500	11	W
3./NJG1	Hptm Manfred Meurer	Wellington	Coesfeld	01.14	5300	38	W
2./NJG1	Ofw Bruno Eikmeier	Wellington	2km SW Vechel	01.17	4600	4	W
1./NJG5	Hptm Siegfried Wandam	Halifax	10km S Jülich	01.18	4600	7	W
StI/NJG1	Maj Werner Streib	Halifax	2km SW Rheinberg	01.20	5800	52	W
1./NJG1	Ofw Reinhard Kollak	Lancaster	3km W Poortvliet	01.24	6000	22	W
StII/NJG1	Oblt Wilhelm Telge	Stirling	3km S Aubel	01.35	3600	5	W
3./NJG1	Hptm Manfred Meurer	Halifax	Zutphen	01.40	4200	39	W
6./NJG1	Hptm E-W. von Bonin	Halifax	Oyeplage, NE Luttich	01.47	5700	15	W
4./NJG1	Oblt Walter Barte	Halifax	4km S Bladel	01.52	6200	12	W
3./NJG1	Hptm Manfred Meurer	Halifax	7km W Xanten	01.55	4800	40	W
StI/NJG1	Maj Werner Streib	Halifax	3km N Mook	01.55		53	W
./NJG1	Oblt Helmut Bergmann	Lancaster	Zandbergen	02.02	5000	13	W
2./NJG1	Fw Heinz Oloff	Lancaster	5km N Gouda	02.05	5900	5	W
StI/NJG1	Maj Werner Streib	Lancaster	18km SW Nijmegen	02.16		54	W
10./NJG1	Ofw Karl-Heinz Scherfling	Lancaster	50km W Bergenlaan Ze	02.19	3300	12	W
StI/NJG1	Maj Werner Streib	Halifax	3km W Sambeek	02.22	5900	55	W
1./NJG1	Lt Werner Baake	Halifax	6km N Gorinchem	02.43	5000	2	W
7./NJG1	Oblt Werner Rapp	Lancaster	5km NE Appeldoorn	02.44	6000	7	W
12./NJG1	Lt Heinz Grimm	Wellington	50km NW Vlieland	02.46	3700	17	W
10./NJG1	Uffz Karl Pfeiffer	Wellington	Schager	02.49	4500	6	W
St ./NJG4	Maj Kurt Holler	Wellington	W Urk	02.55		13	W
./NJG1	Oblt Werner Husemann	Wellington		02.55	3300	3	W
7./NJG1	Lt Hans-Heinz Augenstein	Lancaster	8km NE Deventer	03.00	6200	7	W
St ./NJG4	Maj Kurt Holler	Stirling	25km SW Den Helder	03.04		14	W
10./NJG1	Ofw Karl-Heinz Scherfling	Lancaster	40km W Alkmaar	03.06	4200	13	W

Hauptmann Eckardt-Wilhelm von Bonin of 5./NJG 1 scored his 15th combat victory on 11/12th June by shooting down Halifax JD168 of 77 Squadron at Oyeplage (left). One man from Sergeant Endicott's crew died.

12-13 June 1943

10(N)./ZG1	Oblt Egon Mangold	TB-4	Russia	20.47		3	E
10(N)./ZG1	Oblt Egon Mangold	DB-3F	Russia	21.20		4	E
12./NJG1	Lt Heinz Grimm	Lancaster	Vollenhoven	01.05	6100	18	W
3./NJG1	Hptm Manfred Meurer	Lancaster	Beaumetz	01.14	6200	41	W
III./NJG1	Hptm E. Lippe-Weissenfeld	Lancaster	3km NW Burgsteinfurt	01.22	5000	40	W
10./NJG1	Hptm Rudolf Sigmund	Lancaster	Grafhorst	01.23	5400	16	W
8./NJG1	Oblt Dietrich Schmidt	Lancaster	1km W Ahaus	01.30	6200	5	W
StI/NJG1	Maj Werner Streib	Lancaster	45km S Doetinchen	01.30		56	W
3./NJG3	Ofw Willi Schmale	Halifax	1km N Burgsteinfürt	01.31	6500	3	W
III./NJG1	Hptm E. Lippe-Weissenfeld	Lancaster	6km N Nienberg	01.34	5400	41	W
11./NJG1	Maj Rolf Leuchs	Halifax	50km NW Den Helder	01.35	3200	2	W
3./NJG3	Ofw Willi Schmale	Halifax	1km W Metelen	01.35		4	W
7./NJG1	Oblt Werner Rapp	Lancaster	6km E Hoogeveen	01.50	6300	8	W
7./NJG1	Lt Hans-Heinz Augenstein	Halifax	5km W Gildehaus	01.56	6200	8	W
7./NJG1	Lt Hans-Heinz Augenstein	Halifax	N-E Eldenzaal	02.03	6200	9	W
8./NJG1	Oblt Rolf Bokemeyer	Halifax	Rheine	02.04	5800	3	W
7./NJG1	Oblt Werner Rapp	Halifax	5km SE Meppen	02.06	6000	9	W
10./NJG1	Hptm Rudolf Sigmund	Lancaster	35km NW Vlieland	02.11	5300	17	W
11./NJG1	Maj Rolf Leuchs	Halifax	Texel	02.13	3500	3	W
12./NJG1	Oblt Ernst-Georg Drunkler	Lancaster	N Schagen	02.20	1600	1	W
./NJG1	Oblt Werner Husemann	Lancaster	Tubbergen	02.20	1100	4	W
1./NJG1	Ofw Hermann Sommer	Lancaster	W Zandvoort	02.20	5800	15	W
./NJG1	Oblt Helmut Bergmann	Lancaster	4km SE Tubergen	02.21	6500	14	W
1./NJG1	Ofw Hermann Sommer	Lancaster	W Zandvoort	02.30	4800	16	W
7./NJG1	Oblt August Geiger	Lancaster	7km NNE Raalte	02.37	4300	30	W
11./NJG1	Hptm Hans-Joachim Jabs	Lancaster	10km W Zandvoort	02.39	6200	12	W
12./NJG1	Uffz Georg Kraft	Lancaster	22km NE Den Helder	02.50	3700	11	W
12./NJG1	Oblt Ernst-Georg Drünkler	Lancaster	45km W Alkmaar	02.51	5500	2	W
11./NJG1	Hptm Hans-Joachim Jabs	Lancaster	50km W Zandevoort	03.04	6500	32	W
10./NJG1	Ofw Karl-Heinz Scherfling	Halifax	35km SW Den Helder	03.35	6500	14	W

Many Luftwaffe nightfighter pilots were, by mid-1943, building up impressive victory totals, but none more so than Major Helmut Lent, Kommandeur IV./NJG 2. He had been a Bf110 pilot with 3./ ZG 76 and had taken part in the Wellington slaughter of 18th December 1939. By the time he joined the nightfighter force in June 1940, had at least seven daylight victories credited. On the night of 25/26th June 1943 he claimed a Wellington west of Urk. He had been awarded the Ritterkreuz in August 1941, the Eichenlaub in June 1942 and further high honours were to follow.

13 June 1943

3./NJG3	Ofw Willi Schmale	B-17	Oldenbrock	09.57	4000	4	W
6./NJG3	Oblt Willi Elstermann	B-17	7km SW Tideroog	10.12	8000	2	W

14-15 June 1943

2./NJG1	Uffz Rudolf Frank	Lancaster	8km N Antwerp	00.53	6000	9	W
2./NJG1	Hptm Hans-Dieter Frank	Lancaster	14km SE s'Hertog'bosch	00.59	6200	31	W
3./NJG1	Hptm Manfred Meurer	Lancaster	Hushoven	01.02	6500	42	W
3./NJG1	Hptm Manfred Meurer	Lancaster	ESE Sittard	01.11	6500	43	W
StII/NJG1	Oblt Wilhelm Telge	Lancaster	23km NE Hasselt	01.12	6100	6	W
2./NJG1	Hptm Hans-Dieter Frank	Lancaster	1km S Hien	01.13	6400	33	W
3./NJG1	Hptm Manfred Meurer	Lancaster	4km WSW Bracht	01.15	6500	44	W
StII/NJG3	Maj Günther Radusch	Lancaster	10km S Venlo	01.15	6100	15	W
StII/NJG3	Maj Günther Radusch	Lancaster	20km WSW Venlo	01.20	6800	16	W
5./NJG1	Ofw Fritz Schellwat	Lancaster	Nr. Düsseldorf	01.32	6000	13	W
7./NJG1	Oblt August Geiger	Lancaster	7km SE Terlet	01.44	2000	31	W
3./NJG1	Hptm Wilhelm Herget	Lancaster	1km N Oijen	01.55	7000	17	W
11./NJG5	Hptm Walter Fenske	Stirling	12km SE Rennes	02.15	2500	10	W
2./NJG1	Hptm Hans-Dieter Frank	Lancaster	Nr. Amsterdam	02.20	6300	34	W

15-16 June 1943

10(N)./ZG1	Ofw Müller	SB-2	Russia	20.52		1	E

16-17 June 1943

10(N)./ZG1	Ofw Josef Kociok	Boston	Russia	21.00	4500	11	E
10(N)./ZG1	Ofw Josef Kociok	Boston	Russia	21.20	3000	12	E
2./NJG1	Uffz Rudolf Frank	Lancaster	Lierre	00.45	5800	10	W
3./NJG1	Hptm Manfred Meurer	Lancaster	SW Ndr. Krüchen	00.55	6400	45	W
1./NJG1	Lt Werner Baake	Lancaster	Vechel	00.59	7000	3	W
2./NJG1	Uffz Rudolf Frank	Lancaster	Moorleghem	01.00	50	11	W
StII./NJG1	Oblt Wilhelm Telge	Lancaster	4km SW Meeuwen	01.13		7*	W
StII/NJG1	Maj Walter Ehle	Lancaster	SSW Jülich	01.18	5000	25	W
3./NJG1	Hptm Manfred Meurer	Lancaster	2km W Braunsrath	01.20	5900	46	W
3./NJG1	Hptm Heinrich Wohlers	Lancaster	1.5km S Broich	01.20	4500	13	W
2./NJG1	Oblt H-M. Hadeball	Beaufighter		01.25	5400	12	W
3./NJG1	Hptm Manfred Meurer	Lancaster	S Boerdonk	01.41	6700	47	W

Another victim to Oberleutnant Wilhelm Telge of Stab II./NJG 1. This was ED861, a Lancaster of 57 Squadron that crashed near Lantin on 24/25th June. One man from Sergeant Fellows's crew survived to be captured. (E.Nonnenmacher)

3./NJG1	Maj Günther Radusch	Lancaster	NE Neerpelt	01.45	6500	17	W
2./NJG1	Uffz Rudolf Frank	Lancaster	NE Breskens	01.46	7800	12	W
5./NJG5	Uffz Herbert Hubatsch	Lancaster	Vorst	01.51	6500	2	W
3./NJG1	Hptm Manfred Meurer	Lancaster	4km N Bortel	01.55	7400	48	W
2./NJG1	Lt Heinz Strüning	Halifax	10km NW Euddorp	02.06	6600	30	W
StI/NJG1	Oblt Rath	Lancaster		02.08	6200	1	W

19-20 June 1943

11./NJG5	Oblt Gollasch	Lancaster	30km W Angers	02.01	2800	5	W

20-21 June 1943

NJS LF6	Oblt Günter Bertram	Il-4	Russia	23.30		10*	E

21-22 June 1943

6./JG54	Ofw Alfred Humann	MBR-2	Russia	23.52	20	2	E
StII/NJG1	Lt H-W. Schnaufer	Stirling	1km NE Aerschot	01.13	4300	13	W
2./NJG1	Hptm Walter Milius	Lancaster	12km SW Tilburg	01.14	4200	6	W
St ./NJG4	Maj Kurt Holler	Wellington	W Stavenisse	01.18	6000	15	W
2./NJG1	Hptm Hans-Dieter Frank	Lancaster	36km NE Eindhoven	01.23	6000	35	W
3./NJG1	Hptm Manfred Meurer	Lancaster	2km W Oeffelt	01.23	6100	49	W
2./NJG1	Uffz Rudolf Frank	Wellington	Gramersweil	01.24	5700	13	W
3./NJG1	Hptm Wilhelm Herget	Lancaster	2km S Asten	01.27	6700	18	W
1./NJG1	Lt Werner Baake	Wellington	3km SW Bergeyk	01.30	5000	4	W
StI/NJG1	Maj Werner Streib	Stirling	35km NW Venlo	01.30	7000	57	W
1./NJG1	Oblt Gerhard Raht	Stirling	12km W Antwerp	01.32	3800	5	W
3./NJG1	Maj Günther Radusch	Halifax	16km NNW Venlo	01.34	5500	18	W
1./NJG5	Hptm Siegfried Wandam	Wellington	4km N Roermond	01.34	5400	8	W
1./NJG1	Lt Werner Baake	Halifax	5km NNE Bergeyk	01.37	3000	5	W

2./NJG1	Hptm Hans-Dieter Frank	Halifax	500m E Kaakhoven	01.37	5500	36	W
3./NJG1	Hptm Wilhelm Herget	Wellington	20km. WSW Venlo	01.38	5200	19	W
2./NJG1	Lt Wilhelm Johnen	Halifax	5km E Rousendaal	01.43	5300	3	W
1./NJG5	Hptm Siegfried Wandam	Stirling	500m E Achel	01.45	3600	9	W
2./NJG1	Hptm Hans-Dieter Frank	Halifax	1km W Vechel	01.48	6400	37	W
3./NJG1	Hptm Manfred Meurer	Halifax	½ km SW Wamel	01.48	5500	50	W
6./NJG1	Hptm E-W. Bonin.von	Stirling	18km SE Eindhoven	01.50	2300	16	W
2./NJG1	Lt Heinz Strüning	Lancaster		02.02	5200	31	W
2./NJG1	Hptm Hans-Dieter Frank	Halifax	4km NW Boxtel	02.09	6600	38	W
3./NJG1	Hptm Heinrich Wohlers	Halifax	6km E Utrecht	02.12	5000	14	W
3./NJG1	Hptm Wilhelm Herget	Lancaster	Leitler-Heide	02.15	5900	20	W
12./NJG1	Hptm Franz Buschmann	Stirling	15km NW Den Helder	02.20	3000	2	W
2./NJG1	Hptm Paul Förster	Wellington	2km N Eudenbosch	02.21	3900	1	W
2./NJG1	Hptm Hans-Dieter Frank	Halifax	2km SW Zeist	02.24	5500	39	W
2./NJG1	Hptm Hans-Dieter Frank	Halifax	½km N Yaarsfeld	02.30	4800	40	W
3./NJG1	Fw Heinz Wilberg	Halifax	Harskamp	02.30	5400	2	W
2./NJG1	Lt Wilhelm Johnen	Wellington	NW Haamstede	02.33	1500	4	W
10./NJG1	Hptm Rudolf Sigmund	Wellington	25km W Den Helder	02.35	5400	18	W
12./NJG1	Oblt Ernst-Georg Drünkler	Stirling	1½km W Makkum	02.39	5000	3	W
StII/NJG1	Maj Walter Ehle	Halifax	4km NW Bortel	02.39	5000	26	W
10./NJG1	Hptm Rudolf Sigmund	Lancaster	60km W Egmont	02.40	4000	19	W
11./NJG1	Hptm Hans-Joachim Jabs	Stirling	8km WSW Hoorn	02.44	1800	14	W
10./NJG1	Hptm Rudolf Sigmund	Wellington	60km W Egmont	02.49	4000	20	W
StIV/NJG1	Maj Helmut Lent	Lancaster	70km W Vlieland	03.09	3200	64	W
10./NJG1	Hptm Rudolf Sigmund	Halifax	140km W Den Helder	03.24	2500	21	W

22-23 June 1943

1./NJG1	Lt Werner Baake	Lancaster	15km SSE Utrecht	00.55	6800	6	W
12./NJG1	Uffz Georg Kraft	Lancaster	Utrecht	00.55		13	W
12./NJG1	Uffz Georg Kraft	Halifax	1km E Zutphen	01.20	5000	14	W
10./NJG1	Uffz Karl Pfeiffer	Halifax	2km NE Scherwerhorn	01.22	5400	7	W
7./NJG1	Oblt August Geiger	Wellington	9km N Wesel	01.33	5600	32	W
6./NJG4	Oblt Hans Autenrieth	Stirling	2km NW Kettinich	01.35	2800	12	W
7./NJG1	Oblt August Geiger	Wellington	3km N Rees	01.35	5800	33	W
3./NJG1	Hptm Heinrich Wohlers	Stirling	2km S Appeldorn	01.35	2800	15	W
1./NJG1	Lt Werner Baake	Lancaster	W Nijmegen	01.38	6000	7	W
./NJG1	Hptm Günter Friedrich	Halifax	2km E Zutphen	01.40	5700	3	W
1./NJG1	Lt Werner Baake	Halifax	6km NW Utrecht	01.58	5500	8	W
3./NJG1	Hptm Wilhelm Herget	Halifax	1km NW Vechel	02.01	5300	21	W
StIV/NJG1	Maj Helmut Lent	Halifax	7km W Urk	02.06	300	65	W
9./NJG1	Hptm Wilhelm Dormann	Stirling	5km SW Hoenderloo	02.07	2800	11	W
1./NJG1	Hptm H. Sayn-Wittgenstein	Stirling		02.09	4400	28	W
6./NJG4	Oblt Hans Autenrieth	Wellington	12km NE Hasselt	02.10	3800	13	W
3./NJG1	Hptm Heinrich Wohlers	Halifax	10km E s'Hertogenbosch	02.12	3000	16	W
11./NJG1	Hptm Hans-Joachim Jabs	Stirling	12km NW Bergen	02.19	2700	15	W
11./NJG1	Fw Heinz Vinke	Stirling	8km E Eosterland	02.26	2400	27	W
1./NJG1	Oblt Viktor Sorko	Stirling		02.29	3200	2	W
StII/NJG1	Oblt Wilhelm Telge	Halifax	W Mechelen	02.30	5200	8	W
2./NJG1	Ofw Bruno Eikmeier	Lancaster		02.36	3400	5	W
2./NJG1	Lt Heinz Strüning	Halifax	16km E Utrecht	02.39	4000	32	W

The pathetic remains of Pilot Officer Tay's Halifax HR838, L-London, of 51 Squadron. This went down to Leutnant Franz Günther of 5./NJG 1 east-south-east of Antwerp at 02.22 hours on 28/29th June. Just twelve minutes earlier, Franz had claimed a Stirling at Netersel.

III./NJG1	Hptm E. Lippe-Weissenfeld	Stirling	2km S Markelo	02.47	4300	42	W
III./NJG1	Hptm E. Lippe-Weissenfeld	Stirling		02.55		43	W
1./NJG1	Oblt Viktor Sorko	Halifax		03.25	200	3	W

23-24 June 1943

St./NJG1	Oblt Werner Husemann	Halifax	6km NNW Amsterdam	03.17	50	5	W

24-25 June 1943

10(N)./ZG1	Ofw Müller	DB-3	Russia	21.18		2	W
NJS LF6	Oblt Günter Bertram	Li-2	Russia	22.47		11*	E
NJS LF6	Oblt Günter Bertram	An-6	Russia	22.59		12*	E
1./NJG1	Oblt Gerhard Raht	Lancaster	W Woensdrecht	00.25	3100	8	W
StI/NJG1	Maj Werner Streib	Halifax	Kempen	00.51	6000	58	W
StI/NJG1	Maj Gunther Radusch	Wellington	Wellen	00.57	5900	19	W
2./NJG1	Lt Wilhelm Johnen	Halifax	3km N Rosenburg	01.01	3800	5	W
2./NJG1	Hptm Hans-Dieter Frank	Halifax	Acht	01.05	3800	41	W
3./NJG1	Hptm Heinrich Wohlers	Halifax	1km N Welldorf	01.07	4500	17	W
1./NJG1	Lt Werner Baake	Wellington	Driel	01.10	6000	9	W
7./NJG4	Ofw Reinhard Kollak	Halifax	2km S Roermond	01.11	3200	23	W
StI/NJG1	Maj Günther Radusch	Wellington	Jasmund	01.15	5700	20	W
StII/NJG1	Oblt Wilhelm Telge	Lancaster	6km NE Lüttich	01.20	6200	9	W
4./NJG5	Hptm Werner Hoffmann	Lancaster	3km SW Erkelenz	01.21		5*	W
4./NJG5	Hptm Werner Hoffmann	Lancaster	PQ 6278	01.24	5300	6	W
StI/NJG1	Maj Günther Radusch	Lancaster	NW Venlo	01.24	5500	21	W
2./NJG1	Hptm Walter Milius	Halifax	1km N Zetten	01.30	1800	7	W
7./NJG4	Ofw Reinhard Kollak	Stirling	9km SS- Leuven	01.39	4100	24	W

5./NJG5	Uffz Herbert Hubatsch	Stirling	9km E Honnef	01.50	4800	3	W
4./NJG5	Hptm Werner Hoffmann	Wellington	1km S Brasel	01.54	4700	7	W
6./NJG4	Oblt Hans Autenrieth	Stirling	4km WSW Diest	01.59	4000	14	W
StI/NJG4	Lt Fritz Graef	Halifax	6km NE Barvaux	02.17	5700	4	W
1./NJG1	Hptm H. Sayn-Wittgenstein	Lancaster	30km W Katwijk-am-S	02.25	3200	29	W
2./NJG1	Lt Heinz Strüning	Lancaster	30km W Schouwenh	02.49	4800	33	W
1./NJG1	Hptm H. Sayn-Wittgenstein	Stirling	40km W Scheveningen	02.50	250	30	W
2./NJG1	Lt Heinz Struning	Halifax	20km WNW Schouwen	02.56	2300	34	W
StII/NJG1	Lt H-W. Schnaufer	Wellington	11km S Lowen	02.58	4700	14	W
III./NJG4	Oblt Helmut Bergmann	Stirling	N point of Walcheren	03.02		15	W
1./NJG1	Ofw Kurt Karsten	Stirling		03.02	3300	4	W
2./NJG1	Lt Heinz Strüning	Stirling	25km W Schouwen	03.02	2500	35	W
1./NJG1	Hptm H. Sayn-Wittgenstein	Stirling	20km W Katwijk-aan-Z	03.09	850	31	W
1./NJG1	Hptm H. Sayn-Wittgenstein	Wellington		03.35	200	32	W

25-26 June 1943

11./NJG1	Maj Rolf Leuchs	Halifax	1.5km S Bergen-a-S	00.51	5500	4	W
10./NJG1	Ofw Karl-Heinz Scherfling	Lancaster	20km N Amsterdam	00.54	6400	15	W
2./NJG1	Oblt H-M. Hadeball	Hurricane	Westkapelle	01.00	2300	13	W
2./NJG5	Oblt Hans Leickhardt	Lancaster	20km NNW H'derwijk	01.06	4000	1	W
./NJG1	Oblt Werner Husemann	Halifax	E Zevenaar	01.13	6000	6	W
StIV/NJG1	Maj Helmut Lent	Wellington	W Urk	01.13	6000	66	W
./NJG1	Oblt Werner Husemann	Stirling	15km WNW W'terwijk	01.17		7	W
1./NJG4	Oblt Ludwig Meister	Stirling	3km SW Aalter	01.23	3200	13	W
7./NJG1	Oblt August Geiger	Stirling	Empe	01.26	5600	34	W
9./NJG1	Hptm Wilhelm Dormann	Halifax	7km NNW Coesfeld	01.28	5800	12	W
StI/NJG1	Maj Günther Radusch	Halifax	20km S Nijmegen	01.58	5700	22	W
7./NJG1	Oblt August Geiger	Lancaster	10km SW Steenwilk	02.00	4200	35	W
9./NJG1	Hptm Wilhelm Dormann	Halifax	Gersel	02.01	6000	13	W
7./NJG1	Oblt August Geiger	Lancaster	25km W Volendam	02.09	5600	36	W
11./NJG5	Oblt Gollasch	Lancaster	18km NW Augers	02.11	3200	6	W
12./NJG1	Hptm Franz Buschmann	Lancaster	20km NE Harderwijk	02.17	3000	3	W
2./NJG1	Hptm Hans-Dieter Frank	Wellington	14km NNW Gouda	02.17	5300	42	W
./NJG1	Oblt Werner Husemann	Lancaster	13km S Haarlem	02.19		8	W
10./NJG1	Hptm Rudolf Sigmund	Stirling	50km WNW Den H'der	02.22	3000	22	W
12./NJG1	Uffz Baldischweiler	Stirling	8km E Texel	02.25	2300	2	W
10./NJG1	Hptm Rudolf Sigmund	Stirling	3km S Hoorn	02.31	6000	23	W
11./NJG1	Hptm Hans-Joachim Jabs	Halifax	NW Alkmaar	02.35	5400	16	W
2./NJG1	Hptm Hans-Dieter Frank	Wellington	11km ENE Gouda	02.37	3800	43	W
11./NJG1	Oblt Martin Drewes	Lancaster	S Texel	02.40	5000	5	W
11./NJG1	Hptm Hans-Joachim Jabs	Halifax	EK5-8	02.40	4200	17	W
10./NJG1	Hptm Rudolf Sigmund	Wellington	20km ESE Den Helder	02.43	4800	24	W
12./NJG1	Hptm Franz Buschmann	Halifax	60km W Den Helder	02.45	900	4	W
11./NJG1	Fw Heinz Vinke	Lancaster	50km W Den Helder	02.47	2400	28	W
10./NJG1	Ofw Karl-Heinz Scherfling	Lancaster	W Wijklaan see	02.51	5500	16	W
11./NJG1	Hptm Hans-Joachim Jabs	Lancaster	50km WNW Den H'der	02.56	5000	18	W
10./NJG1	Hptm Rudolf Sigmund	Lancaster	NW Makkum	03.01	4300	25	W
10./NJG1	Ofw Karl-Heinz Scherfling	Lancaster	15km S Alkmaar	03.08	4000	17	W

Ostfront Experten. Officially credited with 44 victories, Alois Lechner (far right), after serving with NJG 2 in the west, went to the eastern front, where he would be credited with 36 Russian aircraft destroyed, bring his tally to 45 before being killed in action. Josef Kociok flew with 4./SKG 210 and then moved to 4(N)./ZG 1, achieving a total of 33 victories, including 21 at night. He would also be killed in action, when flying with NJG 200 in September.

27-28 June 1943

4./NJG2	Fw Fritz Hobusch	Wellington	KK8-3	01.42	1000	5	W
11./NJG5	Oblt Gollasch	Lancaster	25km WNW Angers	02.27	2800	7	W

28-29 June 1943

StII/NJG1	Lt H-W. Schnaufer	Lancaster	9km SE Verviers	01.25	5900	15	W
4./NJG1	Lt Johannes Hager	Halifax	1km NW Maastricht	01.27	5600	3	W
3./NJG1	Hptm Wilhelm Herget	Lancaster	1½km SW Roermond	01.30	5400	22	W
4./NJG3	Hptm Hans Baer	Wellington	6km SE Julich	01.32	5800	6	W
4./NJG1	Lt Johannes Hager	Halifax	1km E Volkenburg	01.35	5200	4	W
StII/NJG1	Lt H-W. Schnaufer	Halifax	7km NE Lüttich	01.45	5300	16	W
StII/NJG1	Lt H-W. Schnaufer	Halifax	Nr. Volken	01.55	5300	17	W
5./NJG1	Ofw Fritz Schellwat	Halifax	8km W Euskirchen	02.10	6200	14	W
5./NJG1	Lt Günter Franz	Stirling	Netersel	02.14	4600	1	W
StI/NJG1	Maj Günther Radusch	Halifax	Waalre	02.15	7100	23	W
3./NJG1	Hptm Heinrich Wohlers	Halifax	S Volkel	02.17	5500	18	W
4./NJG5	Hptm Werner Hoffmann	Stirling	3km NW Lommel	02.18	2000	8	W
1./NJG3	Ob;t Paul Zorner	Wellington	SE Antwerp	02.20		7*	W
5./NJG1	Lt Günter Franz	Halifax	30km ESE Antwerp	02.22	3900	2	W
1./NJG3	Oblt Paul Zorner	Wellington	9km NE Lowen	02.22	4900	8	W
StI/NJG1	Maj Günther Radusch	Halifax	7km S Eindhoven	02.24	6200	24	W
StII/NJG1	Maj Walter Ehle	Stirling	40km WSW Diest	02.33	4200	27	W
StI/NJG1	Maj Günther Radusch	Halifax	5km N Helmond	02.34	4800	25	W
10./NJG1	Ofw Karl-Heinz Scherfling	Lancaster	5km W Zandvoort	02.36	2400	18	W
3./NJG1	Hptm Heinrich Wohlers	Stirling	12km W Utrecht	02.40	2300	19	W
2./NJG1	Hptm Hans-Dieter Frank	Lancaster	12km SE Rotterdam	02.47	3700	44	W
2./NJG1	Lt Werner Hopf	Stirling		02.47	4200	2	W
2./NJG1	Fw Heinz Oloff	Lancaster	8km E Maastricht	02.53	5500	6	W
StI/NJG1	Maj Günther Radusch	Lancaster	5km NW Eindhoven	02.56	6700	26	W
./NJG1	Oblt Werner Husemann	Lancaster		03.07	2300	9	W

1./NJG1	Oblt Viktor Sorko	Lancaster		03.26	4000	4	W
10./NJG1	Ofw Karl-Heinz Scherfling	Lancaster	Wognum	24.00	2400	19	W

30 June-1 July 1943

3./JG302	Uffz Werner Dienst	Halifax	W Munich	00.52	4000	1	W

1-2 July 1943

10(N)./ZG1	Ofw Josef Kociok	PS-84	SE Taman	21.00	1000	13	E

3-4 July 1943

2./NJG1	Oblt H-M. Hadeball	Stirling	1km NW Zaamslag	00.46	3500	14	W
StII/NJG1	ObLt H-W. Schnaufer	Wellington	1km N Grooten Haak	00.48	4500	18	W
1./NJG5	Hptm Siegfried Wandam	Wellington	8km W St.Truiden	00.48	5400	10	W
7./NJG4	Ofw Reinhard Kollak	Wellington	9km N Liege	00.55	5500	25	W
4./NJG1	Lt Johannes Hager	Wellington	SE Liege	01.01	5600	5	W
StII/NJG1	ObLt H-W. Schnaufer	Stirling	Nr. Geetbets	01.01	4400	19	W
2./NJG1	Uffz Rudolf Frank	Halifax	Huissen	01.06	5300	14	W
2./NJG1	Hptm Walter Milius	Halifax	Aerschot 9 km N Diest	01.08	4200	8	W
1./NJG3	Oblt Paul Zorner	Wellington	SW St.Truiden	01.09	5600	9	W
4./NJG3	Hptm Hans Baer	Wellington	N Jackerath	01.18	1500	7	W
7./NJG4	Ofw Reinhard Kollak	Halifax	5km NW Liege	01.28	5500	26	W
KdoHerr	Oblt F-K. Muller	Halifax	Nr. Cologne	01.28	4500	1	W
5./NJG3	Oblt Paul Szameitat	Stirling	30km NW Terschelling	01.28	700	12	W
1./NJG5	Hptm Siegfried Wandam	Halifax	10km NE Tongeren	01.29	5300	11	W
KdoHerr	Maj Hajo Herrmann	Lancaster	Nr. Mehlen	01.30		1	W
KdoHerr	Fw Ernst Muller	Four-eng	SE Cologne	01.31	3500	1	W
4./NJG1	Lt Johannes Hager	Wellington	NE Malmedy	01.35	5300	6	W
5./NJG5	Uffz Herbert Hubatsch	Stirling	Winterscheid	01.46	5500	4	W
KdoHerr	Ofw Heinrich Lonnecker	Four-eng	Cologne	01.55	5000	1	W
KdoHerr	Hptm Ewald Janssen	Four-eng	Cologne	02.05	5000	1	W
KdoHerr	Ofw Hofer	Four-eng	Koln area	02.05		1*	W
12./NJG1	Hptm Franz Buschmann	Wellington	30km W Terschelling	02.22	1000	5	W
StI ./NJG4	Hptm Wilhelm Herget	Wellington	Solre-sur-Sambre	02.23	4000	23	W
7./NJG4	Ofw Reinhard Kollak	Halifax	22km SE Brussels	02.30	5100	27	W
1./NJG4	Oblt Ludwig Meister	Halifax	Nr. Rance	02.40	4800	14	W
5./NJG1	Oblt Wilhelm Telge	Lancaster	ESE Namur	02.42	6100	10	W
3./NJG4	Oblt Fritz Söthe	Wellington	10km E Avernes	02.45	5000	1	W
5./NJG1	Oblt Wilhelm Telge	Lancaster	NW Namur	03.05	4500	11	W
9./NJG4	Uffz Dichaus	Halifax	1km E Toulaine	03.08	2800	1	W

4-5 July 1943

12./NJG5	Oblt Hans Gref	PS-84	E Karatschev	00.16	2400	3	E

5-6 July 1943

11./NJG1	Oblt Hermann Greiner	Stirling	20km NW Leeuwarden	02.31	900	7	W

8-9 July 1943

12./NJG4	Oblt H-M. Hadeball	Lancaster		00.40	6700	15	W
6./NJG1	Lt O-F. Erlinghagen	Lancaster	E Tongeren	01.05	5200	1	W
1./NJG3	Oblt Paul Zorner	Lancaster	W Malmedy	01.53	6500	10	W

5./NJG4	Hptm H. Niebelschutz.von	Lancaster	6km S Calamuer	02.20	6200	2	W
StII/NJG1	ObLt H-W. Schnaufer	Lancaster	Nr. Grobbendonk	02.33	6000	20	W
StI/NJG4	Oblt Fritz Graef	Lancaster	S Hirson	02.50		5	W

9-10 July 1943

4./KG3	Uffz Kuper	Pe-2	Russia	01.15		1	E
StI./NJG1	Hptm Hans-Dieter Frank	Lancaster	8km ENE Enschede	01.22		45*	W
7./NJG1	Oblt August Geiger	Halifax	Eprave, SE Dinant	02.40		37*	W
2./NJG4	Lt Wilhelm Schneiderwind	Lancaster	10km SW Schouwen	03.41	5000	1	W

11-12 July 1943

StIV/NJG5	Hptm H. Sayn-Wittgenstein	DB-3	Russia	21.50	2300	33	E
5./NJG2	Oblt Walter Riedlberger	Douglas.39	Russia	22.58	150	1	W

13-14 July 1943

10./NJG1	Ofw Karl-Heinz Scherfling	Halifax	4km SE Utrecht	01.23	5700	20	W
StI/NJG1	Hptm Hans-Dieter Frank	Halifax	1km S Helmond	01.26	5900	46	W
1./NJG1	Lt Werner Baake	Halifax	3km NW Utrecht	01.30	5000	10	W
StI/NJG1	Hptm Hans-Dieter Frank	Halifax	N Leuith	01.39	5400	47	W
1./NJG1	Lt Rolf Bussmann	Halifax	10km N Tilburg	01.40	6100	15	W
3./NJG1	Oblt Heinz Strüning	Halifax	Venlo railway stn	01.40	6400	36	W
5./NJG1	Oblt Wilhelm Telge	Halifax	NE Geilenkirchen	01.40	5700	12	W
IV./NJG4	Hptm Heinrich Wohlers	Stirling	2km NE Rothem	01.48	4000	20	W
5./NJG1	Oblt Wilhelm Telge	Halifax	Nr. Weissweiler	01.51	5500	13	W
1./NJG6	Ofw Richard Launer	Halifax	1km W Scharperweert	02.05		2	W
7./NJG1	Oblt August Geiger	Lancaster	Forstes-Veneurs	02.10		5*	W
3./NJG1	Lt Josef Nabrich	Halifax	25km ESE Dordrecht	02.00		38*	W
2./NJG4	Oblt Rudolf Altendorf	Halifax	10km S Mabeuge	02.19	5000	15	W
3./NJG4	Hptm Hubert Rauh	Lancaster	18km SW Cambrai	02.28	5400	6	W
1./NJG4	Oblt Ludwig Meister	Halifax	Vossigny la Victoire	02.30		15*	W
StI ./NJG4	Hptm Wilhelm Herget	Wellington	NW Givet	02.35	5200	24	W
2./NJG4	Oblt Rudolf Altendorf	Halifax	1NE Froid-Chapelle	02.36	5800	16	W
StI ./NJG4	Hptm Wilhelm Herget	Halifax	NE Givet	02.40	6000	25	W

14-15 July 1943

StIV/NJG5	Hptm H. Sayn-Wittgenstein	DB-3	E Etrada	21.47	4800	34	E
StIV/NJG5	Hptm H. Sayn-Wittgenstein	TB-7	W Poldsow	22.05	4800	35	E
10./NJG5	Hptm Alois Lechner	DB-3	Russia	01.13	2600	9	E

15-16 July 1943

5./NJG4	Hptm Pich	Halifax	500m E Sacquenay	01.30	2500	-	W
6./NJG4	Hptm Hubert Rauh	Halifax	500m E Sacquenay	01.30		7*	W
6./NJG4	Uffz Philipp	Stirling	Besancon	01.36	3000	1	W
5./NJG4	Fw Fritz Späte	Lancaster	Recey-sur-Ourthe	02.25	2000	1	W

16-17 July 1943

10./NJG5	Oblt Landau	R-5	Russia	21.54	800	1	E
10./NJG5	Oblt Landau	R-5	Russia	21.58	700	2	E
10./NJG5	Oblt Landau	R-5	Russia	00.12	1100	3	E
10./NJG5	Lt Josef Pützkuhl	R-5	Nr. Molkev	00.36	2000	2	E

17-18 July 1943

StIV/NJG5	Hptm H. Sayn-Wittgenstein	DB-3	30km NE Erel	21.26	2000	36	E
10./NJG5	Oblt Landau	R-5	Russia	22.16	1400	4	E
10./NJG5	Uffz Artenhauer	DB-3	Russia	22.55	2900	-	E
10./NJG5	Hptm Alois Lechner	E/A	Russia	00.18		10	E
10./NJG5	Egfr Christian Rohrs	R-5	Russia	00.18	1000	1	E
12./NJG5	Oblt Hans Gref	TB-3	Russia	00.36	1900	4	E
12./NJG5	Oblt Hans Gref	TB-3	Russia	00.58	2800	5	E
10./NJG5	Ofw Helmut Dahms	SB-	Russia	01.43	3300	1	E
10./NJG5	Hptm Godbersen	pe-2	Russia	01.54	1500	1	E

18-19 July 1943

12./NJG5	Lt Robert Weiss	DB-3	Russia	21.07	1800	1	E
StIV/NJG5	Hptm H. Sayn-Wittgenstein	Mitchell	Russia	21.39	2500	37	E
StIV/NJG5	Hptm H. Sayn-Wittgenstein	DB-3	Russia	21.47		38	E
./JG54	Oblt Jank	U-2	Russia	23.38	700	2	E
12./NJG5	Lt Robert Weiss	U-2	Russia	00.19	1200	2	E
12./NJG5	Lt Robert Weiss	PS-84	Russia	01.34		3	E

19-20 July 1943

StIV/NJG5	Hptm H. Sayn-Wittgenstein	DB-3	N Erel	21.10	3000	39	E
StIV/NJG5	Hptm H. Sayn-Wittgenstein	Boston	N Erel	21.20	3000	40	E
1./NJG100	Oblt Landau	E/A	Russia	21.47	1800	5	E
StIV/NJG5	Hptm H. Sayn-Wittgenstein	DB-3	N Erel	22.10	3500	41	E
10./NJG5	Lt Josef Pützkuhl	PS-84	Russia	22.31	2300	3	E
StIV/NJG5	Hptm H. Sayn-Wittgenstein	DB-3	N Erel	01.30	4000	42	E

20-21 July 1943

StIV/NJG5	Hptm H. Sayn-Wittgenstein	DB-3	15km N-E Erel	21.08	3500	43	E
StIV/NJG5	Hptm H. Sayn-Wittgenstein	Mitchell	10km N-E Erel	21.15	3500	44	E
StIV/NJG5	Hptm H. Sayn-Wittgenstein	Mitchell	Russia	21.20	3500	45	E
12./NJG5	Oblt Hans Gref	PS-84	Russia	21.30	1800	6	E
StIV/NJG5	Hptm H. Sayn-Wittgenstein	TB-7	10km N-E Erel	21.38	4000	46	E
StIV/NJG5	Hptm H. Sayn-Wittgenstein	TB-7	15km N-E Erel	21.50	4000	47	E
StIV/NJG5	Hptm H. Sayn-Wittgenstein	TB-7	10km N-E Erel	21.55	4000	48	E
11./NJG5	Fw Karl Maisch	DB-3	Russia	21.56	3000	2	E
10./NJG5	Fw Heintze	TB-3	Russia	22.24	1300	3	E
NJS LF6	Oblt Günter Bertram	Li-2	Russia	23.35		13*	E
1./NJG100	Hptm Kurt Bonow	R-5	Russia	00.18	2500	1	E
12./NJG5	Lt Waldemar Rathke	DB-3	Russia	01.24	2500	1	E
StIV/NJG5	Hptm H. Sayn-Wittgenstein	Mitchell	Erel	01.30	5000	49	E
12./NJG5	Lt Waldemar Rathke	DB-3	Russia	01.47	3700	2	E

21-22 July 1943

StIV/NJG5	Hptm H. Sayn-Wittgenstein	TB-7	Russia	22.25	4000	50	E
10./NJG5	Hptm Alois Lechner	DB-3	Russia	22.42	4000	11	E

Chapter Six

Hamburg and After

24th July - 17th November 1943

As early as mid-May 1943, Sir Arthur Harris had decided up the next step of the strategic bomber offensive. The great port and ship-building facility of Hamburg had survived, thus far, relatively untouched and the production of *U-boats* continued to worry the naval hierarchy. The new 'down-looking' H^2S radar, that supplied a moving picture of the ground features beneath the aircraft was, Harris believed, ideal for such a task, for the coastline would show up well and there were several prominent landmarks to aid bombing accuracy. However, the destruction of the industrial facilities was not to be the prime objective. Harris wanted the destruction of the city, together with the thousands of workers that provided the factories and shipyards with manpower.

Harris also had a 'secret weapon' to employ; metallic strips, cut to the precise wavelength of the *Lichtenstein SN-2* airborne radar, were to dropped by the thousand, to confuse the *Luftwaffe* radar operators by swamping their screen with false echoes.

The first attack took place on the night of 24th July, when 791 aircraft took part with a loss of only twelve aircraft. The metallic strips, code-named *Window*, had worked perfectly. Three nights later, 787 bombers returned, to add to the severe damage already caused and resulted shocked even the planners. A firestorm had developed, incinerating 16,000 apartment buildings and killing at least 40,000 people. Again, the losses to Bomber Command were low. Follow-up raids on 29th July and 2nd August were hardly necessary, for over 1,000,000 people had fled the city after the firestorm. However, the Germans had begun to counter *Window*, the radar operators learning to follow only targets that moved quickly and ignoring the slower-drifting clouds of foil.

With the destruction of Hamburg, the whole concept of the night air war, and particularly the night defence of Germany, had changed. The old close-controlled box system of *Himmelbett* had virtually gone. Now nightfighter crews were scrambled to orbit radio beacons, awaiting instruction from ground control as to the position, course and probable target of the attackers. This system became known as *Zahme Sau* (tame boars), while a new concept was also tried, the *Wilde Sau*, or wild boars. This had been advocated by *Oberstleutnant* Hajo Hermann, a former bomber pilot. He was allowed to establish *Versuchskommando Hermann*, using FW190s and Bf109Gs from *JG 1* and *JG 11*, as point defence of targets. With *Flak* given a height restriction, the *Wilde Sau* cruised above, picking out targets against the glare of fires and then storming in to attack. This tactic was to be employed for more than a year against Bomber Command.

In July 1943, the *Luftwaffe* strength was as follows:

Stab NJG 1	*Obst* Streib	Deelen	Bf110
I./NJG 1	*Hptm* Frank	Venlo	Bf110/He219
II./NJG 1	*Maj* Ehle	St. Trond	Bf110/Do217
III./NJG 1	*Maj* Priz zur Lippe Weissenfeld	Twente	Bf110
IV./NJG 1	*Maj* Lent	Leeuwarden	Bf110
Stab NJG 2	*Obstlt* Hulshoff	Aquino*	Ju88C
I./NJG 2	*Maj* Jung	Parchim	Ju88C
II./NJG 2	*Hptm* Sewing	Parchim	Ju88C
** returning from Middle East in August*			
Stab NJG 3	*Maj* Schalck	Stade	Bf110
I./NJG 3	*Hptm* Peters	Schleswig	Ju88G
II./NJG 3	*Hptm* Radusch	Schleswig	Bf110/Do217/Ju88C

III./NJG 3	*Hptm* Milius	Stade	Bf110
IV./NJG 3	*Maj* Simon	Grove	Ju88C
Stab NJG 4	*Maj* Stoltenhoff	Metz	Bf110
I./NJG 4	*Maj* Herget	Florennes	Bf110/Do217/Bf109G
II./NJG 4	*Hptm* Griese	Coulommieres	Bf110
III./NJG 4	*Hptm* Kamp	Juvincourt	Bf110/Do217
IV./NJG 4	*Maj* Wohlers	Mainz-Finthen	Bf110*
**renamed I./NJG 6 on 1.8.1943*			
Stab NJG 5	*Maj* Schaffer	Doberitz	Bf110
I./NJG 5	*Maj* Hoffmann	Stendal	Bf110
II./NJG 5	*Hptm* Schoenert	Parchim	Bf110
III./NJG 5	*Maj* Borchers	Neuruppin	Bf110
IV./NJG 5	*Maj* Prinz Zu Sayn Wittgenstein	Insterburg	Bf110
IV./NJG 6	*Maj* Lutje	Otopeni, Zilistea	Bf110
Stab NJG 101	*Obstlt* Stoltenhoff	Schleissheim	Bf110
I./NJG 101	*Hptm* Rossiwall	Ingolstadt	Bf110/Bf109G
II./NJG 101	*Maj* Knoetzsch	Lechfeld	Bf110
III./NJG 101	*Hptm* Thimmig	Echterdingen	Bf110
IV./NJG 101	*Maj* Fehling	Kitzingen	Bf110
Stab JG Hermann	*Obstlt* Hajo Hermann	Deelen	FW190A
I./JG Hermann	*Obstlt* Ewald Janssen	Bonn-Hangelar	Bf109G
II./JG Heermann	Unknown	Rheine	FW190A
III./JG Hermann	*Maj* Buchwald(?)	Oldenburg	Bf109G

There were several new formations during August; on 1st, *I./NJG 6* was renamed from *IV./NJG 4* at Mainz-Finthen, commanded by *Major* Wohlers and equipped with Bf110s; *I./NJG 100* was formed at Brjansk under *Major* Prinz Zu Sayn Wittgenstein, with Ju88s, Do217s, Bf110s and Ju87s. On 11th, *V./NJG 5* was formed at Insterburg with Bf110s under command of *Hauptmann* Peters and, on 15th· *III./NJG 2* was re-established at Stuttgart with Ju88Cs under Major Paul Semrau.

On 15th September, *Major* Schaffer moved from *NJG 5* to command the new *Stab NJG 6* at Schleissheim and *II./NJG 6* was formed at Neuburg/Donau under *Major* Leuchs , both units equipped with Bf110s. A new *Wilde Sau Geschwader* had commenced formation in August with the establishment of *JG 301*, commanded by *Oberstleutnant* Helmut Weinreich. *Stab* and *I./JG 301* formed at Neubiberg, with *Hauptmann* Richard Kamp leading *I Gruppe*, *II Gruppe* at Altenburg under *Hauptmann* Graf Resugier and, in November, *III Gruppe* was established at Zerbst, commanded by *Hauptmann* Manfred Mössinger. All were equipped with Bf109Gs.

24-25 July 1943

12./NJG1	Oblt Ernst-Georg Drünkler	Halifax	Kiel Bay	00.54	1600	4	W
6./NJG3	Fw Hans Meissner	Stirling	8km NNW Neumunster	01.11	4500	3	W
5./NJG3	Lt Gotthard Sachsenberg	Lancaster	10km W St.Peter	01.15	5800	1	W
6./NJG3	Lt Karl Böttinger	Stirling	8km NE Schleswig	01.20	3000	1	W
9./NJG3	Uffz Walter Rohlfing	Halifax	S Hamburg	01.20	5400	1	W
11./NJG3	Oblt Gunther Köberich	Halifax	4km NW Sonderborg	01.24		3	W
5./NJG3	Ofw WilhemZiegler	Halifax	7km E Schleswig	01.27	6500	1	W
9./NJG3	Uffz Walter Rohlfing	Wellington	2km E Buchholz	01.40	2500	2	W
7./NJG3	Uffz Berg	Lancaster	4km NE Dühnen	02.00	7500	3	W
11./NJG1	Oblt Hermann Greiner	Halifax	30km SW Den Helder	02.50	2000	8	W
1./NJG3	Oblt Paul Zorner	Halifax	11km NE Gröningen	03.54	5600	11	W
7./NJG1	Hptm August Geiger	Lancaster	10km E Cloppenburg			39*	W

The architect of the Wilde Sau operations was Hajo Hermann (right), a former bomber pilot. Friedrich-Karl 'Nose' Müller (left) became one of the leading exponents of this tactic. A former blind-flying instructor, he was instrumental in training and personally claimed thirty night victories with JG300 and later with NJG 10 and NJG 11. Both men survived the war.

25 July 1943

12./NJG1	Oblt Ernst-Georg Drünkler	B-17	25km NW Vlieland	18.22		5	W
12./NJG1	Lt Gerhard Dittmann	B-	75km WNW Vlieland	18.30		1	W

25-26 July 1943

10./NJG1	Hptm Rudolf Sigmund	Halifax	15km NW Vlieland	23.57	5700	26	W
12./NJG1	Fw Georg Kraft	Halifax	2km N Midoen Lake	00.08	5800	15	W
10./NJG1	Uffz Busch	Halifax	6km E Nieuwe Sluss	00.17	6300	1	W
./NJG1	Maj Werner Streib	Stirling	11km NE Eindhoven	00.28	5500	59	W
7./NJG1	Lt Richard Philipp	Stirling	12km SW Apeldoorn	00.31	5400	3	W
II./KdoHerr	Maj Kurt Kettner	Stirling	Essen	00.36	6200	1	W
9./NJG1	Hptm Wilhelm Dormann	Stirling	5km WNW Disterwijk	00.40	5000	14	W
./NJG1	Maj Werner Streib	Lancaster	10km N Helmond	00.46	5000	60	W
StI/NJG1	Hptm Hans-Dieter Frank	Lancaster	10km SE Nijmegen	00.56	6400	48	W
6./KdoHerr	Lt Werner Treusch	Halifax	Essen	00.58	4500	1	W
5./KdoHerr	Uffz Wolfgang Knobloch	Four-eng	Essen	01.15	7500	1	W
5./KdoHerr	Fw Horst Heinstein	Four-eng	Essen	01.16	4600	1	W
3./KdoHerr	Fw Hermann Wischnewski	Four-eng				1*	W
./NJG1	Maj Werner Streib	Halifax	6km SE Osterwijk	01.20	5700	61	W
StI/NJG1	Hptm Hans-Dieter Frank	Wellington	500m . Cülemburg	01.30	5500	49	W
1./NJG1	Lt Rolf Bussmann	Halifax		01.36	4500	16	W
./NJG1	Maj Werner Streib	Lancaster	17km NE s'Herto'bosch	01.42	6000	62	W
2./NJG1	Hptm Paul Förster	Halifax	2km E Zierikzee	01.43	2500	2	W
1./NJG1	Oblt Wolfgang Knieling	Wellington		01.52	3500	1	W
I./NJG1	Lt Ludwig Wirtz	Wellington	Culemborg rw. Stn	02.30		1	W

26 July 1943

12./NJG1	Lt Gerhard Dittmann	B-17	40km SE Gröningen	12.00		2	W
StI/NJG3	Oblt Johan Miklar	B-17	1km E Diepholz	12.55	8500	1	W

26-27 July 1943

12./NJG3	Lt Hermann Stock	Halifax	10km S Arhaus	02.39	20	1	W

27-28 July 1943

10./NJG200	Ofw Josef Kociok	R-5	Krimskaya	21.20	1200	14	E
StII/NJG3	Maj Günther Radusch	Lancaster	S Vollerwiek	00.26	6000	27	W
StII/NJG3	Maj Günther Radusch	Lancaster	Wöhrden	00.43	5900	28	W
5./NJG3	Ofw Kair	Wellington	Nr Torheimbül	00.49	6000	1	W
6./NJG3	Fw Hans Meissner	Halifax	1km SW Hobeck	01.01	5500	4	W
9./NJG3	Gfr Koenig	Halifax	3km NE Meerzen	01.15	6000	1	W
3./NJG3	Ofw Willi Schmale	Lancaster	S-W Bremen	01.25	500	6	W
StII/NJG1	Maj Walter Ehle	Lancaster	WSW Bremervörde	01.26	6000	28	W
II./KdoHerr	Uffz Friedrich Hagios	Four-eng	Hamburg	01.30	5000	1	W
II./KdoHerr	Ofw Heinrich Lonnecker	Four-eng	Hamburg	01.38	5000	2	W
8./NJG3	Fw Wilhelm Kurreck	Halifax	8km N Heide	01.38	5500	3	W
./JG300	Lt Nöll	Lancaster	Hamburg	01.40	5000	1	W
9./NJG3	Uffz Adolf Loschner	Halifax	NE Osterholz	01.53	4800	1	W
10./NJG5	Hptm Alois Lechner	DB-3	Russia	02.29	350	12	E
StIV/NJG1	Ofw Kubisch	Lancaster	Off Terschelling	02.37	4000	1	W
11./NJG1	Hptm Hans-Joachim Jabs	Halifax	10km W Vlieland	02.50	3000	19	W

28-29 July 1943

11./NJG5	Hptm Walter Fenske	DB-3	Russia	22.44	2000	11	E

29-30 July 1943

10./NJG5	Hptm Alois Lechner	PS-84	Russia	21.30	2400	13	E
10./NJG5	Hptm Alois Lechner	PS-84	Russia	22.55	400	14	E
4./NJG3	Oblt Gerhard Raht	Four-eng	Hanover	23.10	4800	7	W
StII/NJG3	Oblt Gerhard Raht	Lancaster	20km SE Heide	00.23	6000	8	W
2./NJG5	Uffz Hans Krepp	Lancaster	30km W St. Peter	00.24	6000	1	W
StII/NJG3	Maj Günther Radusch	Lancaster	1km N Hemminger	00.25	6000	29	W
8./NJG3	Oblt Günther Stege	Lancaster	30km NE Heligoland	00.30	5300	1	W
StII/NJG3	Oblt Gerhard Raht	Halifax	20km SE Heide	00.40	5900	9	W
II./KdoHerr	Uffz Edmund Rayzner	Lancaster	Hamburg	00.45	5000	1	W
5./KdoHerr	Uffz Kurt Röder	Lancaster	Hamburg	00.47	4500	1	W
II./KdoHerr	Uffz Walter Kauderer	Lancaster	SW Hamburg	00.50	5000	1	W
StII/NJG3	Oblt Gerhard Raht	Halifax	25km NE Heide	00.52	6000	10	W
7./KdoHerr	Uffz Brinkmann	Four-eng	Hamburg	00.55	4500	1	W
II./KdoHerr	Oblt Otto Ertel	Lancaster	Hamburg	00.55	5000	1	W
7./KdoHerr	Uffz Brinkmann	Four-eng	Hamburg	00.57	4700	2	W
5./NJG3	Lt Gotthard Sachsenberg	Halifax	3km W Todesfelde	00.57	5700	2	W
5./KdoHerr	Uffz Kurt Röder	Stirling	Hamburg	00.58	4000	2	W
II./KdoHerr	Lt G. A. Bärsdorf	Lancaster	Hamburg	01.04	3000	1	W
II./KdoHerr	Oblt Otto Ertel	Lancaster	Hamburg	01.05	4000	2	W
8./NJG3	Oblt Gustav Tham	Halifax	2km from Tarterfeld	01.07	7000	3	W
9./NJG3	Uffz Adolf Loschner	Halifax	25km NE Hamburg	01.08	5100	2	W
III./KdoHerr	Maj H-J Buchwald von	Four-eng	Hamburg	01.10	5800	1	W
II./KdoHerr	Fw Alfred Riediger	Four-eng	SE Hamburg	01.12	4500	1	W
9./NJG3	Uffz Walter Rohlfing	Stirling	3km SE Ahrenswohlre	01.12	3500	3	W
5./KdoHerr	Uffz Kurt Roder	Lancaster	Hamburg	01.15	3000	3	W
7./NJG1	Hptm August Geiger	Lancaster	2km SE Ahrensfelde	01.25		40	W
I./KdoHerr	Lt Franz Rübsam	Four-eng	Hamburg	01.25	4500	1	W
9./NJG3	Uffz Walter Rohlfing	Stirling	SE Ahrenswohlre	01.26	38000	4	W

Oberleutnant Walter Barte of II./NJG 1 (left) pictured in June/July 1943 at St.Trond. He was soon to transfer to take command of III./NJG 3 with the rank of Hauptmann, quickly rising to Major. His final victory tally is credited as 23. (S.Hall)

II./KdoHerr	Hptm F-W. Angermann	Four-eng	Hamburg South	01.30	6000	1	W
2./KdoHerr	Uffz Horst John	Stirling	S Hamburg	01.31	6000	1	W
8./NJG3	Fw Wilhelm Kürreck	Halifax	2km NE Walstedt	01.33	4500	4	W
III./NJG1	Hptm E. Lippe-Weissenfeld	Lancaster	Hägbluer Holz	01.40	6800	44	W
./NJG1	Maj Werner Streib	Halifax	25km NW Stade	01.45		63	W
./NJG2	Lt Bruno Heilig	Stirling	70km NE Schierm'koog	01.54		2	W
3./KdoHerr	Ofw Heinz Wohlecke	Halifax	SW Hamburg	02.00	5800	1	W
3./KdoHerr	Fw Willi Rüllkotter	Four-eng	Hamburg	02.02	4000	1	W
StIV/NJG1	Maj Helmut Lent	Lancaster	25km N Ameland	02.10	3800	67	W

30-31 July 1943

St./KdoHerr	Fw Ernst Schäfer	Halifax	Nr. Remscheid		4600	1	W
2./NJG4	Lt Wilhelm Schneiderwind	Stirling	16km S Givet	00.43	3900	2	W
4./NJG1	Lt Johannes Hager	Halifax	Regen	00.55	5600	7	W
V./KG2	Unknown crew	Single-eng	England	00.57		-	W
6./NJG1	Hptm E-W. Bonin.von	Halifax	Mariaweiler	00.57	6200	17	W
2./KdoHerr	Uffz Horst John	Stirling	SW Remscheid	01.01	6000	2	W
II./KdoHerr	Lt G. A. Bärsdorf	Lancaster	Remscheid	01.09	3000	2	W
./JG300	Fw Heinz Radloff	Lancaster	Remscheid	01.10	6000	1	W
5./KdoHerr	Hptm Rudolf Scharfenberg	Lancaster	Wuppertal	01.14	5000	1	W
II./KdoHerr	Uffz Hubert Engst	Stirling	SE Wuppertal	01.16	4000	1	W
3./KdoHerr	Fw Hurdler	Four-eng	N Remscheid	01.18	4000	1	W
3./KdoHerr	Oblt Berg.von	Four-eng	Remscheid	01.19	5000	1	W
II./KdoHerr	Oblt Hans Silberg	Stirling	Wuppertal	01.20	2000	1	W
5./KdoHerr	Uffz Steinhagen	Stirling	NW Remscheid	01.22	2800	1	W
3./KdoHerr	Ofw Heinz Wohlecke	Four-eng	Remscheid	01.27	4000	2	W
2./KdoHerr	Fw Willi Rüllkotter	Four-eng	W Remscheid	01.29	2000	2	W
5./KdoHerr	Uffz Ernst Müller	Lancaster	Remscheid	01.30	4500	2	W
I./JG300	Ofw Kurt Becker	Four-eng	Remscheid	01.35	4000	1	W

3./KdoHerr	Fw Hermann Wischnewski	Stirling	W Remscheid	01.40	4500	1	W
E./NJG2	Fw Ternieden	Stirling	2km SW Dinteloord	01.46	5300	2	W
II./KG53	Unknown crew	Mosquito	Orleans/Bricy airfield	02.40		-	W

31 July-1 August 1943

StIV/NJG5	Hptm H. Sayn-Wittgenstein	PS-84	Russia	21.13	3000	51	E
10./NJG200	Ofw Josef Kociok	R-5	Krimskaya	21.15	1200	15	E
10./NJG200	Ofw Josef Kociok	R-5	Krimskaya area	22.00		16	E
10./NJG200	Ofw Josef Kociok	R-5	Krimskaya area	22.05		17	E
12./NJG5	Uffz Herbert Simross	DB-3	10km NE Roslavl	23.17	2200	1	E

1 August 1943

10./NJG6	Uffz Hugo Knobel	B-24	Russia	14.31	10	1	W
10./NJG6	Uffz Hugo Knobel	B-24	Russia	14.38	10	2	W
10./NJG6	Ofw Georg Klever	B-24	Russia	14.43	10	3	W

1-2 August 1943

StIV/NJG5	Hptm H. Sayn-Wittgenstein	R-5	Russia	21.40	1800	52	E
StIV/NJG5	Hptm H. Sayn-Wittgenstein	PS-84	Russia	21.47	2700	53	E
StIV/NJG5	Hptm H. Sayn-Wittgenstein	R-5	Russia	23.01	2600	54	E
11./NJG5	Fw Georg Bargel	DB-3	S Jelenovka	23.21	3800	2	E
10./NJG5	Unknown crew	R-5	Russia	23.30	500	-	E
StII./NJG5	Hptm Rudolf Schoenert	Halifax	Russia	02.23		26*	W

2-3 August 1943

StIV/NJG5	Hptm H. Sayn-Wittgenstein	R-5	5km NNW Djatkovo	21.27	2500	55	E
StII/NJG3	Maj Günther Radusch	Lancaster	40km ESE Helgoland	01.22	5000	30	W
5./NJG3	Fw Krauter	Stirling	5km NNE Cuxhaven	01.42	1600	1	W
2./NJG3	Ofw Erich Heitmann	Lancaster	5km S Spiekeroog	01.55	50100	2	W
StII/NJG3	Maj Günther Radusch	Halifax	20km NNW Helgoland	02.14	5500	31	W
III./NJG1	Hptm E. Lippe-Weissenfeld	Halifax	20km S Stade	02.26		45	W
11./NJG1	Oblt Hermann Greiner	Lancaster	20km NW Schier'koog	03.06	2100	9	W
StIV/NJG1	Hptm Hans-Joachim Jabs	Halifax	3km NW Hoorn	03.10	5500	20	W
2./NJG3	Lt Hermann Leube	Stirling	20km NW Juist	03.28	4200	1	W
StIV/NJG1	Hptm Hans-Joachim Jabs	Stirling	22km W Harlington	03.29	2300	21	W
10./NJG1	Ofw Karl-Heinz Scherfling	Lancaster	In sea	03.33	5100	21	W
4./NJG3	Lt Georg Führer	Halifax	30km NW Cuxhaven	03.40	4600	1	W
11./NJG1	Oblt Hermann Greiner	Wellington	20km SW Schier'koog	03.48	500	10	W
StIV/NJG1	Hptm Hans-Joachim Jabs	Stirling	20km NW Tersch'ling	03.52	2100	22	W

3-4 August 1943

StIV/NJG5	Hptm H. Sayn-Wittgenstein	DB-3	Sossnewka	20.58	4600	56	E
12./NJG5	Ofw Karl Strohecker	DB-3	30km N Karatschev	21.00	3800	1	E
StIV/NJG5	Hptm H. Sayn-Wittgenstein	DB-3	N Karatschew	21.12	2000	57	E
StIV/NJG5	Hptm H. Sayn-Wittgenstein	DB-3	7km N-W Star	22.09	4100	58	E
12./NJG5	Oblt Hans Gref	DB-3	25km NE Karatschew	03.04	2800	7	E
1./NJG100	Hptm Alois Lechner	E/A	Russia	23.30		15	E

5-6 August 1943

1./NJG100	Hptm Alois Lechner	Mitchell	27km NE Brjansk	22.16	5200	16	E
StIV/NJG5	Hptm H. Sayn-Wittgenstein	Mitchell	Manjeschino	22.44	5300	59	E

In August 1943, NJG 100, consisting of several fully independent Staffeln, was formed in Russia and continued into the winter.

6-7 August 1943

12./NJG5	Oblt Hans Gref	Mitchell	15km SW Orel	01.56	3200	8	E
12./NJG5	Oblt Hans Gref	Mitchell	70km NE Karatschev	02.58	1200	9	E

7-8 August 1943

5./NJG4	Hptm H. Niebelschutz.von	Lancaster	Vendenesse	02.57	49000	3	W
10./NJG5	Fw Rudi Duding	DB-3	Russia	21.25	4300	1	E

8-9 August 1943

StIV/NJG5	Hptm H. Sayn-Wittgenstein	DB-3	3km NE Star	21.50	2000	60	E
12./NJG5	Lt Waldemar Rathke	TB-3	7km SE Krasskoje	23.32	1400	3	E

9-10 August 1943

StII./NJG6	Oblt Johannes Engels	Halifax	Wolmenbach			1*	W
StI/NJG6	Oblt H. J. Birkenstock	Lancaster	Nr. Erfenbach			1	W
2./NJG4	Lt Norbert Pietrek	Halifax	20km SSE Neufchat'	00.43	4300	3	W
2./NJG4	Lt Norbert Pietrek	Lancaster	10km NW St.Hubert	01.31	4200	4	W

10-11 August 1943

./JG54	Oblt Jank	PS-84	Russia	23.20	1800	3	E
./JG54	Oblt Jank	E/A	Russia	23.45	1500	4	E
./JG54	Oblt Jank	DB-3	Russia	23.50	1300	5	E
I./NJG6	Maj Heinrich Wohlers	Halifax	Spessbach			21*	W
StI/NJG6	Oblt H. J. Birkenstock	Lancaster	Nr. Alsenborn			2	W
3./NJG4	Oblt Fritz Söthe	Halifax	Wambrechies	00.03	5200	2	W
StII/NJG1	ObLt H-W. Schnaufer	Lancaster	Hahnlein	00.32	5400	21	W
2./KdoHerr	Uffz Horst John	Halifax	Mannhein	00.36	5500	3	W
I./NJG6	Maj Heinrich Wohlers	Halifax	Bad Kreuznach	00.37		22*	W
I./KdoHerr	Hptm F-K. Muller	Halifax	35km NNE Biblis	00.44	4200	2	W
I./KdoHerr	Hptm F-K. Muller	Halifax	area N Heidelberg	00.45	4200	3	W
3./NJG4	Uffz Kustusch	Halifax	12km W Mons	02.17	1600	1	W
5./NJG1	Fw Otto Fries	Lancaster	SW Charleroi	02.56	3500	1	W
StII/NJG1	Lt Johannes Hager	Stirling	1km SE Doïsche	03.25	3700	8	W

11-12 August 1943

./JG54	Oblt Jank	DB-3	Russia	23.24	1000	6	E
1./NJG100	Fw Rudi Duding	R-5	Russia	23.57	2400	2	E
1./NJG100	Hptm Alois Lechner	PS-84	Russia	01.40	2900	17	E

12-13 August 1943

2./JG2	Uffz Herbert Pedreth	Halifax	La Café	23.10	3000	1	W
2./JG2	Lt Detlef Grossfüss	Halifax	1km NNW Verneuil	23.17	3000	1	W
2./JG2	Lt Detlef Grossfüss	Halifax	bei Beaune	23.17	3000	2	W
2./NJG100	Hptm Ulrich Meien.von	PS-84	7km SE Nowkovka	00.16	3000	1	E
2./NJG100	Hptm Ulrich Meien.von	PS-84	5km E Nowkovka	00.27	2500	2	E
5./NJG4	Hptm H. Niebelschutz.von	Lancaster	Labuissière	02.42	3500	4	W

13-14 August 1943

2./NJG100	Hptm Ulrich Meien.von	DB-3	22km NE. Poltava	00.24	2900	3	E

14-15 August 1943

3./NJG100	Fw Walter Scheinbein	DB-3	Schatalovka	21.23	2600	2	E
5./NJG4	Hptm H. Niebelschutz.von	Lancaster	Laisenay	23.53	4000	5	W

15-16 August 1943

2./JG2	Fw German Elflein	Lancaster	6km SE Chartres	23.20	4300	1	W
2./JG2	Lt Detlef Grossfüss	Halifax	3km SSW Rugles	23.20	4000	3	W
2./JG2	Ofw Josef Bigge	Lancaster	Courville	23.23	4300	5	W
2./JG2	Uffz Herbert Benz	Lancaster	Grigny	23.26	4200	1	W
12./NJG1	Lt Heinz Grimm	Wellington	45km NNW Ters'ling	00.07	700	19	W
12./NJG1	Lt Heinz Grimm	Wellington	45km NNW Ters'ling	00.36	80	20	W
11./NJG1	Hptm Fritz Buchholtz.von	Hampden	25km W Egmont	01.15	100	1	W
2./JG2	Uffz Albert Kreuzer	Halifax	Chartainviliers	03.52	5200	2	W
2./JG2	Lt Detlef Grossfüss	Lancaster	3km SE Orbec-en-Auge	04.03	4200	4	W
2./JG2	Lt Detlef Grossfüss	Lancaster	500m S Houlgate	04.19	3000	5	W

16-17 August 1943

1./NJG100	Hptm Alois Lechner	PS-84	Russia	21.35	3600	18	E
1./NJG100	Hptm Alois Lechner	PS-84	Russia	21.48	3700	19	E
StI/NJG100	Hptm Rudolf Schoenert	TB-7	Russia	22.08	2400	27	E
StI/NJG100	Hptm Rudolf Schoenert	R-5	Russia	23.36	1500	28	E
2./JG2	Oblt Karl Haberland	Halifax	6km E Alencon	23.38	3500	1	W
StI/NJG100	Hptm Rudolf Schoenert	R-5	Russia	23.56	1900	29	E
II./NJG4	Oblt Heinz Hamacher	B-17				1	W
9./NJG4	Oblt Hans Kulow	Stirling	65km NE Lyon	00.43	3400	1	W
1./NJG100	Hptm Alois Lechner	DB-3	Russia	02.32	5200	20	E
2./JG2	Lt Detlef Grossfüss	Four-eng	25km N Ornemündung	04.20	3700	6	W

17 August 1943

1./NJG1	Oblt Rolf Bussmann	B-17	S-W Genck	11.30	3800	17	W
4./NJG1	Oblt Walter Barte	B-17	E Liege	11.35		13*	W
1./NJG1	Oblt Rolf Bussmann	B-17	4km SW Genck	11.37	3800	18	W
4./NJG1	Lt Hans Witzke	B-17	Montzen	11.40	5000	1	W
6./NJG1	Hptm E-W. Bonin.von	B-17	10km S-W Aachen	11.45		18	W
II./NJG1	Lt Johannes Hager	B-17	S Wertheim	12.26		9	W
3./NJG101	Lt Ernst Terborg	B-17		13.46	5500	1	W
3./NJG101	Uffz Werner Dotzauer	B-17		13.47	5500	1	W
3./NJG101	Uffz Werner Dotzauer	B-17		13.49	5500	2	W
6./NJG1	Oblt Herbert Kleinschmidt	B-17	Esch	15.00	2500	1	W
6./NJG1	Hptm E-W. Bonin.von	B-17	3km NE Aachen	15.07		19	W

Although outclassed by more nimble single-engined fighters in daylight, The Messerschmitt Bf110 became a formidable opponent at night. The punch of the nose armament on the G-4 version was supplemented by a belly-pack and even the drag and weight of the nose-mounted Lichtenstein SN-2 radar array could not detract from its total superiority over the cumbersome heavy bombers of the Royal Air Force.

10./NJG101	Fw Nagel	B-17	15km W Bad Kissengen	16.10	500	3	W
7./NJG101	Lt Josef Springer	B-17	7km S Ebrach	16.17	400	1	W
10./NJG101	Fw Nagel	B-17	Eifershausen	16.24	200	4	W
2./NJG6	Fw Erich Becker	B-17	Nr. Wetz	16.44	5500	1	W
10./NJG101	Hptm Krzywo	B-17		17.25	40000	1	W
3./NJG1	Ofw Werner Hellenbrucks	B-17	NNW Antwerp	17.30	7300	1	W
3./NJG1	Lt Josef Nabrich	B-17				3	W

17-18 August 1943

10./NJG1	Lt Hans Kroner	Wellington	NW Alkmaar	23.30	150	1	W
4./JG300	Uffz Edmund Rayzner	Mosquito	W Berlin	00.17	5500	1	W
4./NJG3	Uffz Gunther Liersch	Halifax	300m W Westerland af	00.36	4200	1	W
V./KG2	Unknown crew	Sing-eng		00.57	5500	-	W
7./NJG3	Fw Paul Bahr	Lancaster	15km SW Esbjerg	01.03		1	W
10./NJG1	Ofw Karl-Heinz Scherfling	Halifax	Greifwald	01.28	2800	22	W
4./NJG1	Oblt Walter Barte	Lancaster	SW Peenemünde	01.35	2000	14	W
I./KdoHerr	Lt Walter Harjes	B-17	Swinemünde	01.40	2000	1	W
E./NJG2	Lt Helmut Perle	Fighter	10km N Peenemünde	01.42		1*	W
StII/NJG1	Maj Walter Ehle	Lancaster	NNW Peenemünde	01.44	3000	29	W
./JG300	Hptm F-K. Müller	Lancaster	Swinemünde	01.45	1000	4	W
StII/NJG1	Maj Walter Ehle	Lancaster	NNW Peenemünde	01.46	2800	30	W
./NJG5	Maj Gunther Radusch	Four-eng	Peenemünde	01.46	1500	32	W
5./NJG1	Lt Dieter Musset	Four-eng	Peenemünde	01.47	2800	1	W

9./NJG1	Lt Erwin Ernst	Lancaster	7km SW Peenemünde	01.50	1800	1	W
5./NJG5	Lt Peter Ehrhardt	Lancaster	Peenemünde	01.52	2400	3	W
5./NJG1	Lt Dieter Musset	Four-eng	Peenemünde	01.52	2700	2	W
7./NJG3	Oblt Paul Zorner	Lancaster	Baltic Sea	01.53	4400	12	W
./NJG5	Maj Günther Radusch	Four-eng	Peenemünde	01.55	2500	33	W
5./NJG5	Lt Peter Ehrhardt	Lancaster	Peenemünde	01.57	2300	4	W
5./NJG1	Ofw Fritz Schellwat	Lancaster	Peenemünde	01.57	3800	15	W
./KdoHerr	Hptm F-K. Müller	Lancaster	10km NW Swinemünde	01.58	1000	5	W
5./NJG1	Lt Dieter Musset	Lancaster	Peenemünde	01.58	2000	3	W
5./NJG1	Lt Dieter Musset	Four-eng	Peenemünde	01.58	2700	4	W
4./NJG1	Oblt Walter Barte	Halifax	SW Peenemünde	01.59	2500	15	W
4./NJG3	Uffz Heinz Oberheide	Halifax	12km W Peenemünde	02.00	2500	1	W
./NJG5	Maj Günther Radusch	Four-eng	Peenemünde	02.00	2000	34	W
5./NJG5	Lt Peter Spoden	Lancaster	Forsthagen	02.00	3000	1	W
5./NJG1	Lt Dieter Musset	Four-eng	Peenemünde	02.01	2000	5	W
9./NJG5	Uffz Baader	Halifax	Peenemünde	02.02	500	1	W
5./NJG5	Lt Peter Ehrhardt	Lancaster	Peenemünde	02.02	3000	5	W
5./NJG5	Uffz Holker	Halifax	Peenemünde	02.02	3000	1	W
7./NJG3	Oblt Paul Zorner	Lancaster	Peenemünde	02.03		13	W
7./NJG1	Hptm August Geiger	Halifax	15km NW Greifswald	02.08	2400	41	W
5./NJG5	Lt Peter Ehrhardt	Lancaster	Peenemünde	02.12	3000	6	W
7./NJG1	Hptm August Geiger	Lancaster	2km S Reinberg	02.13	2400	42	W
5./NJG5	Uffz Holker	Lancaster	Peenemunde	02.17	4000	2	W
6./NJG3	Fw Hans Meissner	Lancaster	N coast of Alsen	02.56	4500	5	W
StII/NJG3	Oblt Gerhard Raht	Lancaster	3km E Flensburg	02.57	3500	11	W
6./NJG3	Fw Hans Meissner	Lancaster	2km NW Barsmark	03.07	3700	6	W
6./NJG3	Fw Hans Meissner	Lancaster	7km SW Hadersleben	03.11	3500	7	W
10./NJG3	Uffz Benno Gramlich	Stirling	55km N Juist	03.15		1	W
4./NJG3	Uffz Günther Liersch	Halifax	30km NNW Helgoland	03.16	3200	2	W

21-22 August 1943

1./NJG100	Hptm Alois Lechner	DB-3	Russia	21.42	1200	21	E
1./NJG200	Oblt Jank	DB-3F	Russia	22.33		1	E
1./NJG200	Oblt Jank	DB-3F	Russia	22.40		2	E

22-23 August 1943

1./NJG100	Lt Josef Pützkuhl	PS-84	Russia	20.46	3000	4	E
1./NJG100	Oblt Adolf Kaiser	DB-3	Russia	21.02	3700	1	E
1./NJG100	Oblt Adolf Kaiser	PS-84	Russia	22.01	2500	2	E
1./NJG100	Lt Josef Pützkuhl	R-5	Russia	22.45	2000	5	E
3./NJG100	Lt Waldemar Rathke	PS-84	Russia	23.57	4000	4	E
II./NJG1	Lt Johannes Hager	Halifax	ENE Aachen	00.41		10	W
4./NJG1	Lt Hans Witzke	Halifax	5km W Maastricht	00.42	5400	2	W
II./NJG1	Lt Johannes Hager	Lancaster	20km S Utrecht	00.51		11	W
3./JG300	Lt Gerhard Stamp	Four-eng	Berlin	01.30		1*	W
3./NJG1	Oblt Heinz Strüning	Stirling	20km E Eindhoven	01.43	6300	37	W
8./NJG1	Lt Heinz Bock	Lancaster	5km E Meerkerk	01.46	5000	1	W

23-24 August 1943

1./NJG100	Hptm Kurt Bonow	DB-3	Russia	20.50	2300	2	E
1./NJG200	Oblt Jank	MBR-2	Russia	22.05		3	E

Sergeant Harrison's 434 Squadron Halifax never made it to Leverkusen. The smoking wreckage of EB255, shot down near Maastricht on 22/23rd August by Leutnant Hans Witzke of 4./NJG 1. On 17th August, Witzke had charged into a daylight raid by the 8th Air Force and had been credited with a B-17 destroyed, but on the night of 30th August he and his crew died in a crash near St. Trond.

1./NJG100	Hptm Alois Lechner	SB-	Russia	22.15	2500	22	E
StI/NJG1	Hptm Hans-Dieter Frank	Lancaster	7km SE Emmen	23.40	5400	50	W
12./NJG1	Lt Heinz Rolland	Lancaster	E Bremen			1	W
./JG300	Hptm F-K. Müller	Halifax	Gassen, nr Gentheim	00.15	4600	6	W
2./JG300	Fw F. Laubenheimer	Halifax	W Berlin	00.18	5500	1	W
StII/NJG3	Hptm H. Sayn-Wittgenstein	Halifax	NNW Neuhalde'leben	00.18	5500	61	W
3./NJG6	Uffz Benno Kratz	Halifax	W Berlin	00.35	5500	1	W
1./NJG2	Oblt Heinz Rökker	Lancaster	20km SW Berlin	00.35		7	W
3./NJG6	Fw Gunther Bahr	Halifax	Berlin	00.40		2	W
1./NJG5	Ofw Johannes Scholz	Lancaster	SW Berlin	00.46	3200	1	W
3./NJG6	Fw Günther Bahr	Stirling	N Berlin	00.47	5000	3	W
6./JG300	Uffz Anton Gaissmayer	Stirling	Berlin	00.48	5000	1	W
8./NJG3	Oblt Günther Stege	Lancaster	Zossen	00.48	6400	2	W
9./JG300	Uffz Werner Dienst	B-24	NW Berlin	00.50	5000	2	W
./NJG1	Oblt Werner Husemann	Halifax	Biesenthal	00.50	5200	10	W
1./NJG2	Lt Heinz Rökker	Halifax	15km SSW Berlin	00.50	5200	8	W
3./JG300	Ofw Ernst Wick	Lancaster	NW Berlin	00.50	5000	1	W
3./NJG3	Oblt Heinz Ferger	Four-eng	SW Berlin	00.51	3000	1	W
1./JG300	Uffz Fischer	Four-eng	Berlin	00.53	5000	1	W
2./NJG1	Oblt Lenz Finster	Halifax	Schonerlinde	00.54	5500	1	W
./NJG3	Maj Helmut Lent	Stirling	Berlin-Lichtenberg	00.56		68	W
StII/NJG5	Hptm Manfred Meurer	Lancaster	east part of Berlin	00.58	5600	51	W

./JG300	Hptm F-K. Müller	Stirling	Berlin	00.58	4500	7	W
StII/NJG3	Oblt Gerhard Raht	Halifax	1km SE Thomsdorf	00.58	5000	12	W
5./NJG5	Oblt Wilhelm Johnen	Stirling	NW Berlin	01.02	4200	6	W
II./NJG1	Lt Johannes Hager	Stirling	Löhne	01.03		12	W
2./NJG3	Uffz Rudolf Frank	Stirling	NW Berlin	01.05	3500	15	W
8./JG300	Lt Robert Plewa	Lancaster	Berlin	01.05	5000	1	W
St/JG300	Hptm Max Langer	Four-eng	Berlin	01.05		1*	W
5./NJG5	Oblt Wilhelm Johnen	Stirling	NW Berlin	01.07	4500	7	W
2./NJG1	Oblt Lenz Finster	Halifax	W Trebitz	01.12	5000	2	W
5./NJG5	Lt Peter Spoden	Halifax	W Berlin	01.13	4500	2	W
StIII/NJG5	Hptm Walter Borchers	Stirling	Helenau	01.14	5300	5	W
./JG300	Hptm F-K. Müller	Stirling	Berlin	01.15	4200	8	W
8./NJG4	Ofw Reinhard Kollak	Four-eng	NW Berlin	01.16	3000	28	W
./NJG3	Maj Helmut Lent	Halifax	Berlin-Mahlsdorf	01.16	4500	69	W
./NJG3	Maj Helmut Lent	Halifax	Berlin-Spandau			70*	W
2./NJG4	Oblt Rudolf Altendorf	Halifax	N Berlin	01.18	3000	17	W
8./JG300	Lt Robert Plewa	Halifax	Nr. Berlin	01.18	6000	2	W
5./NJG1	Ofw Fritz Schellwat	Stirling	S Lanke	01.18	3800	16	W
StI ./NJG4	Hptm Wilhelm Herget	Lancaster	1km SE Thomsdorf	01.23	5000	26	W
9./NJG3	Lt Hans Raum	Lancaster	Kustrin	01.23	3800	1	W
./NJG1	Maj Werner Streib	Halifax	Heiligen Lake	01.23	6000	64	W
8./JG300	Lt Robert Plewa	Halifax	Berlin	01.28	5000	3	W
5./NJG3	Oblt Ruprecht Panzer	Halifax	N Berlin	01.30	3500	1	W
3./JG300	Lt Hermann Stumpe	Four-eng	Berlin	01.30	4000	1	W
2./NJG1	Uffz Josef Brunner	Halifax	N Templin	01.35	2800	1	W
II./NJG1	Lt Johannes Hager	Stirling	NE Berlin	01.35		13	W
4./NJG3	Uffz Heinz Oberheide	Four-eng	10-20km NNE Berlin	01.35	4500	2	W
StII/NJG5	Hptm Manfred Meurer	Lancaster	Eranienburd/Furst'rg	01.38	4000	52	W
10./NJG1	Oblt Hermann Greiner	Halifax	NW Berlin	01.43	4600	11	W
1./NJG100	Hptm Kurt Bonow	DB-3	Russia	01.52	2000	3	E
5./NJG3	Hptm Paul Szameitat	Lancaster	10km E Kappen	02.03	5500	13	W
StIV/NJG1	Hptm Hans-Joachim Jabs	Stirling	10km S Enkhuizen	02.14	2400	23	W
10./NJG3	Oblt Hans-Hermann Müller	Stirling	5km SE Hornum	02.15	5500	2	W
1./NJG1	Lt Werner Baake	Lancaster	S-E Wittenberg	02.18	5200	11	W
II./NJG1	Hptm Walter Rupprecht	Four-eng	NW Berlin	02.21		1	W
6./NJG3	Lt Karl Böttinger	Halifax	15km W Ribe	02.35	4500	2	W
12./NJG1	Lt Heinz Rolland	Stirling	45km N-W Vlieland	03.31	2300	2	W
12./NJG1	Lt Heinz Grimm	Lancaster	25km N Ameland	03.40	3900	21	W
V./KG2	Ofw Gräber	Four-eng	N Cambridge	04.08		1	W

24-25 August 1943

1./NJG100	Hptm Alois Lechner	DB-3	Russia	20.34	2100	23	E

25-26 August 1943

1./NJG100	Hptm Alois Lechner	DB-3	Russia	21.08	3000	24	E
1./NJG100	Hptm Alois Lechner	DB-3	Russia	21.18	4000	25	E
1./NJG100	Hptm Kurt Bonow	DB-3	Russia	21.20	3200	4	E

26-27 August 1943

1./NJG100	Ofw Helmut Dahms	R-5	Russia	22.00	2100	2	E

Martin Drewes served as a Bf110 pilot in daylight operations in the spring of 1941 with 6./ZG 76 in Iraq, where he claimed a Gladiator destroyed. In August, leading a training mission over the North Sea, the Bf110s were attacked by Spitfires of 19 Squadron, of which two were destroyed without loss, one by Drewes. He then transferred to NJG 1 and ended the war as Kommandeur of III./NJG 1, the Ritterkreuz mit Schwerten and a total of 49 confirmed victories, including four American heavy bombers in daylight. On 27/28th August he claimed a Halifax near Huy, his 6th confirmed victory.

27-28 August 1943

StI/NJG100	Hptm Rudolf Schoenert	TB-7	Krotovko	20.20	5300	30	E
StI/NJG100	Maj Harold Lange	TB-7	Russia	20.27	4300	1	E
StI/NJG100	Maj Harold Lange	PS-84	Russia	20.49	3100	3	E
3./NJG100	Oblt Hans Gref	PS-84	58km SW Newel	21.55	1300	10	E
12./NJG3	Lt Karl Rechberger	Wellington	100km NW Alborg	23.41	5100	1	W
8./NJG4	Ofw Reinhard Kollak	Lancaster	15km SW Verdun	00.42	5500	29	W
3./JG300	Ofw Ernst Wick	Halifax	NW Berlin	00.50		1	W
8./NJG4	Uffz Herbert Ter.Steegen	Stirling	6km E Saarburg	00.53	3000	1	W
6./NJG1	Lt Johannes Hager	Stirling	SW Saarburg	00.55	3200	14	W
8./NJG4	Oblt Helmut Bergmann	Halifax	SW Echternacht	01.04	5400	16	W
3./JG300	Hptm Max Langer	Four-eng	Nr. Nuremberg	01.09		2	W
6./NJG5	Lt Josef Kraft	Lancaster	22km W Nurnberg	01.20		1	W
StI/NJG6	Hptm Heinrich Wohlers	Halifax	Eppenheim	01.33	5500	23	W
1./NJG4	Oblt Ludwig Meister	Lancaster	Wolpertshausen	01.35	5000	16	W
3./JG300	Lt Hermann Stumpe	Four-eng	Nuremberg	01.40		2	W
2./NJG4	Ofw Walter Eckel	Lancaster	NW Nuremberg	01.45	4000	1	W
StI/NJG6	Hptm Heinrich Wohlers	Stirling	Eisenheim/Asbach	01.46	5400	24	W
6./NJG1	Lt Johannes Hager	Lancaster	Nr. Nuremberg	01.47	4000	15	W
4./JG300	Uffz Hubert Engst	Lancaster	Nr. Nuremberg	01.50	5000	2	W
3./NJG3	Oblt Heinz Ferger	Stirling	NW Nuremberg	01.50	4800	2	W
StI ./NJG4	Maj Wilhelm Herget	Halifax	Nuremberg	01.50	4000	27	W
6./JG300	Lt Klaus Bretschneider	Stirling	Nuremberg	01.55		1	W
1./NJG4	Oblt Ludwig Meister	Four-eng	Nr. Nuremberg	01.57	4200	17	W
2./NJG5	Uffz Hans Krepp	Lancaster	Nr. Nuremberg	01.58	4500	2	W
StII/NJG1	Maj Walter Ehle	Stirling	NW Nuremberg	02.00		31	W
6./NJG1	Lt Johannes Hager	Halifax	NW Nuremberg	02.00		16	W
5./JG300	Uffz Kurt Röder	Stirling	E Nuremburg	02.02	4500	4	W
2./JG300	Fw Horst John	Stirling	Nuremburg	02.04		4	W
3./NJG1	Lt Josef Nabrich	Halifax	S Nuremberg	02.08	4000	4	W
StI/NJG5	Oblt Manfred Tischtau	Stirling	Nr. Nuremberg	02.08	2200	1	W
3./JG300	Hptm Max Langer	Four-eng	W Nuremberg	02.09		3*	W

1./NJG1	Lt Werner Baake	Stirling	W Barnberg	02.10	3500	12	W
10./NJG101	Uffz Rupp	Halifax	N Nuremberg	02.10	4300	1	W
1./NJG6	Oblt Albert Walter	Stirling	25km NE Kitzingen	02.10		1	W
2./NJG4	Oblt Rudolf Altendorf	Halifax	Nuremberg	02.12		18	W
3./NJG3	Oblt Heinz Ferger	Four-eng	15-25km NW Nurem'rg	02.12	3500	3	W
./JG300	Hptm F-K. Müller	Lancaster	10-15km W Nuremberg	02.12	3200	9	W
StI/NJG6	Hptm Heinz Reschke	Halifax	NW Nuremberg	02.13	4000	1	W
1./NJG1	Lt Werner Baake	Stirling	20km NW Nuremberg	02.15	3500	13	W
5./JG300	Oblt Grisal	Stirling	Nuremberg	02.15	4300	1	W
5./JG300	Oblt Grisal	Four-eng	Nuremberg	02.15	4800	2	W
5./JG300	Oblt Otto Ertel	Four-emg	Nuremburg	02.15		3*	W
7./NJG1	Hptm August Geiger	Stirling	Gerlachhausen	02.20		43	W
StI/NJG6	Hptm Heinrich Wohlers	Lancaster	Jungholzhausen	02.53		25	W
1./NJG1	oblt Rolf Bussmann	Wellington	W Marienstadtfeld	03.15		19*	W
./NJG1	Oblt Werner Husemann	Wellington	NW Nuremberg	03.15	4000	11	W
11./NJG1	Oblt Martin Drewes	Halifax	Huy	03.19	5300	4	W
3./NJG4	Hptm Fritz Söthe	Lancaster	Estinnes	03.36	3800	3	W
8./NJG4	Ofw Reinhard Kollak	Four-eng	22km NNE Amiens	03.46	5600	30	W
12./NJG1	ObLt H-W. Schnaufer	Halifax	Jemeppe sur Sambre	03.59	4000	22	W
7./NJG1	Lt Hans-Heinz Augenstein	Stirling	20km NW Nuremburg			10*	W
7./NJG1	Lt Hans-Heinz Augenstein	Four-eng	Nuremburg-Lohe			11*	W

29-30 August 1943

2./NJG100	Uffz Paul	DB-3	15km W Debelzewo	21.36	2500	1	E

30-31 August 1943

9./NJG1	Oblt Fritz Lau	Stirling	SSE Enschede	23.50		1*	W
StI/NJG1	Hptm Hans-Dieter Frank	Stirling	Mönchengladbach	03.18	4700	51	W
3./NJG1	Oblt Heinz Strüning	Halifax	20km W M'gladbach	03.20	5500	38	W
StI/NJG1	Hptm Hans-Dieter Frank	Wellington	Siggerath	03.30	4300	52	W
4./NJG1	Lt Johannes Hager	Halifax	Niederkruchten	03.30	5200	17	W
4./NJG1	Lt Wilhelm Henseler	Stirling	4km SW Rheydt	03.30	4500	1	W
E./NJG2	Lt Helmut Perle	Stirling	10-15km SW M'gladbach	03.33	3800	2	W
StI/NJG1	Hptm Hans-Dieter Frank	Lancaster	Bruggen	03.35	5700	53	W
E./NJG2	Lt Helmut Perle	E/A	15km SW M'gladbach	03.35	4000	3	W
4./NJG1	Lt Johannes Hager	Halifax	Mönchengladbach	03.40	5200	18	W
3./NJG1	Ofw Werner Hellenbrucks	Lancaster	W Mönchengladbachh	03.40	5700	2	W
9./NJG4	Ofw Heinz Macke	Halifax	Mönchengladbach	03.40	4800	3	W
2./NJG3	Uffz Rudolf Frank	Halifax	Mönchengladbach	03.43		16	W
StII/NJG1	Maj Walter Ehle	Lancaster	Giesenkirchen	03.45	5200	32	W
3./NJG1	Oblt Heinz Strüning	Halifax	W Mönchengladbach	03.45	5700	39	W
StI/NJG1	Lt Walter Schön	Halifax	35km WNW M-gladbach	03.50	4200	6	W
3./NJG1	Oblt Heinz Strüning	Halifax	60km WSW M-gladbach	03.50	5700	40	W
12./NJG1	ObLt H-W. Schnaufer	Wellington	St.Trond	03.53	4000	23	W
2./NJG1	Ofw Bruno Eikmeier	Stirling	25km SE Eindhoven	03.56	3200	6	W
11./NJG1	Oblt Martin Drewes	Halifax	Maastricht	03.58	4800	5	W
12./NJG1	ObLt H-W. Schnaufer	Wellington	Lozen, N Bree	04.00	4000	24	W
9./NJG4	Hptm Kurt Fladrich	Stirling	6km W Weert	04.01	4200	7	W
10./NJG1	Lt Friedrich Potthast	Wellington	4km S Delft	04.03	5500	3	W
II./NJG1	Hptm Walter Rupprecht	Halifax	3km SW Westmalle	04.03	3000	2	W

By September 1943, the Luftwaffe night Experten were multiplying. Wilhelm Herget of Stab I./NJG 4 (right) claimed his 42nd victory on 31st August/1st September, he would end the war flying with JV 44, flying the Me262, credited with 73 victories. Herbert Lutje (centre) claimed his 30th on 3/4th September. He too would survive the war, as Kommodore or NJG 6 with 50 victories credited. Josef Kraft (left), was a relatively late-starter, claiming his third kill on the night of 31st August. He too survived the war, leading 12./NJG 1 with 56 victories credited to him.

31 August-1 September 1943

StI/NJG100	Hptm Rudolf Schoenert	Mitchell	Russia	22.01	2900	31	E
StI/NJG100	Hptm Rudolf Schoenert	Mitchell	Russia	22.21	3000	32	E
StI/NJG100	Hptm Rudolf Schoenert	Mitchell	Russia	22.37	2600	33	E
12./NJG1	ObLt H-W. Schnaufer	Halifax	2km SE Kuinre	22.41	4500	25	W
StI/NJG100	Hptm Rudolf Schoenert	Mitchell	Russia	23.00	2300	34	E
3./NJG1	Oblt Werner Baake	Halifax	NE Lemgo	23.25	4500	14	W
2./NJG3	Uffz Rudolf Frank	Lancaster	E Diepholz	23.28	5200	17	W
7./NJG1	Hptm August Geiger	Stirling	9km SSW Enschede	23.28	4700	44	W
3./NJG1	Oblt Werner Baake	Halifax	Neu-Rebstock	23.30	4500	15	W
6./NJG1	Hptm E-W. Bonin.von	Halifax		23.30	4500	20	W
12./NJG1	Lt Heinz Grimm	Lancaster	Meppen	23.48	4800	22	W
12./NJG1	Lt Heinz Grimm	Stirling	Bielefeld	24.00	3800	23	W
12./NJG1	Lt Heinz Grimm	Four-eng	40km E Hanover	00.13	3400	24	W
StIII/NJG3	Hptm Rudolf Sigmund	Halifax	Nr. Brandenburg	00.14	4200	27	W
5./NJG1	Hptm Wilhelm Telge	Stirling	ENE Gottingen	00.30	5800	14	W
6./JG300	Fw Hans Kugler	Lancaster	Middle of Berlin	00.35	4500	1	W
4./JG300	Ofw Richard Heemsoth	Stirling	SE Berlin	00.40	4500	1	W
6./NJG5	Lt Josef Kraft	Lancaster	25km SSW Berlin	00.47	4200	2	W
StIII/NJG5	Hptm Walter Borchers	Stirling	S Berlin	00.50	4500	6	W
3./NJG3	Uffz Klemenz Koch	Stirling	NW Berlin	00.50	3000	1	W
3./NJG5	Uffz Ernst Reitmeyer	Stirling	E Berlin	00.52		1	W
6./NJG1	Lt Johannes Hager	Halifax	SW Berlin	00.53		19	W
./NJG3	Maj Helmut Lent	Halifax	SW Berlin	00.54	4200	71	W
4./JG300	Ofw Richard Heemsoth	Four-eng	SW Berlin	00.55	5200	2	W
6./NJG5	Lt Josef Kraft	Stirling	E Berlin	00.55		3	W
8./NJG3	Oblt Günther Stege	Halifax	30km SSW Berlin	00.55	4900	3	W

StI ./NJG4	Maj Wilhelm Herget	Halifax	SW Trebbin	00.56	4000	28	W
4./JG300	Uffz Edmund Rayzner	Lancaster	middle of Berlin	00.56	4200	2	W
6./JG300	Fw Hans Kugler	Stirling	middle of Berlin	00.57	4500	2	W
5./NJG5	Oblt Wilhelm Johnen	Lancaster	20km SW Berlin	00.58	4200	8	W
5./JG300	Hptm Rudolf Scharfenberg	Lancaster	NNW Berlin	00.58	5000	2	W
6./JG300	Lt Werner Treusch	Lancaster	W Berlin	00.58	4300	2	W
3./NJG1	Oblt Werner Baake	Lancaster	Wustermark	00.59	4200	16	W
9./NJG3	Hptm Siegfried Hahn	Stirling	S Berlin	00.59	4200	2	W
3./NJG3	Uffz Wilhelm Brockerhoff	Stirling	ESE Berlin	01.00	2000	1	W
StII/NJG5	Hptm Manfred Meurer	Stirling	20km WNW Berlin	01.00	3000	53	W
5./NJG3	Uffz Heinrich Wortmann	Stirling	WNW Berlin	01.02	4500	1	W
./NJG3	Maj Helmut Lent	Halifax	E Berlin	01.03	3800	72	W
6./NJG1	Lt Johannes Hager	Four-eng	SW Berlin	01.04	5400	20	W
4./JG300	Uffz Friedrich Hagios	Lancaster	10km S Berlin	01.04	2000	2	W
12./NJG1	Lt Heinz Rolland	Stirling	S Potsdam	01.05	4000	3	W
3./NJG1	Oblt Heinz Strüning	Halifax	Nrandenburg	01.05	4500	41	W
1./NJG6	Oblt Albert Walter	Halifax	W Berlin	01.05	5400	2	W
II./NJG1	Uffz Hubert Bergen.von	Stirling		01.06	5300	1	W
6./NJG1	Lt Johannes Hager	Stirling	S Berlin	01.06		21	W
3./NJG6	Oblt H-M. Hadeball	Halifax	N Kothen	01.07	5500	16	W
5./NJG5	Fw Radtke	Stirling	SW Berlin	01.07	3000	1	W
StI ./NJG4	Maj Wilhelm Herget	Lancaster	NE Trebbin	01.11	3800	29	W
5./NJG5	Lt Erich Metz	Stirling	5-10km NW Berlin	01.12	3500	1	W
StI/JG300	Hptm Max Langer	Halifax	Brandenburg/Berlin	01.15	4500	2	W
9./NJG3	Uffz Adolf Löschner	Halifax	S Brandenburg	01.15		3	W
4./JG300	Ofw Richard Heemsoth	Stirling	SW Berlin	01.20	4200	3	W

3-4 September 1943

StI/NJG100	Hptm Rudolf Schoenert	R-5	Poltava	22.34	2100	35	E
8./NJG1	Hptm Herbert Lütje	Lancaster	22km NE Rheine	23.25	6000	30	W
1./NJG2	Uffz Hans Kurz	Halifax	N Viersen	23.41	3900	2	W
7./NJG1	Hptm August Geiger	Stirling	1km SW Hannover	23.50		45*	W
2./NJG1	Oblt Lenz Finster	Four-eng	Nr. Glienicke	00.28	5800	3	W
1./NJG5	Ofw Johannes Scholz	Lancaster	10km E Neuruppin	00.30	4000	2	W
./JG300	Uffz Brinkmann	Lancaster	Berlin	00.31	4500	3	W
StI/NJG5	Hptm Werner Hoffmann	Lancaster	middle of Berlin	00.32	6000	9	W
1./NJG5	Ofw Johannes Scholz	Lancaster	E Neuruppin	00.42	3500	3	W
StIII/NJG5	Hptm Walter Borchers	Lancaster	20km SSE Neuruppin	00.45	4300	7	W
St/NJG1	Oblt Werner Husemann	Wellington	Nr. Strassbourg	00.50	2500	12	W
4./NJG5	Ofw Heinz Säwert	Lancaster	Berlin	00.50	4800	7	W
10./NJG3	Uffz Heinz Koppe	Lancaster	15km S Skagen	02.15	1500	1	W
12./NJG3	Lt Karl Rechberger	Lancaster	15km S Skagen	02.32	4300	2	W
14./KG2	Lt Baak	Twin-eng	S Lincoln	03.51	200	1	W

5-6 September 1943

2./NJG100	Hptm Ulrich Meien.von	DB-3	55km from Poltava	21.21	2200	4	E
6./JG300	Fw Hans Kugler	Lancaster	2km SE Mannheim	00.05	5000	3	W
./JG300	Hptm F-K. Müller	Halifax	Nr. Mannheim	00.07	4600	10	W
I./JG300	Fw Horst John	Four-eng	Nr. Mannheim	00.10	4300	5	W
3./JG300	Fw Hans Graf	Four-eng	S Mannheim	00.13	4300	1	W
1./NJG6	Ofw Richard Launer	Halifax	5-10km SW Mannheim	00.14	5500	6	W

A rare picture of a Do217 nightfighter in flight. This is from 5./NJG 3. Note the radar array. (H.Scholl)

3./NJG1	Oblt Werner Baake	Halifax	E Germersheim	00.15	4500	17	W
StI/NJG1	Hptm Hans-Dieter Frank	Lancaster	NE Pirmasens	00.15	5600	54	W
3./JG300	Fw Ullrich Veh	Four-eng	Mannheim	00.15		1*	W
StI/NJG6	Hptm Heinrich Wohlers	Stirling	Pirmasens	00.16	4800	26	W
4./JG300	Ofw Richard Lofgen	Stirling	20km S Mannheim	00.18	3800	1	W
3./JG300	Fw Hans Graf	Four-eng	Nr. Mannheim	00.20	4000	2	W
II./NJG1	Lt Heinz-Wolfgang Hoppner	Stirling	Mannheim	00.20	4300	1	W
6./JG300	Lt Werner Treusch	Lancaster	Nr. Mannheim	00.20		3	W
3./JG300	Fw Ullrich Veh	Four-eng	S Mannheim	00.20		2*	W
6./JG300	Lt Klaus Bretschneider	Lancaster	S Mannheim	00.20		2	W
7./NJG1	Hptm August Geiger	Lancaster	Oppen, Saarland	00.24	5000	46	W
4./JG300	Uffz Friedrich Hagios	Four-eng	25km SSE Mannheim	00.24	2000	3	W
1./NJG6	Ofw Richard Launer	Stirling	10km WSW Mannheim	00.25	5200	7	W
StII/JG300	Maj Kurt Kettner	Stirling	SW Mannheim	00.28	5500	2	W
5./NJG4	Hptm H. Niebelschutz.von	Stirling	Mannheim	00.29	5000	6	W
StI/NJG6	Hptm Heinrich Wohlers	Stirling	Jagersbuch	00.29	5200	27	W
7./JG300	Fw Kurt Emler	Lancaster	Munich	00.32	2200	1	W
7./JG300	Ofw Seeler	Four-eng	Mannheim	00.32	2200	1	W
StIII/NJG1	Maj E. Lippe-Weissenfeld	Stirling	7km SE Hassloch	00.36	5500	46	W
./JG300	Hptm F-K. Muller	Stirling	10km N Speyer	00.37	4800	11	W
7./JG300	Fw Kurt Emler	Halifax	Munich	00.44	3500	2	W
9./NJG1	Lt Hans-Heinz Augenstein	Stirling	Schönborn	00.45	5500	12	W
7./NJG1	Hptm August Geiger	Halifax	Rheinhausen	00.45	4900	46	W
1./JG302	Ofw Nunuff	Lancaster	Munich	00.47	4500	1	W
7./JG300	Ofw Paul Streuff	Four-eng	N Mannheim	00.47	4500	1	W
3./NJG6	Oblt H-M. Hadeball	Halifax	2km SW Elmsheim	00.47	5400	17	W
2./NJG3	Fw Rudolf Frank	Halifax	SSE Mannheim	00.50	4800	18	W
II./JG300	Oblt Rudi Hoffmann	Lancaster	Nr. Mannheim	00.51	4500	1	W
6./JG300	Fw Kofler	Lancaster	Munich	00.52	4200	1	W
3./JG302	Fw Herbert Herre	Lancaster	Munich	00.53	4300	1	W
2./NJG6	Hptm Karl Floitgraf	Halifax	Sandhofen	00.57		2*	W

2./NJG6	Hptm Gerhard Friedrich	Halifax	Nr. Mannheim-S'hofer	00.57	3500	9	W
2./NJG1	Oblt Lenz Finster	Halifax	Lugwigshafen	00.58	4800	4	W
3./NJG6	Oblt H-M. Hadeball	Halifax	SE Lindenberg	00.59	5300	18	W
5./NJG4	Hptm H. Niebelschutz.von	Halifax	Mannheim	00.59	4500	7	W
9./NJG1	Lt Hans-Heinz Augenstein	Lancaster	5km NW Landau	01.07	4900	13	W
2./NJG1	Oblt Lenz Finster	Halifax	32km ESE Reims	01.10	2800	5	W
4./NJG1	Lt Wilhelm Henseler	Halifax	Hornissenwiesen	01.15	4800	2	W
3./NJG1	Ofw Werner Hellenbrucks	Lancaster	Oranburgweiler	01.30	3800	3	W
3./NJG1	Oblt Werner Baake	Lancaster	Kaiserslautern	02.00	4500	18	W

6 September 1943

9./NJG101	Uffz Konty	B-17		11.15	6000	1	W
9./NJG101	Uffz Busch	B-17		11.28	700	1	W

6-7 September 1943

StI/NJG100	Hptm Rudolf Schoenert	R-5	Russia	21.03	900	36	E
1./NJG100	Hptm Alois Lechner	R-5	Russia	21.15	1000	26	E
1./NJG100	Hptm Alois Lechner	Mitchell	Russia	22.36	1400	27	E
5./JG300	Lt Klaus Bretschneider	Four-eng	Munich area			3*	W
StI/NJG5	Hptm Werner Hoffmann	Four-eng	E Munich	00.24	5800	10	W
StI/NJG6	Hptm Heinrich Wohlers	Halifax	SW Munich	00.29	4500	28	W
7./NJG101	Oblt Jakob Bender	Halifax	SW Munich	00.30	4500	4	W
./JG300	Hptm F-K. Müller	Lancaster	15km SE Munich	00.37	4300	12	W
5./NJG101	Lt Schulze	Stirling	SE Munich	00.37	5600	1	W
9./JG300	Fw Nachtigall	B-24	Munich	00.38	4500	1	W
StII/NJG101	Oblt Alfred Günther	Lancaster	SW Munich	00.40	7000	1	W
StII/NJG5	Hptm Manfred Meurer	Lancaster	S Munich	00.40	5500	54	W
6./NJG1	Hptm E-W. Bonin.von	Halifax	SW Munich	00.45	4500	21	W
StI/NJG5	Hptm Werner Hoffmann	Lancaster	S Munich	00.45	5000	11	W
./NJG1	Oblt Werner Husemann	Halifax	30km S Munich	00.45	4000	13	W
8./NJG3	Oblt Günther Stege	Stirling	W Munich	00.45	5200	4	W
6./NJG1	Hptm E-W. Bonin.von	Halifax	SW Munich	00.47	4600	22	W
2./JG300	Fw Horst John	Halifax	E Munich	00.48	4600	6	W
./JG300	Hptm F-K. Müller	Lancaster	Munich	00.48	5000	13	W
StII/NJG5	Hptm Manfred Meurer	Lancaster	25km S Munich	00.50	4000	55	W
9./JG300	Uffz Werner Dienst	Halifax	W Munich	00.52	4000	3	W
I./JG300	Fw Schäfer	Halifax	SSW Munich	00.55	5000	1	W
6./NJG101	Ofw Jeschke	Halifax	5-10km S Munich	01.10	500	1	W

7 September 1943

3./NJG100	Lt Waldemar Rathke	PS-84	E Newel	10.50	3500	5	E

7-8 September 1943

StI/NJG100	Hptm Rudolf Schoenert	DB-3	E Slobodka	20.58	4000	37	E
1./NJG100	Ofw Helmut Dahms	R-5	S Rabanitza	21.08	1600	3	E
1./NJG100	Ofw Helmut Dahms	U-2	W Ljuboshna	21.30	1300	4	E
StI/NJG100	Hptm Rudolf Schoenert	DB-3	Brünischlschi	21.48	2600	38	E
3./NJG100	Oblt Hans Gref	PS-84	W Newel	22.12	2400	11	E
StI/NJG100	Hptm Rudolf Schoenert	PS-84	Grischany	22.17	2100	39	E
StI/NJG100	Hptm Rudolf Schoenert	PS-84	Mischaly	22.29	4500	40	E
1./NJG100	Ofw Helmut Dahms	R-5	Russia	23.11	1000	5	E

The Wilde Sau pilots were out in force on the night of 22nd September. Feldwebel Anton Benning of 2./JG 300 (left) and Oberfeldwebel Heinz Gossow of 1./JG302 (right) each claimed two Lancasters and Feldwebel Hermann Wischnewski also from 3./JG 300 (centre) claimed a Halifax. All would survive the war, winning the Ritterkreuz. Benning and Wischnewski would each be credited with 28 victories and Gossow twelve, achieved in both day and night operations.

3./NJG100	Lt Waldemar Rathke	PS-84	NE Newel	01.27	3500	6	E
3./NJG100	Lt Waldemar Rathke	PS-84	NE Newel	01.57	2200	7	E

9-10 September 1943

1./NJG200	Oblt Jank	DB-3	N Oranienbaum	21.18		4*	E

10-11 September 1943

1./NJG2	Oblt Albert Schulz	Wellington	SW Emden	23.45		12	W

12-13 September 1943

3./NJG100	Oblt Hans Gref	PS-84	SW Newel	20.15	2800	12	E
1./NJG6	Ofw Kurt Karsten	Lancaster	SW Stuttgart	23.29		5	W
3./NJG100	Lt Waldemar Rathke	R-5	S Newel	24.00	1200	8	E
3./NJG100	Lt Waldemar Rathke	R-5	NE Newel	00.16	1100	9	E

13-14 September 1943

3./NJG100	Lt Waldemar Rathke	PS-84	NE Newel	02.18	2900	10	E
1./NJG3	Uffz Rudolf Franz	Wellington	10km NE Hardenberg	03.10	3600	1	W
2./NJG3	Uffz Kutter	Wellington	25km NE Siegen	03.45		1	W

14-15 September 1943

11./NJG6	Uffz Karl Grundmann	Il-4	5km S Kaschina	20.43	1700	1	E
5./NJG2	Uffz Unuke	Wellington	over the Sea	23.48		1	W
1./NJG200	Oblt Jank	TB-7	Russia	21.36		5	E
2./JG2	Fw Herbert Benz	Halifax	L'Aigle	23.30	4800	2	W

16-17 September 1943

11./NJG3	Oblt Johannes Hiedlmayr	Halifax	SE Aurich	22.22	350	1	W
2./JG2	Unknown pilot	Stirling	12km S Lisieux	03.20	3500	-	W
2./JG2	Fw Herbert Benz	Stirling	19km SW Lisieux	03.22	3200	3	W
11./NJG3	Lt Richard Burdyna	Halifax	W Ringsted	04.45		1	W

17 September 1943

5./NJG200	Ofw Josef Kociok	Boston	Russia	14.00	2000	18	E
5./NJG200	Ofw Josef Kociok	Boston	Russia	14.01	2000	19	E
5./NJG200	Ofw Josef Kociok	MiG-1	Russia	14.04	2000	20	E

18 September 1943

5./NJG200	Ofw Haufe	IL-2	Russia	07.03	200	1	E
5./NJG200	Ofw Josef Kociok	LaGG-	Russia	10.05	50	21	E

18-19 September 1943

2./NJG100	Ofw Maisch	DB-3	Saporotnya	21.47	2500	3	E

19-20 September 1943

./NJG1	Maj Werner Streib	Halifax	Zenderen SE Eindhoven	01.01	100	65	W

20-21 September 1943

3./NJG100	Oblt Hans Gref	PS-84	NW Newel	21.47	3000	13	E
3./NJG100	Lt Waldemar Rathke	DB-3	SW Newel	23.10	3800	11	E
StI/NJG100	Hptm Rudolf Schoenert	DB-3	Kislyaki	23.27	5400	41	E
StI/NJG100	Hptm Rudolf Schoenert	DB-3	W Mirgorod	23.57	4500	42	E

21-22 September 1943

4./NJG3	Ofw Karl Haisch	Wellington		23.38	1400	9	W
4./NJG3	Ofw Karl Haisch	Wellington		00.18	2300	19	W
StII/NJG3	Maj Günther Radusch	Wellington		01.01	1800	35	W

22-23 September 1943

II./NJG3	Lt Hans Wallrath	Halifax	35km S Hanover			1	W
7./JG300	Fw Kurt Emler	Lancaster	SE Bremen	21.45		3	W
7./JG300	Fw Heinz Gossow	Lancaster	S Bremen	21.45		1	W
3./NJG3	Oblt Heinz Ferger	Lancaster	10-15km SS- Bremen	22.15	4600	4	W
II./JG302	Fw Pietsch	Halifax	Hanover	22.25		1	W
2./NJG3	Uffz W. Golzenleuchter	Lancaster	Scharmbeck b. Bremen	22.26	5200	1	W
8./JG301	Fw Erich Teubner	Lancaster	40km W Hanover	22.26		1	W
9./NJG3	Hptm Siegfried Hahn	Halifax	S Rotterdam	22.27	5600	3	W
8./JG300	Fw Anton Benning	Lancaster	15-20km SE Bremen	22.29		1	W
1./JG302	Ofw Heinz Gossow	Lancaster	S Bremen	22.30		2	W
12./NJG1	Lt Heinz Grimm	Lancaster	Braunstedt	22.30	3800	25	W
8./JG301	Fw Erich Teubner	Lancaster	NW Hanover	22.30		2	W
II./JG302	Fw Pietsch	Halifax	Hanover	22.35		2	W
2./JG302	Fw Anton Benning	Lancaster	NE Bremen	22.35		2	W
7./JG300	Fw Kurt Emler	Lancaster	SE Bremen	22.35		4	W
7./NJG1	Hptm August Geiger	Stirling	Gestorf	22.35	4000	47	W
3./JG300	Lt Hermann Stumpe	Four-eng	Hanover	22.37		3	W
3./JG300	Fw Hermann Wischnewski	Halifax	Hanover	22.37		2	W

Helmut Lent had taken command of NJG 3 on 1st August and was visited at Stade by Reichs Marschall Herman Goering. Goering was no technocrat and had little time for the latest developments; "Radar? It's all boxes with wires - and I hate boxes with wires". His understanding of air fighting in any respect was limited to World War combats and he demonstrated little understanding or respect for the fighter aircrews that now attempted to stem a growing avalanche of day and night Allied air attacks. (left) Lent, (centre) Goering, (right background) General Josef 'Beppo' Schmid. (via H.Scholl)

2./NJG3	Fw Rudolf Frank	Halifax	W. edge of Hanover	22.39	5700	19	W
8./NJG1	Lt Heinz Bock	Halifax	Strohnen	22.40	4500	2	W
./NJG3	Maj Helmut Lent	Stirling	Kröhnsberg	22.41	4200	73	W
6./NJG3	Oblt Helmut Niklas	Halifax	Hanover	22.44	5600	3	W
9./NJG1	Lt Hans-Heinz Augenstein	Lancaster	Hüppede	22.45	5400	14	W
6./NJG3	Oblt Helmut Niklas	Halifax	Hanover	22.47	5500	4	W
8./NJG1	Uffz Horst Becker	Halifax	area S Hameln	22.50	5200	1	W
I./NJG1	Lt Walter Schön	Halifax		22.50	5200	7	W
3./NJG5	Lt Robert Wolf	Stirling	E. edge of Hanover	22.52	5500	1	W
2./JG302	Ofw Artur Gross	Lancaster	NE Hanover	22.54		1	W
5./JG300	Lt Ernst-Erich Hirschfeld	Stirling	Hanover	22.55		1	W
3./JG300	Oblt Hermann Bertram	E/A	20km NW Bremen	22.57		1	W
5./NJG3	Hptm Paul Szameitat	Stirling	S-W Hanover	23.02	40000	14	W
1./NJG5	Ofw Richard Tripke	Halifax	40km SSW Hanover	23.02	5000	1	W
9./JG300	Fw Herbert Chantelow	Lancaster	N-W Hanover	23.03	5000	1	W
5./JG301	Ofw Kurt Welter	Lancaster	Hanover	23.04	4300	1	W
5./JG301	Ofw Kurt Welter	B-24	Hanover	23.12	2000	2	W
3./JG300	Ofw Ullrich Veh	Stirling	Hanover	23.20	4800	3	W
10./NJG1	Oblt Hermann Greiner	Halifax	Bergfeine	23.28	4300	12	W
7./JG300	Ofw Ernst Fleischmann	Mosquito	Oldenburg	24.00	3500	2	W
5./JG300	Lt Klaus Bretschneider	Lancaster				4*	W
14./KG2	Maj Wolf Dietrich Meister	Four-eng	S Lincoln	01.34	800	1	W

23-24 September 1943

3./NJG6	Oblt H-M. Hadeball	Halifax	16km SW Munich	22.39	5400	19	W
3./JG300	Fw Ullrich Veh	Four-eng	S Mannheim	22.50	3800	4	W
9./NJG1	Lt Hans-Heinz Augenstein	Lancaster	3km N Speyer	22.50	5000	15	W
2./NJG1	Uffz Josef Brunner	Lancaster	S Dürkheim	22.50	5300	2	W
6./NJG1	Lt Johannes Hager	Halifax	Mannheim	22.50		22	W
8./NJG1	Hptm Herbert Lütje	Halifax		22.50	5200	31	W
./JG300	Hptm Ewald Janssen	Four-eng	Mannheim	22.51		2	W
8./NJG1	Lt Heinz Bock	Lancaster	10km W Ludwigshafen	22.55	5500	3	W

2./NJG1	Uffz Josef Brunner	Lancaster	8km W Koblenz	22.56	3500	3	W
3./NJG1	Lt Josef Nabrich	Lancaster	5km W Ludwigshafen	22.58	4800	5	W
6./JG300	Uffz Anton Gaissmayer	Lancaster	15km NW Mannheim	22.59		2	E
9./NJG1	Lt Hans-Heinz Augenstein	Lancaster	3km N Kirchhelmb'den	23.00	4800	16	W
12./NJG1	ObLt H-W. Schnaufer	Stirling	5km S Kirchheim-B'dn	23.00	4600	26	W
2./JG300	Ofw Günther Migge	Stirling	Mannheim	23.02		1	W
2./JG300	Fw Willi Rüllkotter	Four-eng	Mannheim	23.03	8000	3	W
9./NJG1	Uffz Heinz Amsberg	Stirling	Edsheim	23.07	3500	1	W
StIV/NJG1	Lt Heinz Grimm	Lancaster	5km E Mannheim	23.07	4900	26	W
II./JG300	Oblt Rudi Hoffmann	Lancaster	W Mannheim	23.07	5500	2	W
9./JG300	Oblt Robert Plewa	Halifax	W Mannheim	23.10		4	W
1./NJG3	Oblt Martin	Lancaster	35km SS- Mannheim	23.13	5500	1	W
1./JG300	Ofw Walter Mackens	Four-eng	Mannheim	23.14	5000	1	W
2./JG302	Ofw Artur Gross	Lancaster	Mannheim	23.15	2200	2	W
3./NJG3	Fw Günther	Lancaster	20km NW Mannheim	23.15	5500	1	W
8./JG300	Oblt Gerhard Stamp	Lancaster	10km W Mannheim	23.15	2200	2	W
2./NJG3	Ofw Erich Heitmann	Stirling	W Ludwigshafen	23.16	5300	3	W
2./NJG1	Oblt Lenz Finster	Lancaster	Mannheim	23.20	5500	6	W
5./JG300	Ofw Norbert Graziadei	Lancaster	Mannheim	23.20	4500	1	W
6./JG300	Fw Hans Kugler	Lancaster	5km SE Mannheim	23.20	3900	4	W
5./JG300	Ofw Kurt Greineisen	Lancaster	Mannheim	23.20		1*	W
2./NJG6	Ofw Martin Becker	Lancaster	Nr. Speyer	23.21	5000	1	W
11./NJG5	Uffz Josef Sallmütter	Lancaster	6km N-E Mannehein	23.25	600	1	W
11./NJG1	Ofw Heinz Vinke	Lancaster	10km S	23.25	4600	29	W
2./NJG3	Lt Hermann Leube	Stirling	SW Ludwigshafen	23.26	5600	2	W
2./NJG3	Lt Hermann Leube	Lancaster	SW Ludwigshafen	23.28	5200	3	W
10./NJG1	Oblt Hermann Greiner	Lancaster	10km S Lampertheim	23.30	5900	13	W
2./NJG3	Fw Rudolf Frank	Lancaster	Mannheim	23.42	5500	20	W
5./JG300	Lt Klaus Bretschneider	Four-eng				5*	W

24 September 1943

5./NJG200	Ofw Josef Kociok	MBR-2	Russia	11.48	1500	22	E
2./NJG100	Ofw Dr..er	Yak-7	Russia	16.20	50	-	E

24-25 September 1943

7./NJG2	Lt Karl Heinz Brandes	Lancaster		02.13	1900	3	W

26-27 September 1943

5./NJG200	Ofw Josef Kociok	MBR-2	NE Newel	20.20		23	E
3./NJG100	Lt Waldemar Rathke	PS-84	NE Cherkassy	22.19	400	12	E
1./NJG100	Lt Josef Pützkuhl	PS-84	Russia	23.03	1500	6	E

27 September 1943

2./NJG100	Uffz Fuchs	R-5	Russia	10.15	40	1	E

27-28 September 1943

3./NJG100	Hptm Kurt Bonow	PS-84	Felenjuka	20.54	1200	5	E
7./NJG1	Hptm August Geiger	Lancaster	Zuiderzee	22.10	5500	48	W
3./NJG1	Oblt Werner Baake	Lancaster	Quakenbruck	22.40	4000	19	W
3./NJG1	Ofw Werner Hellenbrucks	Halifax	nr Neuenkirchen	22.47	4900	4	W
I./JG300	Oblt Robert Plewa	Stirling	Hannover	22.58	3000	5*	W

The British had known of the German developments in airborne radar and, in 1942, began to formulate countermeasures. In a courageous action, Pilot Officer Jordan of 192 Squadron, flying a specially equipped Wellington, deliberately exposed his aircraft to fighter attack, the on-board detection equipment pin-pointing the exact frequency of the Lichtenstein transmissions. He brought his badly damaged aircraft home and the information obtained formed the basis of Serrate. This detected German airborne radar and, when fitted to Beaufighters of 141 Squadron, was used to 'home' onto Luftwaffe nightfighters. On the night of 29th September, Hauptmann August Geiger of 7./NJG 1, a holder of the Ritterkreuz and with 54 victories credited, was caught over the Ijsselmeer by Wing Commander J.R.D.Braham. He was shot down and killed, with his Bordfunker Feldwebel Koch. Many years later, the wreckage of his aircraft was found, still showing the victory bars on the rudder.

9./NJG1	Lt Hans-Heinz Augenstein	Halifax	W Lubbecke	23.00	6200	17	W
2./NJG3	Ofw Erich Heitmann	Lancaster	1km E Hanover	23.00	5500	4	W
9./NJG1	Lt Kraft	Halifax	S Hildesheim	23.00	4500	1	W
3./JG302	Fw Herbert Herre	Stirling	Gross-Piesen	23.03	4600	2	W
5./JG300	Lt Ernst-Erich Hirschfeld	Lancaster	Hanover	23.04	4300	2	W
7./JG300	Fw Gnerern	Stirling	S Hanover	23.06	4600	1	W
11./NJG5	Uffz Josef Sallmutter	Four-eng	Hanover	23.07	5000	2	W
1./JG302	Oblt Karl-Heinz Seeler	Halifax	Geidingen	23.07		1*	W
II./JG301	Uffz Andreas Hartl	B-24	WSW Hanover	23.08	5500	1	W
2./NJG6	Oblt Heinz Knigge	Stirling	20km NNE Hanover	23.08		2	W
./JG300	Hptm F-K. Müller	Lancaster	Hildesheim	23.09	3800	14	W
1./JG300	Ofw Walter Mackens	Four-eng	WSW Hanover	23.10	4000	2	W
9./NJG3	Hptm Siegfried Hahn	Lancaster	10-15km NW Brauns'g	23.10		4	W
4./NJG3	Oblt Gerhard Raht	Lancaster	Hanover	23.10	4800	13	W
7./NJG2	Uffz Hans Werthner	Stirling	Hanover	23.12	4800	1	W
9./NJG3	Ofw Noack	Four-eng	Hannover-Braunschweig	23.14		1	W
I./JG300	Hptm Graf Karl Resseguier	B-17	W Hanover	23.15	5500	1	W
5./JG300	Lt Ernst-Erich Hirschfeld	Stirling	W Hanover	23.15	4500	3	W
4./JG300	Ofw Hans Wegener	Lancaster	NW Braunschweig	23.17	4500	1	W
5./NJG2	Oblt Walter Riedlberger	Stirling	SW Edge Hanover	23.17	4000	2	W
StI/NJG5	Hptm Werner Hoffmann	Lancaster	Hanover	23.18	4500	12	W
1./NJG2	Lt Willi Reisgies	Stirling	15km SSE Hanover	23.19	3500	1	W
3./JG300	Ofw Heinz Wohlecke	Four-eng	W Hanover	23.19		3	W
3./JG300	Ofw Ullrich Veh	Four-eng	NE Hannover	23.20		5*	W
3./JG300	Ofw Lothar Sachs	Four-eng	NE Hanover	23.20	4300	1	W
9./NJG3	Gfr Koenig	Wellington	NW Hanover	23.20	4000	2	W
StII/NJG5	Hptm Manfred Meurer	B-17	S Hanover	23.20	6000	56	W
5./NJG3	Uffz Heinrich Wortmann	Halifax	SW Hanover	23.20	5000	2	W
II./JG301	Oblt Sauter	Lancaster	WSW Hanover	23.23	9500	1	W
II./JG301	Uffz Andreas Hartl	B-24	30km SW Hanover	23.24	4500	2	W
./NJG5	Maj Günther Radusch	Lancaster	SW edge Hanover	23.24	3600	36	W
5./NJG2	Ofw Willi Glitz	Stirling	5-8km SW Hanover	23.25		1	W

3./NJG5	Uffz Ernst Reitmeyer	Stirling	24km S Hanover	23.25	4000	1	W
6./NJG5	Lt Josef Kraft	Lancaster	Hanover	23.26		4	W
I./JG300	Oblt Robert Plewa	Stirling	Hannover	23.26	4500	6	W
7./JG300	Fw Kurt Emler	Lancaster	S Hanover	23.27	7000	5	W
7./NJG5	Uffz Hermann Schönthier	Lancaster	7km NNE Hanover	23.27	5700	3	W
7./NJG1	Hptm August Geiger	Halifax	Nr. Munich	23.30		49	W
8./NJG1	Uffz Emil Heinzelmann	Stirling	5km NW Hameln	23.30	5400	5	W
12./NJG1	ObLt H-W. Schnaufer	Halifax	Nr. Stemmen	23.31	4500	27	W
3./NJG6	Hptm H-M. Hadeball	Halifax	13km NNE Minden	23.35	5200	20	W
8./NJG1	Hptm Herbert Lütje	Halifax	Neustadt	23.40	5100	32	W
11./NJG1	Oblt Martin Drewes	Halifax	Stadthagen	23.50	4400	6	W
7./NJG1	Hptm August Geiger	Halifax	S Papenburg	00.01	5500	50	W
14./KG2	Oblt Rudolf Abrahamczik	Four-eng	Cromer	01.12	350	1	W

29-30 September 1943

5./NJG200	Fw Pies	DB-3	Russia	21.00		1	E
5./NJG200	Fw Pies	PS-84	Russia	21.30		2	E
9./NJG1	Oblt Fritz Lau	Lancaster	4km N Raalte	21.37	4500	2	W
III./NJG1	Hptm E. Lippe-Weissenfeld	Halifax	7km S Hengelo	21.44	6000	47	W
III./NJG1	Hptm E. Lippe-Weissenfeld	Halifax	2km NW Legden	21.55	6000	48	W
1./JG300	Lt Otto Brüning	Lancaster	Bochum	22.12		1	W
I./JG300	Fw Arthur Gross	Stirling	Bochum	22.16		1	W
11./NJG1	Uffz Hans Gienanth.von	Lancaster	Deventer	22.42	5300	1	W
10./NJG3	Oblt Werner Speidel	Lancaster	E Dortmund	01.43	2300	1	W

30 September 1943

2./NJG100	Uffz Fuchs	R-5	Russia	10.40	150	2	E

30 September-1 October 1943

2./JG302	Uffz Schneider	Four-eng	Berlin			1	W

1 October 1943

2./NJG100	Uffz Fuchs	R-5	Russia	12.05	100	3	E

2-3 October 1943

11./NJG3	Uffz Eduard Florian	Halifax	Skaggerak	22.49	2200	1	W
StIII/NJ101	Maj Wolfgang Thimmig	Lancaster	2km W Kirchheim	23.05	6200	19	W
./JG300	Hptm F-K. Müller	Halifax	middle of Munich	23.34	5000	15	W
2./JG300	Fw Willi Rüllkotter	Stirling	S Munich	23.38		4	W
1./NJG3	Lt Reinhard Keib	Halifax	Karolinenfeld	23.45	6500	1	W
6./NJG101	Fw Ludwig	E/A	SSW Munich	23.54	5200	1	W
5./NJG101	Ofw Hans Kreutzberg	Four-eng	SSW Munich	00.05	3600	1	W
StIII/NJG3	Hptm Rudolf Sigmund	Lancaster	Walchensee.S Munich	00.05	5600	28	W

3-4 October 1943

11./NJG1	Ofw Heinz Vinke	Lancaster		21.12		30	W
9./NJG1	Oblt Fritz Lau	Stirling	3km N Osnabrück	21.49	4000	3	W
12./NJG1	ObLt H-W. Schnaufer	Halifax	8km N Minden	21.50	5300	28	W
2./NJG3	Fw Rudolf Frank	Halifax	15km SW Oldenburg	21.56		21	W
7./JG300	Fw Kurt Emler	Stirling	NW Hanover	22.04	4200	6	W
4./NJG3	Hptm Hans Baer	Four-eng	NW Kassel	22.06	4500	8	W

A Ju88C-6 of 7./NJG 5, with Lichtenstein SN-2 radar array. (H.Scholl)

5./JG300	Lt Ernst-Erich Hirschfeld	Stirling	S Kassel	22.27	4500	4	W
./NJG3	Maj Helmut Lent	Stirling	Harleshausen	22.27	6300	74	W
5./JG301	Ofw Kurt Welter	Halifax	SW Kassel	22.29	5000	3	W
5./NJG2	Oblt Walter Riedlberger	Halifax	2-5km NNW Kassel	22.33	5600	3	W
6./JG300	Lt Werner Treusch	Stirling	S Kassel	22.34	4300	4	W
9./NJG1	Lt Hans-Heinz Augenstein	Halifax	7km E Kassel	22.37	4200	18	W
4./NJG5	Hptm Ernst Zechlin	Halifax	40-50km WS- Kassel	22.39	3500	1	W
./JG300	Hptm F-K. Müller	Four-eng	10-15km SE Kassel	22.41	4000	16	W
5./JG301	Ofw Kurt Welter	Halifax	ESE Kassel	22.41	2000	4	W
5./NJG5	Hptm Leopold Fellerer	Lancaster	40km S Kassel	22.57	4800	13	W
1./NJG6	Ofw Wilhelm Engel	Halifax	1.5km WSW St.Vith	23.15	5100	6	W
11./NJG1	Ofw Heinz Vinke	Halifax	W Terschelling	23.22		31	W
5./JG300	Lt Klaus Bretschneider	Halifax				6*	W
2./NJG100	Ofw Maisch	R-5	N Neu-Augustinovka	00.12	1000	4	E

4th October 1943

11./NJG101	Hptm Franz Evers (shared)	B-17	Nr. Koblenz	12.00		1	W
4./NJG4	Lt Jakob Schaus	B-17	Juvincourt	12.55	500	3	W

4-5 October 1943

1./NJG100	Oblt Adolf Kaiser	PS-84	Prowski	18.58	3300	3	E
1./NJG100	Oblt Adolf Kaiser	PS-84	Kanewski	20.07	3400	4	E
StI ./NJG4	Maj Wilhelm Herget	Halifax	Givet	20.49		30	W
1./NJG4	Oblt Ludwig Meister	Halifax	Neuerberg	21.02	5000	18	W
1./NJG4	Oblt Ludwig Meister	Halifax	Trier	21.03		19*	W
1./NJG100	Hptm Alois Lechner	E/A	Russia	21.25		28	E
1./NJG100	Ofw Christian Rohrs	PS-84	Mogilev	21.25	3200	2	E
3./NJG6	Hptm H-M. Hadeball	Halifax	20km SE Traben-T'ch	21.27	5200	21	W
StI ./NJG4	Maj Wilhelm Herget	Halifax	WNW Neufchateau	21.30	5000	31	W
./JG300	Hptm F-K. Müller	Halifax	Hulst	21.35	5200	17	W
4./JG300	Ofw	Four-eng		21.35	5000	-	W
StI/NJG6	Oblt H J Birkenstock	Halifax	10km SE Limburg	21.49	5500	3	W
4./JG300	Ofw	Four-eng	Frankfurt-am-Main	22.00	5000	-	W
4./JG300	Lt Otto Brüning	Four-eng	Frankfurt	22.05	4500	2	W
StI ./NJG4	Maj Wilhelm Herget	B-17	Friedberg	22.08	7000	32	W

8./NJG3	Uffz Rudnik	Halifax	SE Kassel	22.25	5000	1	W
3./NJG1	Lt Josef Nabrich	Halifax	Florennes	22.36	5200	6	W
7./JG302	Hptm Helmut Suhr	Stirling	Mannheim	22.43		1	W
5./JG300	Lt Klaus Bretschneider	Lancaster				7*	W

6-7 October 1943

1./NJG100	Hptm Hermskerck.von	DB-3	Polna	18.11	3400	1	E
1./NJG100	Lt Eberhard Kaross	DB-3	Rogatpuko	18.14	1700	1	E
1./NJG100	Hptm Hermskerck.von	PS-84	Repotka	18.54	3200	2	E
3./NJG100	Lt Waldemar Rathke	U-2	Newel	19.09	1400	13	E
1./NJG100	Hptm Hermskerck.von	PS-84	SW Gejasiwicz	19.10	3000	3	E
1./NJG100	Hptm Hermskerck.von	PS-84	Gaschin	19.52	3200	4	E
1./NJG100	Oblt Adolf Kaiser	PS-84	Fernikov	20.33	3200		E

7-8 October 1943

11./NJG1	Ofw Heinz Vinke	Stirling	50km NW Vlieland	20.26	400	32	W
5./NJG4	Oblt Ferdinand Christiner	Lancaster	SW Commercy	23.20	5800	3	W
5./NJG4	Oblt Ferdinand Christiner	Lancaster	W Rachecourt	23.24	3800	4	W
2./NJG3	Fw Rudolf Frank	Wellington	30km N Norderney	00.32	1400	22	W
10./NJG3	Uffz Heinz Koppe	Wellington	N Hannover	00.45	2100	2	W

8 October 1943

5./NJG3	Hptm Paul Szameitat	B-24		16.17	6700	15	W
6./NJG3	Oblt Helmut Niklas	B-24	SSW Helgoland	16.35	200	5	W
II./NJG2	Ofw Ternieden	B-17				3	W

8-9 October 1943

11./NJG1	Ofw Heinz Vinke	Halifax	30km S Gröningen	00.28	5600	33	W
9./NJG1	Lt Kraft	Halifax	W Nienburg	01.08	4500	2	W
2./NJG3	Hptm Martin.Becker	Halifax	1½km S Nassdorf	01.10	6000	1	W
2./NJG3	Uffz W. Golzenleuchter	Halifax	Kiefelhorst	01.11	5200	2	W
12./NJG1	ObLt H-W. Schnaufer	Halifax	Nr. Schwarforden	01.13	5100	29	W
7./JG300	Fw Kurt Emler	Stirling	WNW "Marie"	01.17	6200	7	W
12./NJG3	Lt Karl Rechberger	Lancaster	30km SW	01.20	4500	3	W
6./NJG101	Ofw Ludwig	Stirlng	SE Bremen	01.23		3	W
II./JG301	Lt Kummer	Lancaster	5km NW Hanover	01.24	6200	1	W
4./NJG101	Ofw Fritz Brandt	Four-eng	SE Bremen	01.25		1	W
12./NJG3	Lt Karl Rechberger	Four-eng	Hanover	01.25	5300	4	W
5./NJG3	Hptm Paul Szameitat	Lancaster	SW Stade	01.25	4500	16	W
II./JG301	Oblt Karl-Heinz Seeler	Halifax	2km S Bremen	01.27	2800	2	W
./JG300	Hptm F-K. Müller	Halifax	NW Hanover	01.29	4800	18	W
7./JG300	Fw Kurt Emler	Stirling	N Hanover	01.32	5900	8	W
6./NJG5	Lt Josef Kraft	Stirling	40-50km N Bremen	01.32	4000	5	W
StIV/NJG5	Hptm H. Niebelschutz.von	Lancaster	Hanover	01.33		8	W
StIII/NJG2	Hptm Paul Semrau	Halifax	35km W Den Haag	01.35	5900	15	W
II./NJG2	Uffz Wakhye	Halifax	Bremen	01.40		1	W
9./NJG1	Uffz Heinz Amsberg	Lancaster	Schwarmstedt	01.42	5300	2	W
9./NJG1	Lt Hans-Heinz Augenstein	Stirling	SSW Hanover	01.42	6100	19	W
3./NJG5	Ofw A. Klaiber	Lancaster	15km N-E Hannover	01.42	5200	1	W
12./NJG1	ObLt H-W. Schnaufer	Halifax	Nr. Holtensen	01.42	5300	30	W
1./JG300	Lt Otto Brüning	Four-eng	W Hanover	01.43	3500	3	W

Two more pilots were climbing the ranks of the Experten in October. On the night of 20th October Oberleutnant Gerhard Raht of 4./NJG 3 claimed a 'Viermot' and a Lancaster for his 14th and 15th victories, while Hptm Paul Szameitat of 5 Staffel claimed two Lancasters on 18th/19th for his 17th and 18th. Raht survived the war with 54 credited victories, but Szameitat, his aircraft crippled by return fire, crashed near Buckeburg and was killed. He was

5./NJG3	Uffz Otto Kutzner	Lancaster	S Hanover	01.46	3000	1	W
4./NJG3	Lt Georg Führer	Four-eng	15km SW Hanover	01.48	4500	2	W
9./NJG3	Hptm Siegfried Hahn	Halifax	Langenhagen	01.48	5400	5	W
6./JG300	Lt Werner Treusch (shared)	Lancaster	SW Hanover	01.48	3900	5	W
3./NJG5	Uffz Ernst Reitmeyer	Lancaster	NE Hanover	01.50	5000	2	W
3./NJG5	Maj Günther Radusch	Lancaster	20km W Hanover	01.51	4200	37	W
4./NJG5	Hptm Ernst Zechlin	Lancaster	40km WSW Hanover	01.52	6200	2	W
6./NJG101	Lt Hans-Joachim Beck	Four-eng	SW Hannover	01.54		1	W
4./NJG5	Ofw Heinz Säwert	Lancaster	W Hanover	02.00	5500	8	W
11./NJG5	Oblt Gustav Tham	Lancaster	10km W Hanover	02.04		4	W
2./NJG1	Uffz Josef Brunner	Halifax	NW Hook of Holland	03.04		4	W

9 October 1943

8./NJG5	Ofw Herbert Altner	B-17		12.14	3500	3	W
9./NJG5	Fw Brendel	B-17		12.14	3500	1	W
6./NJG101	Lt Hans-Joachim Beck	B-17	SW Hanover	12.32		2	W
6./NJG3	Lt Meess	B-17		12.32	4500	3	W
6./NJG101	Fw Ludwig	B-17	SE Bremen	12.40		2	W
StII/NJG3	Oblt Rudolf Szardenings	B-17	20km N-W Schleswig	12.40	5000	1	W
StII./NJG3	Oblt Rudolf Szardenings	B-17	N Schleswig	13.10		2*	W
./NJG2	Obstlt Karl Hülshoff	B-17		14,35		6	W
StI/NJG2	Maj Rolf Jung	B-17	Samso Island	14.58	4500	3	W
9./NJG3	Oblt Erhard Peters	B-17	10km SW Fomo	15.30	4500	1	W

10-11 October 1943

1./NJG100	Ofw Christian Rohrs	DB-3	Russia	19.33	2300	3	E
StI/NJG100	Hptm Rudolf Schoenert	U-2	Ditbrovnovo	22.41	1000	43	E

11-12 October 1943

3./NJG100	Uffz Herbert Simross	DB-3	Russia	19.20	2300	2	E
StI/NJG100	Hptm Rudolf Schoenert	PS-84	N Charvovo	20.54	2000	44	E
StI/NJG100	Hptm Rudolf Schoenert	PS-84	SE Sapolye	21.00	2000	45	E
StI/NJG100	Hptm Rudolf Schoenert	PS-84	N Alexejevo	21.37	2000	46	E
1./NJG100	Hptm Kaufmann	R-5	Schklovo	23.22	1000	1	E

12-13 October 1943

2./NJG100	Oblt Hagenow.von	DB-3	S-E Alexandrovka	17.53	2000	2	E
1./NJG100	Oblt Adolf Kaiser	DB-3	E Dabskovka	17.59	1700	6	E
1./NJG100	Hptm Alois Lechner	DB-3	Ujuki	18.38	2700	29	E
3./NJG100	Fw Helmut Konter	Mitchell	E Erscha	18.39	5700	1	E
1./NJG100	Hptm Alois Lechner	DB-3	Russia	18.56	2800	30	E
1./NJG100	Hptm Alois Lechner	Mitchell	Russia	20.48	2000	31	E
StI/NJG100	Hptm Rudolf Schoenert	U-2	Russia	22.43	2000	47	E
StI/NJG100	Hptm Rudolf Schoenert	DB-3	Babinovitski	01.07	2000	48	E

13-14 October 1943

StI/NJG100	Hptm Rudolf Schoenert	DB-3	Berseja	17.46	6000	49	E
1./NJG100	Oblt Adolf Kaiser	Mitchell	Russia	20.37	1300	7	E
StI/NJG100	Hptm Rudolf Schoenert	DB-3	Russia	21.24	4500	50	E
1./NJG100	Hptm Alois Lechner	DB-3	Russia	21.28	4200	32	E
StI/NJG100	Ofw Richter	PS-84	Russia	21.53	2600	1	E

14 October 1943

7./NJG101	Hptm H-J. Thomeschat	B-17	20km SE Kitzingen	14.40	6800	2	W
3./NJG2	Hptm Eduard Jacob	B-17	NE Mergentheim	14.56	5600	1	W
3./NJG4	Hptm Fritz Söthe	B-17	Revigny	14.56		4	W
3./NJG6	Fw Gunther Bahr	B-17	20km SE Schweinfurt	14.57	7000	4	W
1./NJG6	Ofw Wilhelm Engel	B-17	20km SE Schweinfurt	14.57	7000	7	W
StII/NJG6	Maj Rolf Leuchs	B-17	Bad Mergentheim	15.03	5000	5	W
3./NJG101	Hptm H.Renner	B-17		15.15	6500	1	W
3./NJG101	Ofw Helmut Munchinger	B-17		15.16	7600	1	W
9./NJG101	Oblt Kranz (shared)	B-17	Gadheim/Wurzburg	15.28	3000	1	W
3./NJG2	Oblt Berge [sic]	B-17	Gandelsheim	15.39	5000	1	W
10./NJG101	Hptm Krzywo	B-17	2km S Sennfeld	15.40	5000	2	W
4./NJG6	Lt Funder	B-17	9km NW Wurzburg	15.44		1	W
9./NJG101	Uffz Tachezy	B-17	1km S Sillich	15.45	6000	1	W
5./NJG2	Lt Wendelin Breukel	B-17		16.01		1	W
4./NJG4	Fw Fritz Lehmann (shared)	B-17		16.25		1	W

14-15 October 1943

1./NJG100	Hptm Alois Lechner	PS-84	Stupniki	17.52	1200	33	E
1./NJG100	Hptm Alois Lechner	PS-84	Chausy, Mogilev	18.15	1700	34	E
StI/NJG100	Hptm Rudolf Schoenert	DB-3	W Demenino	18.21	1300	51	E
StI/NJG100	Hptm Rudolf Schoenert	DB-3	W Demenino	18.26	1700	52	E
StI/NJG100	Hptm Rudolf Schoenert	DB-3	W Demenino	18.37	1300	53	E
StI/NJG100	Hptm Rudolf Schoenert	DB-3	Naravoki	20.56	1900	54	E

15-16 October 1943

StI/NJG100	Hptm Rudolf Schoenert	PS-84	Haschuki	20.17	2300	55	E
2./JG2	Uffz Gerhard Frass	Halifax	Livarot	00.58		1	W

16-17 October 1943

1./NJG100	Hptm Alois Lechner	PS-84	Usuniki	19.45	1700	35	E
3./NJG100	Oblt Hans Gref	DB-3	Orscha	21.00	1500	14	E

The wreckage of Warrant Officer B.E.E.Allden's Halifax LW301 of 78 Squadron lies in open countryside some 5-10km from Kassel. It is believed that this crew was the 41st to be shot down by Major Gunther Radusch of Stab NJG 5. He had served with the Condor Legion in Spain, where he gained one victory. He was appointed Kommandeur of I./NJG 1 in June 1940 and survived the war with a total of 65 combat successes.

18-19 October 1943

StI/NJG1	Hptm Manfred Meurer	Lancaster	Erichshagen	20.05	5200	57	W
StIV/NJG1	Hptm Hans-Joachim Jabs	Lancaster	35km N-E Diepholz	20.07	5800	24	W
StI/JG300	Hptm Patz	Lancaster	Hannover	20.15	1800	1	W
StI/JG300	Lt Walter Harjes	B-17	Hannover	20.20	4000	5	W
9./NJG3	Lt Hans Raum	Lancaster	S Hanover	20.20	5500	11	W
9./NJG3	Uffz Koenig	Four-eng	Hanover	20.22	6500	3	W
II./JG301	Oblt Hans Gottuck	Four-eng	W Hanover	20.24	4500	3	W
./JG300	Hptm F-K. Muller	Lancaster	Nr. Hildesheim	20.24	5800	19	W
StI/NJG5	Hptm Werner Hoffmann	Lancaster	5-10km NE Hanover	20.25	5900	13	W
12./NJG1	ObLt H-W. Schnaufer	Halifax	NNW Hanover	20.25	5300	31	W
4./JG300	Hptm Spitzer	Lancaster	WNW Hanover	20.26	4500	1	W
5./NJG3	Hptm Paul Szameitat	Lancaster	10km S Hannover	20.30	6000	17	W
5./NJG3	Hptm Paul Szameitat	Lancaster	15km S Hanover	20.34	6000	18	W
StII/NJG3	Hptm H. Sayn-Wittgenstein	Lancaster	Hanover	20.35	6200	62	W
I./JG300	Fw Ernst Schäfer	Halifax	Hannover	20.37	5000	2	W
4./NJG2	Ofw Hans Berschwinger	Lancaster	NW Hanover	20.38	5500	7	W
7./NJG3	Hptm Alfred Heldt	Lancaster	SW Hanover	20.41	5100	2	W
3./NJG6	Fw Günther Bahr	Lancaster	25km W Hamlin	20.43		5	W
2./NJG3	Hptm Martin Becker	Lancaster	15km NE Zwolle	21.18	7100	2	W
9./NJG3	Hptm Siegfried Hahn	Lancaster	ESE Hanover	22.20		6	W

20-21 October 1943

12./NJG1	ObLt H-W. Schnaufer	Lancaster	15km E Assen	19.13	5400	32	W
12./NJG1	Ofw Karl-Heinz Scherfling	Halifax	40km W Bergen-am-Zee	19.24	5500	23	W
12./NJG1	ObLt H-W. Schnaufer	Lancaster	7km NE Sogel	19.26	5400	33	W
4./NJG3	Lt Georg Führer	Four-eng	12km N Gourlaben	20.00	6000	3	W
4./NJG3	Oblt Gerhard Raht	Four-eng	15km NE Stendal	20.15	6000	14	W
8./NJG1	Oblt Dietrich Schmidt	Lancaster		20.45		6	W
3./NJG5	Uffz Ernst Reitmeyer	Halifax	Potsdam	20.50	6000	3	W
7./NJG2	Fw Schulte	Lancaster	E Dessau	21.14	5400	1	W
11./NJG1	Ofw Heinz Vinke	Lancaster	Anreep, SE Assen	21.35	2800	34	W

9./NJG1	Lt Hans-Heinz Augenstein	Lancaster	Ommen	22.24		20	W
11./NJG1	Ofw Heinz Vinke	Lancaster	3km S Nijmegen	22.32	5500	35	W
4./NJG3	Oblt Gerhard Raht	Lancaster	Hagenau	22.35	6000	15	W
9./NJG1	Oblt H-H. Augenstein	Lancaster	Nr. Zwolle	22.45	3200	21	W
12./NJG1	Ofw Karl-Heinz Scherfling	Halifax	20km SW Den Helder	22.52	7000	24	W
12./NJG1	Lt Heinz Rolland	Stirling	DL1-4	22.58	1200	5	W

22-23 October 1943

5./NJG1	Otto Otto Fries	Stirling	SE Schleiden	20.35	5200	2	W
2./JG300	Fw Horst John	Halifax	Kassel	20.48	4600	7	W
./NJG5	Maj Günther Radusch	Halifax	5-10km WNW Kassel	20.57	5000	38	W
I./JG301	Fw Fritz Yung	Four-eng	10km NW Kassel	20.59	5000	1	W
5./NJG3	Oblt Ruprecht Panzer	Halifax	N Kassel	21.00	4800	2	W
7./JG300	Ofw Paul Streuff	Lancaster	N Kassel	21.02	5000	1	W
5./NJG5	Lt Erich Metz	Lancaster	30km NW Kassel	21.02	5900	2	W
11./NJG5	Oblt Gustav Tham	Halifax	NW Kassel	21.02	4000	5	W
12./NJG3	Hptm Eduard Schröder	Lancaster	SW Holzminden	21.03	5000	6	W
./NJG5	Maj Günther Radusch	Halifax	10km NE Kassel	21.04	5500	39	W
II./JG301	Ofw Kurt Welter	Four-eng	4km NW Kassel	21.04	5000	5	W
9./NJG2	Lt Franz Brinkhaus	Lancaster	50km NW Kassel	21.05	5200	3	W
StIV/NJG5	Hptm H. Niebelschutz.von	Halifax	500m E Kassel	21.05	5500	9	W
11./NJG5	Uffz Josef Sallmutter	Halifax	30km NW Kassel	21.05	5800	3	W
II./JG301	Lt Josef Wolfsberger	Lancaster	N Kassel	21.06	4500	1	W
III./NJG2	Lt Horst-Rudiger Blume	Lancaster	15km SE Kassel	21.07		1	W
5./NJG5	Hptm Leopold Fellerer	Lancaster	15km N Kassel	21.07	6000	14	W
4./NJG5	Oblt Helmuth Schulte	Lancaster	NNW Kassel	21.08	5000	1	W
4./NJG3	Uffz Heinz Oberheide	Halifax	15km E Kassel	21.10	5900	3	W
StII/NJG3	Hptm H. Sayn-Wittgenstein	Lancaster	10-20km NNW Kassel	21.10	4500	63	W
IV./NJG3	Lt Achim Woeste	Halifax	25-30km NW Kassel	21.10	5200	1	W
12./NJG5	Fw Günther Nord	Halifax	Nern edge of Kassel	21.12	5500	1	W
I./JG301	Fw Fritz Yung	Four-eng	E Kassel	21.12	4000	2	W
1./NJG2	Uffz Hans Kurz	Halifax	20km S Holzminden	21.13	5500	3	W
5./NJG3	Uffz Otto Kutzner	Lancaster	60km NW Kassel	21.13	6000	2	W
12./NJG5	Fw Günther Nord	Four-eng	WNW Kassel	21.13	5500	2	W
./NJG5	Maj Günther Radusch	Lancaster	20km N Kassel	21.13	4900	40	W
12./NJG5	Hptm Rudolf Altendorf	Halifax	27km E Herford	21.14	4800	19	W
3./JG300	Oblt Hermann Bertram	Four-eng	35km N Kassel	21.14	4200	2	W
StI/NJG5	Hptm Werner Hoffmann	Lancaster	35km N Kassel	21.14	6100	14	W
II./JG301	Ofw Kurt Welter	Four-eng	18km N Kassel	21.14	5800	6	W
I./JG301	Uffz Gerhard Witt	Four-eng	NNW Kassel	21.14	4000	3	W
2./NJG2	Hptm Eduard Jacob	Halifax	2km S Harzberg	21.15	5600	2	W
9./NJG2	Uffz Heinz Misch	Halifax	SW Holzminden	21.15	4000	1	W
12./NJG3	Uffz Klaus Möller	Lancaster	25km W Kassel	21.15	4600	1	W
I./JG301	Fw Viktor Emanuel	Four-eng	W Kassel	21.16	4300	1	W
4./JG300	Ofw Richard Lofgen	Lancaster	NW Kassel	21.16	5000	2	W
4./NJG3	Uffz Heinz Oberheide	Lancaster	5-10km NE Kassel	21.17	5900	4	W
7./NJG3	Hptm Alfred Heldt	Halifax	NW Kassel	21.18	5100	3	W
2./NJG5	Oblt Hans Leickhardt	Lancaster	30-40km N Kassel	21.18	6000	2	W
11./NJG3	Oblt Johannes Hiedlmayr	Lancaster	30km NW Kassel	21.20	6500	2	W
StI/NJG1	Hptm Manfred Meurer	Lancaster	Buhne-Haarbruck	21.20	6000	58	W
3./JG300	Lt Lothar Sachs	Four-eng	30-40km N Kassel	21.20	6000	2	W

Rudolf Schoenert (right), a pre-war Lufthansa pilot, joined the nightfighter force in 1941 and began to score quickly, first against the RAF and later in Russia. On 27th October he claimed his 56th and 57th victories and would increase this total to 65 by the surrender. Werner Baake (left) began nightfighting in 1942. He flew only with I./NJG 1 and became very successful flying the He219. On

2./JG300	Ofw Günther Migge	Four-eng		21.24	5800	2	W
II./JG301	Ofw Kurt Welter	Four-eng	28km N Kassel	21.24	4500	7	W
3./NJG3	Uffz Wilhelm Brockerhoff	Lancaster	Minden	21.25	6000	2	W
9./NJG3	Oblt Erhard Peters	Lancaster	20km SW Kassel	21.25	4500	2	W
10./NJG5	Uffz Erwin Glass	Lancaster	30km N Kassel	21.28	5600	1	W
7./NJG1	Oblt Werner Husemann	Lancaster	20km NW Minden	21.30	6000	14	W
StI/NJG5	Hptm Werner Hoffmann	Lancaster	Buckeburg	21.32	6300	15	W
3./NJG3	Uffz Wilhelm Brockerhoff	Lancaster	5km S Diepholz	21.35	6900	3	W
StII/NJG3	Hptm H. Sayn-Wittgenstein	Lancaster	Paderborn	21.35	6100	64	W
12./NJG1	ObLt H-W. Schnaufer	Lancaster	NE Kassel	21.40	6000	34	W
4./NJG101	Ofw Fritz Brandt	Lancaster	Frankfurt	21.55	5900	2	W
8./NJG3	Fw Karl Marten	Halifax	Nr. Kassel	22.17		1	W
10./NJG3	Oblt Werner Speidel	DC-3	CC7-3	23.01	600	2	W
3./NJG6	Hptm H-M Hadeball	Halifax				22*	W

27-28 October 1943

3./NJG100	Uffz Herbert Simross	DB-3	Schklovo	17.27	3600	3	E
1./NJG100	Hptm Alois Lechner	DB-3	Russia	17.57	4600	36	E
1./NJG100	Hptm Alois Lechner	DB-3	Russia	18.10	4600	37	E
1./NJG100	Hptm Alois Lechner	DB-3	Russia	18.30	3700	38	E
1./NJG100	Hptm Alois Lechner	DB-3	Russia	19.14	3600	39	E
1./NJG100	Hptm Alois Lechner	DB-3	Russia	19.24	3600	40	E
StI/NJG100	Hptm Rudolf Schoenert	DB-3	Warafina	19.37	2000	56	E
1./NJG100	Hptm Alois Lechner	DB-3	Russia	20.07	3400	41	E
1./NJG100	Hptm Alois Lechner	PS-84	Russia	20.25	2500	42	E
StI/NJG100	Hptm Rudolf Schoenert	R-5	Russia	22.37	1800	57	E

3 November 1943

StII/NJG1	Maj Walter Ehle	B-17		14.25		33	W

3-4 November 1943

3./NJG1	Oblt Werner Baake	Lancaster	E Antwerp	19.13	6500	20	W
6./NJG1	Hptm E-W. Bonin.von	Lancaster	SW Gangelt	19.30	5800	23	W
5./NJG1	Lt Otto Fries	Halifax	Nr. Epgrimby	19.30	5700	3	W
StIV/NJG1	Hptm Hans-Joachim Jabs	Lancaster	Gunzdorf	19.34	5800	25	W
3./NJG1	Oblt Werner Baake	Lancaster	N Hasselt	19.36	5800	21	W
StII/NJG1	Maj Walter Ehle	Lancaster	30km WNW Cologne	19.40		34	W
8./NJG1	Oblt Dietrich Schmidt	Halifax	E St.Truiden	19.46	5300	7	W

6./NJG1	Lt Erwin Ernst	Lancaster	Dusseldorf	19.51	5900	2	W
2./NJG6	Ofw Erich Becker	Halifax	5km N Düsseldorf	19.53	6500	2	W
3./NJG6	Fw Gunther Bahr	Halifax	25km W Düsseldorf	19.55	6200	6	W
StIV/NJG1	Hptm Hans-Joachim Jabs	Halifax	Niedermüssen	19.55	5800	26	W
2./NJG3	Oblt Horst-Helmut Orpky	Halifax	SW Düsseldorf	19.55	6500	1	W
4./NJG1	Lt Wilhelm Henseler	Halifax	SE Rheydt	19.57	6100	3	W
1./NJG4	Oblt Ludwig Meister	Halifax	21km WNW Köln	19.57	5800	20	W
2./NJG3	Lt Hermann Leube	Lancaster	W Düsseldorf	20.00	6300	4	W
7./NJG1	Lt Albert Bergerhoff	Lancaster	Mittelfüssen	20.02	5800	1	W
9./NJG1	Oblt H-H. Augenstein	Lancaster	SW Sundern	20.03	5800	22	W
4./NJG1	Oblt Lenz Finster	Lancaster	Wachtendonk	20.10	5400	7	W
StI/NJG1	Hptm Manfred Meurer	Halifax	14km NE Tilburg	20.15	6000	59	W

4 November 1943

3./NJG100	Oblt Hans Gref	DB-3	NE Witebsk	17.10	2300	15	E

4-5 November 1943

12./NJG3	Lt Karl Rechberger	Stirling	JN 9-8	18.53	3200	5	W
12./NJG3	Oblt Gunter Rogge	Halifax	KA 6-8	19.03	3000	2	W
10./NJG3	Oblt Hans-Hermann Müller	Stirling	EU 5-8	19.40	3000	3	W
10./NJG3	Oblt Hans-Hermann Müller	Stirling	EU 8-3	20.02	2200	4	W
10./NJG3	Oblt Werner Speidel	Wellington	HG 3-3	20.18	1800	6	W
11./NJG3	Oblt Johannes Hiedlmayr	Stirling	LG 7	20.46	1500	3	W
11./NJG3	Oblt Johannes Hiedlmayr	Stirling	LQ 7	20.46	1500	4	W
10./NJG3	Oblt Werner Speidel	Stirling	IS 4-3	21.11	1600	4	W

5-6 November 1943

3./NJG100	Ofw Karl Strohecker	DB-3	ESE Welish	18.10	1900	2	E
2./NJG100	Hptm Ulrich Meien.von	DB-3	W Krementschuk	18.46	3600	5	E

11-12 November 1943

11./NJG1	Ofw Heinz Vinke	Lancaster	80km NW Terschelling	20.24	1400	36	W
2./JG2	Uffz Herbert Treff	Halifax	Lisieux	00.58	4500	1	W
8./NJG4	Hptm Helmut Bergmann	Halifax	25km W Chateaudun	01.00	4000	17	W
2./JG2	Uffz Werner Edelmann	Halifax	N. Ste. Marguerite-d-L	01.12	3600	1	W
2./JG2	Uffz Werner Edelmann	Halifax	Nr. Pont L'Eveque	01.25	3300	2	W

16 November 1943

10./NJG3	Oblt Hans-Hermann Müller	B-17		11.40	3600	5	W
10./NJG3	Fw Heinrich Herr	B-17		11.45	3600	1	W

17-18 November 1943

3./NJG6	Fw Günther Bahr	Lancaster	SSW Hochspeyer	20.16	5200	7	W

Chapter Seven

The Battle of Berlin Begins

November 18th - December 31st 1943

Air Marshal Harris had achieved partial success with *Operation Millennium* and the Ruhr campaign and dealt Germany a devastating blow with *Operation Gommorrah* against Hamburg. Now he turned his mind against the greatest, and certainly the most difficult target; Berlin.

Attacks upon the German capital city were very long-range through heavily defended areas. The Berlin *Flak* defences were formidable and these ground defences, coupled with the prospect of continuous fighter attack, made the campaign very dangerous indeed for the Bomber Command crews. Making things even more perilous for the bomber crews, the *Luftwaffe* crews had to a large degree mastered *Window*. The old system of nightfighters loitering at beacons was discontinued and replaced by simple controller vectors to the bomber stream. From that point the nightfighter crew was free to pursue and attack the enemy until fuel or ammunition were exhausted. This inserted many more nightfighters into the bomber stream than had previously been the case with strict radar-controlled interceptions. It became known as *Zahme Sau* (tame boar) nightfighting. For the *Nachtjagd*, the 'Happy Time' was about to begin.

At the start of the battle, the Luftwaffe nightfighter force was as follows:

Stab NJG 1	*Obst* Streib	Deelen	Bf110
I./NJG 1	*Hptm* Meurer	Venlo	Bf110/He219
II./NJG 1	*Maj* v.Bonin	St. Trond	Bf110/Do217
III./NJG 1	*Maj* Priz zur Lippe Weissenfeld	Twente	Bf110
IV./NJG 1	*Hptm* Jabs	Leeuwarden	Bf110
Stab NJG 2	*Obstlt* Hulshoff	Parchim/Tilburg	Ju88C
I./NJG 2	*Maj* Jung	Kassel/Rothwesten	Ju88C/R
II./NJG 2	*Hptm* Sewing	Parchim	Ju88C
III./NJG 2	*Maj* Semrau	Schipol, Venlo, Neuruppin	Ju88C
Stab NJG 3	*Obstlt* Lent	Stade	Bf110
I./NJG 3	*Hptm* Milius	Vechta	Ju88C
II./NJG 3	*Hptm* Prinz zu Sayn-Wittgenstein	Schleswig	Bf110/Do217/Ju88C
III./NJG 3	*Maj* Barte	Stade	Bf110
IV./NJG 3	*Maj* Simon	Grove	Ju88C
Stab NJG 4	*Maj* Thimmig	Metz	Bf110/Do217
I./NJG 4	*Maj* Herget	Florennes	Bf110/Do217Ju88
II./NJG 4	*Hptm* Griese	St Dizier	Bf110/Do217
III./NJG 4	*Hptm* Kamp	Juvincourt	Bf110/Do217/Ju88
Stab NJG 5	*Obst* Radusch	Doberitz	Bf110
I./NJG 5	*Maj* Hoffmann	Stendal	Bf110
II./NJG 5	*Hptm* Baer	Parchim	Bf110/Do217
III./NJG 5	*Maj* Borchers	Neuruppin	Bf110/Do217
IV./NJG 5	*Hptm* v.Niebelschutz	Brandis	Bf110
V./NJG 5	*Hptm* Peters	Insterburg	Bf110

Stab NJG 6	*Maj* Schaffer	Schleissheim	Bf110
I./NJG 6	*Maj* Wohlers	Mainz-Finthen	Bf110
II./NJG 6	*Maj* Leuchs	Neuburg/Donau	Bf110
IV./NJG 6	*Maj* Lutje	Otopeni, Zilistea	Bf110
Stab NJG 100	*Maj* Schoenert	Bryansk	Ju88
I./NJG 100	Unknown	Various (Russia)	Ju88
Stab NJG 101	*Obstlt* Stoltenhoff	Neuburg	Bf110/Do217
I./NJG 101	*Maj* Vohwinkel	Ingolstadt	Bf110/Do217/Ju88
II./NJG 101	*Maj* Knoetzsch	Lechfeld	Bf110/217
III./NJG 101	*Hptm* Floitgraf	Echterdingen	Bf110/Do217
IV./NJG 101	*Maj* Fehling	Kitzingen	Bf110/Do217
NJG 200	Unknown	Various (Russia)	Bf110/Ju88/FW190
Stab JG 300	*Obstlt* Kettner	Deelen	FW190A
I./JG 300	*Hptm* Stamp	Bonn-Hangelar	Bf109G
II./JG 300	*Hptm* Peters	Rheine	FW190A
III./JG 300	*Maj* Buchwald(?)	Erbenheim	Bf109G
Stab/JG 301	*Obstlt* Weinreich	Rhein-Main	Bf109G

Note: LB Staffeln not included. NJG 100 and NJG 200 operated independent Staffeln.

On 15th December *NJG 102* commenced establishment at Kitzingen with Bf110s. The *Geschwader Stab* was commanded by Major Albert Blumensaat and *I Gruppe* by *Major* Ludwig Fehling. This was followed, on 24th, by *II Gruppe* forming at Echterdingen under *Hauptmann* Karl Floitgraf, again with Bf110s. *NJGr.10* was formed at Werneuchen on 1st January 1944, equipped initially with Bf110s and later FW190s, under command of *Major* Rudolf Schoenert. In February 1944 *I./NJG 7* was formed at Munster-Handorf under *Major* Horst Bengsch and equipped with Ju88Gs.

18 November 1943

10./NJG3	Oblt Hans-Hermann Müller	B-24		12.57	3600	6	W
10./NJG3	Uffz Hans Weingartner	B-24		13.05	100	1	W
10./NJG3	Oblt Werner Speidel	B-24		14.00	400	5	W
10./NJG3	Uffz Heinz Koppe	B-24		14.30	1200	3	W
11./NJG3	Fw Herbert Koch	B-24				1*	W

18-19 November 1943

8./NJG4	Hptm Helmut Bergmann	Halifax	NE Chalons s-Marne	19.24	5600	19	W
8./NJG4	Hptm Helmut Bergmann	Stirling	NE Chalons s-Marne	19.49	4200	20	W
3./NJG6	Hptm H-M. Hadeball	Halifax	Polch	19.59		23	W
3./NJG6	Hptm H-M. Hadeball	Halifax	S Treis	20.08		24	W
St./JG300	Maj Helmut Weinreich	E/A	Mannheim	20.35	6000	1	W
2./NJG6	Oblt Martin Becker	Stirling	Nr. Lampertsheim	20.44		2	W
7./JG301	Oblt Siegfried Wegener	B-17	SE Mannheim	20.55	3200	1	W
2./NJG6	Ofw Erich Becker	Halifax	WSW Mannheim	20.56		3	W
StI/NJG6	Oblt H J Birkenstock	Stirling	St.Martin	21.04		4	W
2./NJG6	Oblt Martin Becker	Halifax	Nr. Rimbach	21.05		3	W
2./NJG6	Oblt Heinz Knigge	Halifax	Honensulzen	21.08		1	W
1./NJG6	Oblt Albert Walter	Halifax	3km E Brodenbach	21.17		3	W
8./NJG4	Ofw Reinhard Kollak	Stirling	NE Chalons s-Marne	21.19	3500	31	W
8./NJG4	Ofw Reinhard Kollak	Stirling	NE Chalons s-Marne	21.40	4000	32	W
8./NJG4	Lt Erhard Peters	Halifax	N Laon	22.00	3600	3	W
StI/NJG6	Hptm Franz Evers	Lancaster	Asesse	22.10	5000	2	W

II./NJG 1 lost its Kommandeur, Major Walter Ehle on the night of 17th/18th November. Ehle (left) had claimed his last night victory on 3rd November, bringing his credited total to 39 and he had been awarded the Ritterkreuz in August. He lost his life when, on final approach to St Trond, the airfield illuminations failed. (S. Hall)

8./NJG4	Lt Erhard Peters	Lancaster		22.11	5000	4	W
2./NJG3	Fw Rudolf Frank	Lancaster	E Lingen	23.11		23	W

19-20 November 1943

6./NJG1	Hptm E-W. Bonin.von	Lancaster	NE Ronse	19.30	6000	24	W
5./NJG1	Lt Otto Fries	Stirling	Nr. Horrues	19.59	4100	4	W
2./NJG1	Hptm Paul Förster	B-	N Einhoven	20.24	4500	3	W
8./NJG5	Oblt Peter Ehrhardt	Halifax	NE Düsseldorf	20.24	4500	7	W

23-24 November 1943

11./NJG1	Ofw Heinz Vinke	Lancaster		18.26	4700	37	W
12./NJG1	ObLt H-W. Schnaufer	Lancaster	2km from Ter Apel	18.50	5000	35	W
12./NJG1	ObLt H-W. Schnaufer	Lancaster	Lorup	19.00	5500	36	W
9./NJG3	Hptm Siegfried Hahn	Four-eng	N Brandenburg	19.30	5400	7	W
5./NJG5	Lt Peter Spoden	Lancaster	W Berlin	20.02	6000	3	W
4./NJG3	Oblt Fehre	Four-eng	S area of Berlin	20.05	6000	1	W
StI/NJG5	Hptm Werner Hoffmann	Lancaster	SW Berlin	20.08	5000	16	W
8./NJG3	Oblt Paul Zorner	Four-eng	40km NW Berlin	20.09	6500	14	W
9./NJG3	Lt Hans Raum	Lancaster	Nr. Berlin	20.15	6000	12	W
1./NJG4	Oblt Ludwig Meister	Halifax	Centre Berlin	20.16	5500	21	W
4./NJG5	Ofw Heinz Säwert	Four-eng		20.22	5500	9	W
6./NJG3	Oblt Rudolf Szardenings	Halifax	Berlin	20.25	5700	3	W

25-26 November 1943

5./NJG5	Hptm Leopold Fellerer	Lancaster				15	W
5./NJG5	Hptm Leopold Fellerer	Lancaster				16	W
11./NJG1	Lt Guido Krupinski	Halifax				1	W
12./NJG1	Lt Heinz Rolland	Lancaster				6	W
12./NJG1	Lt Heinz Rolland	Stirling				7	W
1./NJG4	Oblt Hans Autenrieth	Halifax	E Dinant	02.12	4900	15	W
6./NJG1	Hptm E-W. Bonin.von	Halifax	N Prum	02.20		25	W
6./NJG1	Hptm E-W. Bonin.von	Halifax	SE Darmstadt	02.42		26	W
6./NJG1	Hptm E-W. Bonin.von	Lancaster	12km S Darmstadt	02.45	6000	27	W

11./NJG5	Oblt Gustav Tham	Four-eng	Frankfurt-am-Main	02.51	5000	6	W
9./NJG1	Oblt H-H Augenstein	Halifax	Frankfurt-am-Eder	02.55		23	W
StI/NJG3	Oblt Fritz Brinkmann	Halifax	Burgstadt	02.55	5500	1	W
7./NJG5	Fw Hans Kissel	Four-eng	W Frankfurt-am-Main	03.00	5500	1	W
5./NJG6	Uffz Robert Kröhn	Halifax	4km N Schaffenburg	03.25		1	W
4./NJG1	Lt Wilhelm Henseler	Halifax	NE St.Vith	03.35	3800	4	W
StI ./NJG4	Maj Wilhelm Herget	Halifax	4km -E St.Vith	03.35	5800	33	W

26 November 1943

11./NJG1	Oblt Martin Drewes	B-17	W Lemmer	12.47		7	W

26-27 November 1943

6./NJG1	Hptm E-W. Bonin.von	Lancaster	SSW Liege	19.24		28	W
6./NJG1	Hptm E-W. Bonin.von	Halifax	Hermeskel	19.30		29	W
1./NJG6	Oblt Albert Walter	Halifax	1km N Winkel	19.37		4	W
StI ./NJG4	Maj Wilhelm Herget	Lancaster	ESE Trier	19.40	5200	34	W
1./NJG6	Lt Wilhelm Engel	Halifax	Burg Grafenrode	20.05		8	W
StI ./NJG4	Maj Wilhelm Herget	Lancaster	NE Frankfurt-a-M	20.06	5300	35	W
StI ./NJG4	Maj Wilhelm Herget	Lancaster	NE Frankfurt-a-M	20.08		36	W
6./NJG5	Lt Josef Kraft	Lancaster	Berlin	21.52		6	W
8./NJG1	Oblt Dietrich Schmidt	Lancaster	Emlichheim	22.54		8	W
III./NJG1	Oblt Witzleb	Lancaster		22.56		2	W
9./NJG1	Uffz Heinz Amsberg	Lancaster		23.02		3	W

30 November-1 December 1943

3./NJG3	Ofw Willi Schmale	Halifax	6km SSW Cloppenburg	02.37	5500	6	W

1-2 December 1943

4./NJG3	Oblt Gerhard Raht	Stirling	Friesian Islands	17.54		16	W
LBSt 7	Fw Weidling	Bomber	12km N Aschaffenburg	21.15		1	W

2-3 December 1943

5./NJG5	Hptm Leopold Fellerer	Lancaster				17	W
StIII/NJG3	Unknown crew	E/A		19.17		-	W
8./NJG3	Oblt Paul Zorner	Lancaster	Nr. Diepholz	19.24	5800	15	W
6./NJG3	Fw Hans Meissner	Lancaster	50km W Hanover	19.31	6000	8	W
5./NJG3	Hptm Paul Szameitat	Lancaster	NE Hanover	19.35	6000	19	W
5./NJG5	Hptm Leopold Fellerer	Lancaster	Wesendorf	19.36	5500	18	W
5./NJG3	Hptm Paul Szameitat	Lancaster	30km E Celle	19.45	5700	20	W
6./NJG3	Fw Hans Meissner	Lancaster	Hanover	19.55	5800	9	W
./NJG3	Maj Helmut Lent	Lancaster	Magdeburg	19.58	5000	75	W
5./NJG3	Hptm Paul Szameitat	Lancaster	Nr. Stendal	20.02	5700	21	W
4./NJG3	Uffz Dunkerbeck	Lancaster	10-30km N Magdeburg	20.05		1	W
StII./JG302	Lt Alfred Körver	Lancaster	25km W Potsdam	20.11		1	W
2./NJG3	Oblt Horst-Helmut Orpky	Four-eng	Nr. Berlin	20.15	6000	2	W
6./NJG5	Ofw Walter Kammerer	Halifax	SW Berlin	20.12	6000	1	W
5./NJG5	Hptm Leopold Fellerer	Lancaster	Berlin	20.15	5000	19	W
II./JG302	Uffz Andreas Hartl	Four-eng	Berlin	20.20	3400	3	W
./NJG3	Maj Helmut Lent	Stirling	SW Berlin	20.20	4700	76	W
./NJG3	Hptm Erhard Peters	Lancaster	Berlin	20.22	4500	5	W
StI/NJG5	Hptm Werner Hoffmann	Lancaster	Berlin	20.23	6500	17	W

Leopold 'Poldi' Fellerer, originally a bomber pilot, became one of the 'originals' when he joined II./NJG 1 late in 1940. He claimed his first victory in February 1941 and served with 4.NJG 1, 5./NJG 2 (which later became 11./.NJG 1), 5./NJG 5 and finally ended the war as Gruppenkommandeur of III./NJG 6. Awarded the Ritterkreuz in April 1944, he survived the war only to die in the crash of a light aircraft in 1968. On 16th December 1943 he despatched three Lancasters for his 16th - 18th victories.

4./NJG3	Uffz Lorenz Gerstmayer	Halifax	Berlin	20.24	6200	1	W
8./NJG5	Oblt Ulrich Wulff	Lancaster	Berlin	20.24	5200	2	W
1./JG302	Ofw Ernst Haase	Lancaster	Berlin	20.26	4000	1	W
./NJG3	Hptm Erhard Peters	Halifax	Nr. Berlin	20.27	4500	6	W
I./NJG4	Maj Wilhelm Herget	Stirling	Berlin	20.29	5200	37	W
1./JG302	Fw Herbert Herre	Lancaster	Berlin	20.29	6000	3	W
8./NJG3	Oblt Paul Zorner	Lancaster	SW Berlin	20.29	5600	16	W
9./NJG3	Uffz Bahr	Four-eng	Nr. Berlin	20.30	6000	1	W
10./NJG5	Uffz Herbert Ter.Steegen	Lancaster	20km SSW Berlin	20.30	6000	2	W
StI/NJG5	Oblt Manfred Tischtau	Lancaster	Berlin	20.30	5800	2	W
6./NJG5	Lt Josef Kraft	Lancaster	Berlin	20.32	5200	7	W
4./NJG3	Fw Heinz Oberheide	Lancaster	10-35km E Berlin	20.33	6500	5	W
5./JG302	Fw Kurt Becker	Four-eng	Berlin	20.34	3400	1	W
StI/NJG5	Hptm Werner Hoffmann	Lancaster	Berlin	20.34	6500	18	W
II./JG302	Uffz Andreas Hartl	Four-eng	30km SSW Berlin	20.38	4200	4	W
1./JG302	Fw Herbert Herre	Lancaster	Berlin	20.39	3800	4	W
10./NJG5	Uffz Herbert Ter.Steegen	Lancaster	20-30km SSW Berlin	20.39	5500	3	W
1./JG302	Lt Erich Schölta	Stirling	NW Berlin	20.43	8000	1	W
2./NJG3	Ofw Erich Heitmann	Four-eng	Nr. Berlin	20.44	5000	5	W
11./NJG3	Uffz Hugo Fuetscher	Lancaster	NW Berlin	20.45	5500	1	W
StII/NJG4	Oblt Alfred Lelaut	Lancaster	6km NW Zessew	20.45		1	W
2./NJG3	Lt Hermann Leube	Lancaster	W Berlin	20.45	6500	5	W
I./NJG4	Maj Wilhelm Herget	Stirling	Berlin	20.55	7500	38	W
3./JG302	Ofw Egbert Jaacks	Halifax	15km SW Berlin	20.56	4400	1	W
II./JG302	Lt Stoewer	Lancaster	10-20km W Potsdam	21.11	2800	1	W
2./NJG4	Unknown crew	E/A		03.55		-	W

3-4 December 1943

5./NJG3	Hptm Paul Szameitat	Lancaster				22*	W
12./NJG1	Lt Heinz Rolland	Halifax	SW Leipzig	01.49	5200	8	W
./NJG1	Maj Werner Streib	Lancaster	Braunschweig	02.15	6300	66	W
3./NJG3	Uffz Wilhelm Brockerhoff	Halifax	500m from Stalförden	02.44	5500	4	W
1./NJG3	Oblt Martin	Halifax	E Kleinkladen	02.50	5200	2	W

./NJG1	Maj Werner Streib	Lancaster	Nr. Lingen	02.50	6000	67	W
11./NJG3	Oblt Gunther Köberich	Lancaster	N Hanover	03.00	6200	4	W
3./NJG3	Ofw Willi Schmale	Halifax	Achteburg	03.00		8	W
6./NJG3	Fw Hans Meissner	Lancaster	NNE Hanover	03.05	5200	10	W
5./NJG3	Hptm Paul Szameitat	Lancaster	N Hanover	03.18	6200	23	W
5./NJG3	Oblt Eswald Bachmann	Lancaster	E Hanover	03.20	6300	1	W
4./NJG3	Fw Heinz Oberheide	Lancaster	SW Leipzig	03.22	6500	6	W
11./NJG3	Lt Herbert Koch	Lancaster	Braunschweig	03.25	6000	2	W
5./NJG3	Hptm Paul Szameitat	Lancaster	Hamm-Wesendorf	03.25	6000	24	W
11./NJG3	Oblt Gunther Köberich	Lancaster	Nr. Wittenberge	03.31	6000	5	W
5./NJG3	Hptm Paul Szameitat	Lancaster	N Braunschweig	03.36	6300	25	W
IV./NJG3	Hptm Albert Schulz	Stirling		03.37		13	W
5./NJG3	Hptm Paul Szameitat	Lancaster	Nr. Hanover	03.41	6000	26	W
12./NJG1	Fw Karl Pfeiffer	Halifax		03.43	4500	8	W
StIII/NJG3	Lt Achim Woeste	Lancaster	Salzwedel	03.46	5800	2	W
StI/NJG5	Oblt Manfred Tischtau	Lancaster		03.53	6200	3	W
8./NJG5	Oblt Ulrich Wulff	Stirling	Leipzig	04.13	5500	3	W

8-9 December 1943

3./NJG100	Uffz Mayer	PS-84	NW Orscha	18.16		1	E

10-11 December 1943

12./NJG3	Uffz Klaus Möller	Halifax	15km N Ringsted	01.54	400	2	W

12-13 December 1943

StI/NJG1	Hptm Manfred Meurer	Mosquito	W Zaltbommel	19.25	8300	60	W

13 December 1943

2./NJG3	Fw Hans Meissner	B-17	Nordfeld	13.30	6200	11	W
11./NJG3	Uffz Hugo Fuetscher	B-24	S Heligoland	13.55	4000	2	W
12./NJG3	Lt Hermann Stock	B-17	NE Gröningen	14.05	6700	2	W

15-16 December 1943

IV./NJG3	Unknown crew	E/A		21.10		-	W

16-17 December 1943

12./NJG1	ObLt H-W. Schnaufer	Lancaster	Nr. Follega	18.00	5800	37	W
12./NJG1	Lt Heinz Rolland	Lancaster	12km from Alkmaar	18.02	5800	9	W
12./NJG1	ObLt H-W. Schnaufer	Lancaster	Nr. Lemmer	18.11	5700	38	W
12./NJG1	ObLt H-W. Schnaufer	Lancaster	4km SW Wolvega	18.23	6000	39	W
5./NJG2	Fw Gajewski	Lancaster	Eupen	18.45	6400	1	W
8./NJG1	Oblt Dietrich Schmidt	Lancaster	W Cloppenburg	18.45	5100	9	W
12./NJG1	ObLt H-W. Schnaufer	Lancaster	2km SW Wierum	18.45	6000	40	W
III./NJG1	Maj E. Lippe-Weissenfeld	Lancaster	N Ahlhorn	18.50	5500	59	W
7./NJG1	Oblt H-H. Augenstein	Lancaster	S Osnabruck	19.00	5300	24	W
III./NJG1	Maj E. Lippe-Weissenfeld	Lancaster	NW Nordhorn	19.00	5000	50	W
./NJG3	Maj Helmut Lent	Lancaster	Almanbruck	19.04		77	W
StI/NJG3	Oblt Berger	Lancaster	N Munster	19.05		1	W
6./NJG3	Fw Rudolf Frank	Lancaster	E Dummer See	19.15	6300	24	W
I./NJG1	Hptm Manfred Meurer	Lancaster		19.25		61	W
StI/NJG3	Oblt Fritz Brinkmann	Lancaster	N Münster	19.35		2	W

Even during the 'happy time' the Luftwaffe nightfighter crews were in some danger. On the night of 20th December, Hauptmann Helmut Kindor of 6./NJG 6 was stalked over Anbach by a 418 Squadron Mosquito crew. His Bf110 was shot down and he crash-landed injured, almost certainly victim to Flying Officer H.E.Jones. (S.Hall)

1./JG302	Fw Herbert Herre	Lancaster	Berlin	20.04	3500	5	W
StIII/NJG3	Hptm Walter Barte	Lancaster	W Berlin	20.05		16	W
StIII/NJG5	Hptm Walter Borchers	Lancaster	10km NE Berlin	20.08	5800	8	W
8./NJG5	Ofw Herbert Altner	Lancaster	Nr. Berlin	20.09	4500	4	W
1./JG302	Fw Herbert Herre	Lancaster	Berlin	20.12	4500	6	W
I./NJG1	Hptm Manfred Meurer	Lancaster				61	W
5./JG300	Lt Klaus Bretschneider	Lancaster				7*	W

20-21 December 1943

II./NJG1	Oblt Lenz Finster	Lancaster	NW Mainz			8	W
3./NJG1	Oblt Werner Baake	Lancaster	NE Liege	19.00	5500	22	W
1./NJG4	Oblt Ludwig Meister	Halifax	10km S Euskirchen	19.12	5600	22	W
3./NJG4	Hptm Hubert Rauh	Halifax	Mannheim	19.14	5800	8	W
StIII/NJG4	Hptm Hans-Karl Kamp	Halifax	Baugnez, nr Malmedy	19.20	5500	15	W
3./NJG4	Hptm Hubert Rauh	Halifax	24km WSW Koblenz	19.20	5600	9	W
1./.NJG4	Hptm Alfred Haesler	Halifax	Nauenheim	19.21		3	W
3./NJG6	Fw Günther Bahr	Halifax	SSE Mayern	19.24	5800	8	W
6./NJG1	Hptm E-W. Bonin.von	Halifax	Mayen	19.26	6000	30	W
3./NJG6	Ofw Xaver Pflieger	Four-eng	N Niederelbert	19.26	5500	1	W
1./.NJG4	Hptm Alfred Haesler	Halifax	20km W Koblenz	19.27		4	W
I./NJG4	Maj Wilhelm Herget	Halifax	Münstermaifeld	19.27	5300	39	W
5./NJG2	Lt Schröde	Halifax	Wetzlar	19.30	5500	1	W
I./NJG4	Maj Wilhelm Herget	Halifax	Wiesenheim	19.35	5500	40	W
3./NJG6	Uffz Benno Kratz	Halifax	Mergentheim	19.35	5500	2	W
I./NJG4	Maj Wilhelm Herget	Halifax	Florsheim	19.37	5800	41	W
4./NJG2	Hptm Hüschens	Lancaster	Niederdörfelden, Hanau	19.37	5500	2	W
3./NJG6	Fw Günther Bahr	Halifax	50km SW Koblenz	19.38	5000	9	W
2./JG300	Ofw Günther Migge	Halifax	SE Frankfurt-a-M	19.40	4500	3	W
1./JG300	Lt Schröder	Four-eng	Nr. Effenburg	19.40	4200	1	W
I./NJG4	Maj Wilhelm Herget	Four-eng	12km WSW Frankfurt	19.43	5800	42	W
2./NJG2	Lt Edgar Honold	Lancaster	Kelsterbach	19.43	5800	1	W
6./NJG1	Hptm E-W. Bonin.von	Halifax	Dachsenhausen	19.45	5500	31	W
1./NJG2	Ofw Rudolf Winn	Lancaster	ENEE Bad Orb	19.45	5500	1	W
I./NJG4	Maj Wilhelm Herget	Lancaster	NE Hinau	19.47	3500	43	W

1./NJG2	Oblt Heinz Rokker	Lancaster	Rothenbergen	19.47	2000	9	W
2./NJG6	Oblt Martin Becker	Lancaster	Bodenrod	19.50	5000	4	W
2./NJG6	Oblt Martin Becker	Halifax	Weilmünster	19.52	4200	5	W
2./NJG6	Oblt Martin Becker	Halifax	Hintermeilingen	19.55	4000	6	W
6./NJG3	Lt Hermann Leube	Halifax	Nr. Frankfurt-a-M	19.55	6200	6	W
I./NJG4	Maj Wilhelm Herget	Halifax	Hanau	19.57	4500	44	W
1./JG300	Lt Otto Bruning	Halifax	NE Frankfurt-a-M	19.59	7000	4	W
3./NJG6	Fw Gunther Bahr	Halifax	Ruppach	20.00	4400	10	W
8./NJG3	Uffz Hermann Bottcher	Lancaster	Dahlen-Ruppach	20.00	6000	1	W
5./NJG2	Lt Wendelin Breukel	Halifax	nr Habenhausen	20.00	5400	2	W
6./NJG3	Fw Rudolf Frank	Halifax	NE Frankfurt-a-M	20.00	5700	25	W
I./NJG4	Maj Wilhelm Herget	Lancaster	Rossdorf	20.00	4500	45	W
2./JG300	Ofw Gunther Migge	E/A	NNE Frankfurt-a-Mn	20.02	6300	4	W
8./NJG3	Oblt Paul Zorner	Lancaster	Hintetmeiligen	20.02	5400	17	W
4./NJG3	Uffz Lorenz Gerstmayer	Four-eng	ENE Frankfurt-a-M	20.04	5000	2	W
1./NJG6	Oblt Albert Walter	Halifax	Nr. Michelbach	20.04	6700	5	W
8./NJG3	Uffz Werner Röminger	Lancaster	Nidda	20.07		1	W
3./NJG6	Hptm H-M. Hadeball	Lancaster	10km NW Bad Schw'ch	20.10		25	W
5./NJG3	Oblt Rudolf Szardenings	Lancaster	Nr. Frankfurt-a-M	20.12	3100	3	W
I./NJG4	Maj Wilhelm Herget	Lancaster	W Schwalbach	20.15	4000	46	W
LBSt.7	Lt Würtz	Lancaster	Frankfurt	20.15		1	W
LBSt.7	Lt Würtz	Lancaster	Frankfurt	20.20		2	W
9./NJG4	Hptm Kurt Fladrich	Halifax	1km N Merlus	20.29	4200	8	W
2./NJG3	Hptm Paul Szameitat	Halifax	S Bonn	20.30	5200	27	W
4./NJG1	Oblt Wilhelm Henseler	Halifax	Fronville	20.43	5000	5	W
3./NJG1	Oblt Werner Baake	Halifax	NW Eindhoven	21.00	3200	23	W

23-24 December 1943

6./NJG2	Lt Erich Jung	Lancaster	Ober-Sayn	02.30	6500	1	W
1./JG300	Ofw Walter Mackens	Halifax	Frankurt-am-Oder	02.50		3*	W
1./JG300	Lt Förster	Four-eng	near Lindburg	02.52		1	W
7./NJG1	Oblt H-H. Augenstein	Lancaster	Frankfurt-am-Oder	03.00	5500	25	W
8./NJG3	Oblt Paul Zorner	Lancaster	E Giessen	03.02	5500	18	W
2./NJG3	Hptm Paul Szameitat	Lancaster	Lansenheim	03.06	5300	28	W
11./NJG5	Lt Georg Fenk	Lancaster	S Berlin	04.10	6000	1	W
II./JG302	Uffz Andreas Hartl	Four-eng	Berlin	04.15	5000	5	W
12./NJG5	Oblt Etto Karl Klemenz	Halifax	SW Berlin	04.17	5600	1	W
1./JG302	Ofw Kurt Emler	Lancaster	Berlin	04.20		9	W
9./NJG5	Lt Etto Keller	Four-eng	NNE Berlin	04.22	6500	12	W
8./NJG5	Lt Gerhard Wagner	Lancaster	Finow	04.25	5000	1	W
8./NJG3	Oblt Paul Zorner	Lancaster	S Diepholz	05.43	6300	19	W
3./NJG3	Ofw Willi Schmale	Lancaster	Leghorn	05.45	6000	9	W
8./NJG3	Oblt Paul Zorner	Lancaster	Cloppenburg	06.02	6100	20	W
12./NJG1	Ofw Karl-Heinz Scherfling	Lancaster	NE Berlin	06.14	4600	25	W
12./NJG1	Ofw Karl-Heinz Scherfling	Lancaster	Off Alkmaar	06.34	6400	26	W

24-25 December

12./NJG1	Ofw Karl-Heinz Scherfling	Halifax	S Ijssel	22.42	6000	27	W

29-30 December 1943

12./NJG1	ObLt H-W. Schnaufer	Halifax	5km N-E Meppel	18.50	2800	41	W

The Nachtjäger were soon to receive a new fighter; the He219. Powered by two Daimler Benz 603A motors and equipped with a pair of 30mm MK108 and four 20mm MG151 cannon, the big 'Uhu' (Owl) could reach around 385 mph and packed a devastating punch. It was also equipped with ejector seats for both crewmen - the only operational aircraft in WWII to do so. In the spring of 1944 it began to supplement the Bf110Gs of I./NJG 1.

12./NJG1	Ofw Karl-Heinz Scherfling	Halifax		19.15		28	W
12./NJG1	ObLt H-W. Schnaufer	Lancaster	Nr. Lingen	19.15	5600	42	W
1./NJG5	Lt Joachim Hanss	Halifax	5km SE Histedt	19.42		1	W
4./NJG5	Ofw Heinz Säwert	Halifax	Berlin	20.14	5500	10	W
1./JG302	Ofw Otto Pritzel	B-24	Berlin	20.22	6200	1	W
3./NJG5	Uffz Ernst Reitmeyer	Lancaster		21.04	5000	4	W
2./NJG3	Hptm Paul Szameitat	Halifax	3km W Gustrow	21.23	800	29	W
5./NJG1	Lt Otto Fries	Lancaster	Nr. Lingen	22.16	5600	5	W
5./JG300	Lt Klaus Bretschneider	Four-eng				8*	W

30 December 1943

9./NJG4	Fw Gottfried Kurzwernhart	B-24	15km S Charleville	12.37	5300	1	W

31 December 1943-1 January 1944

9./NJG3	Fw Heigl	Stirling		02.02		1	W

Chapter Eight

Berlin, Leipzig and Nurnberg

1st January - 31st March 1944

Prior to the commencement of the Berlin assault in mid-November, Harris had said *"We can wreck Berlin from end to end if the Americans will come in on it. It will cost us five hundred aircraft; it will cost Germany the war."* This was certainly an over-statement, since history has shown that a concerted bombing offensive did nothing but strengthen the resolve of inhabitants of a major city to survive and resist. In the event, the Americans declined the invitation and Bomber Command faced the storm of *Flak* and fighters unaided.

The night battles over Berlin, although the primary aim for Bomber Command crews, paled into almost insignificance following two disastrous raids against Leipzig, on 19th February, and Nürnberg on 30th March. The Leipzig attack, planned with the usual 'spoof raids', diversions, countermeasures and course changes, cost the RAF seventy-eight bombers. This was more than any American daylight attack. However the Nürnberg raid would be worse. Harris decided to abandon the tactic of making his crews fly 'dogleg' courses to hide the true target until the last possible minute. Instead, the Nürnberg attackers would fly directly to the objective. This, Harris reasoned, would itself be regarded as a feint. The night was moonlit and the course took the bomber stream directly over several nightfighter assembly points. The result was carnage. Ninety-five bombers failed to return, with many more classed as write-offs after landing. The raid itself was classed as a failure... .

1-2 January 1944

2./JG302	Ofw Eduard Ries	Four-eng	Berlin		1	W
3./JG300	Fw Hermann Wischnewski	Lancaster			3*	W
8./NJG1	Oblt Dietrich Schmidt	Lancaster	Ramsel	01.58	10	W
Stab./NJG2	Maj H. Sayn-Wittgenstein	Lancaster	Nr. Berlin	02.00	65	W
12./NJG1	Lt Friedrich Potthast	Lancaster	E Coervorden	02.10	4	W
Stab./NJG2	Maj H. Sayn-Wittgenstein	Lancaster	Nr. Berlin	02.18	66	W
Stab./NJG2	Maj H. Sayn-Wittgenstein	Lancaster	Nr. Berlin	02.20	67	W
9./NJG3	Hptm Erhard Peters	Lancaster		02.35	7	W
Stab./NJG3	Unknown	Lancaster		02.49	-	W
Stab./NJG3	Unknown	Lancaster		02.50	-	W
Stab./NJG3	Unknown	E/A		02.50	-	W
2./NJG3	Fw Hans Meissner	Lancaster		02.52	12	W
5./JG301	Ofw Kurt Welter	Four-eng	W Berlin	02.55	8	W
6./JG302	Fw Andreas Hartl	Lancaster	Berlin	02.56	6	W
5./NJG2	Lt Wendelin Breukel	Lancaster		03.00	4	W
Stab./NJG2	Maj H. Sayn-Wittgenstein	Lancaster	Nr. Berlin	03.00	68	W
St.I./NJG5	Hptm Werner Hoffmann	Lancaster	Berlin	03.06	19	W
5./NJG2	Lt Wendelin Breukel	Lancaster		03.10	5	W
Stab./NJG2	Maj H. Sayn-Wittgenstein	Lancaster	Nr. Berlin	03.10	69	W
Stab./NJG2	Maj H. Sayn-Wittgenstein	Lancaster	Nr. Berlin	03.12	70	W
5./NJG3	Oblt Oswald Bachmann	Lancaster		03.18	2	W
1./NJG4	Oblt Ludwig Meister	Lancaster	4km NW St.Pol	05.00	23	W
1./NJG4	Oblt Ludwig Meister	Lancaster	SE Mons	05.15	24	W
I./NJG4	Maj Wilhelm Herget	Lancaster	Grandrieu	05.46	47	W

1./NJG4	Oblt Ludwig Meister	Lancaster	W St Pol	05.51	25*	W
1./NJG4	Oblt Ludwig Meister	Halifax	W St.Pol	06.00	26	W
I./NJG1	Hptm Manfred Meurer	Lancaster			65	W
2-3 January 1944						
3./JG300	Fw Hermann Wischnewski	Lancaster			4*	W
3./JG300	Fw Hermann Wischnewski	Lancaster			5*	W
3 January 1944						
Kom./JG300	Obst Hajo Herrmann	Lancaster				W
11./NJG1	Ofw Heinz Vinke	Lancaster		01.27	38	W
12./NJG1	Lt Kurt Matzak	Lancaster	Quakenbrück	02.00	1	W
2./NJG1	Lt Heinz Oloff	Lancaster		02.25	7	W
9./NJG3	Uffz Walter Rohlfing	Lancaster	10km NNE Rheinsberg	02.38	5	W
2./NJG5	Oblt Hans Leickhardt	Lancaster		02.40	3	W
1./NJG6	Oblt Walter Walter	Lancaster	10km S Potsdam	02.52	6	W
3./NJG5	Lt Robert Wolf	Lancaster		02.55	2	W
5./NJG5	Oblt Wilhelm Johnen	Lancaster	20km SW Berlin	02.56	9	W
11./NJG5	Lt Wilhelm Seuss	E/A	Berlin	03.03	1	W
10./NJG5	Uffz Herbert Ter.Steegen	Lancaster		03.03	4	W
4./NJG3	Uffz Lorenz Gerstmayer	Lancaster		03.04	3	W
St.III./NJG5	Oblt Manfred Tischtau	Lancaster		03.06	4	W
5./NJG5	Fw Herbert Hubatsch	Lancaster		03.08	5	W
6./NJG2	Fw Don	Lancaster		03.10	1	W
8./NJG3	Oblt Paul Zorner	Lancaster	Nr. Lückenwalde	03.10	21	W
5./NJG5	Oblt Wilhelm Johnen	Lancaster	70km SW Berlin	03.12	10	W
St.III./NJG5	Unknown	E/A		03.16	-	W
5./NJG2	Lt Wendelin Breukel	Lancaster		03.18	6	W
5./NJG2	Lt Wendelin Breukel	Lancaster		03.20	7	W
St./JG300	Obst Hajo Hermann	Lancaster			2	W
4 January 1944						
8./NJG5	Lt Gerhard Wagner	B-24			2	W
12./NJG3	Lt Hermann Stock	B-17	Sea, 25km N Juist	12.30	3	W
11./NJG3	Lt Herbert Koch	B-17		12.31	3*	W
St.III./NJG5	Hptm Walter Borchers	B-24		12.45	9	W
5./NJG3	Fw Krauter	B-17		12.50	2	W
2./NJG5	Oblt Hans Leickhardt	B-17		12.50	4	W
8./NJG5	Ofw Herbert Altner	B-24		12.55	5	W
5./NJG3	Uffz Otto Kutzner	B-17		12.55	3	W
1./NJG5	Lt Friedhelm Wustenhagen	B-24		12.55	1	W
5./NJG3	Uffz Otto Kutzner	B-17		22.55	4	W
5 January 1944						
9./NJG5	Fw Brendel	B-24			2	W
5./JG301	Ofw Kurt Welter	Four-eng	W Stettin	03.46	9	W
St.III./NJG5	Hptm Walter Borchers	B-24		11.50	10	W
4./NJG5	Lt Otto Keller	B-24		11.50	1	W
11./NJG1	Oblt Martin Drewes	B-24	100km N Norderney	12.32	8	W
6./NJG3	Oblt Rudolf Szardenings	P-38	N Grove	12.55	5	W
5./NJG3	Ofhr Benno Gramlich	B-17		13.13	2	W

Attrition was high. On 2/3rd January Hauptmann Paul Szameitat (left), Gruppenkommandeur I./NJG 3 and victor in 29 combats, met his match. With his Ju88C-6 severely damaged by return fire, he attempted to force-land near Buckeburg, but crashed and died with his crew. On 5th January, Unteroffizier Alexander Wallbraucht of 8./NJG 5 (right) was killed in action at Elmshorn and on 9th January Leutnant Gerhard Rheinheimer of 4./NJG 1 (centre) perished at Helden, Holland. Neither had opened their score. (S.Hall)

5-6 January 1944

St./NJG10	Hptm Rudolf Schoenert	Four-eng			58*	W

6 January 1944

St.III./NJG5	Hptm Leopold Fellerer	B-24		03.27	20	W
5./NJG5	Oblt Wilhelm Johnen	Lancaster	Nr. Tutos	03.27	11	W
4./NJG5	Hptm Ernst Zechlin	Lancaster		03.31	3	W
4./NJG5	Oblt Josef Kraft	Lancaster	Stettin	03.33	8	W
6./JG302	Fw Andreas Hartl	Lancaster		03.41	7	W
8./NJG3	Oblt Paul Zorner	Lancaster	NW Stettin	03.42	22	W
11./NJG3	Lt Herbert Koch	Lancaster		03.44	4	W
4./NJG5	Oblt Josef Kraft	Lancaster	NW Stettin	03.45	9	W
11./NJG3	Lt Herbert Koch	Lancaster		03.47	5	W
8./NJG3	Oblt Paul Zorner	Lancaster	NW Stettin	03.51	23	W
4./NJG5	Oblt Josef Kraft	Lancaster	NW Stettin	04.01	10	W
5./NJG2	Lt Wendelin Breukel	Lancaster		04.05	8	W
12./NJG3	Lt Hermann Stock	Lancaster	NW Stettin	04.15	4	W

7 January 1944

1./NJG4	Oblt Ludwig Meister	B-17	Cambrai	13.00	27*	W
4./NJG4	Lt Jakob Schaus	B-17	Mainz		4*	W

10 January 1944

1./NJG100	Fw Heinze	R-5		18.01	1	O

11 January 1944

4./NJG3	Fw Bruno Rupp	B-17	Hanover	11.16	1	W
2./NJG5	Oblt Hans Leickhardt	B-17		11.55	5	W
St.I ./NJG5	Unknown	E/A		12.01	-	W
3./NJG5	Oblt Rolf Bokemeyer	B-17		12.10	4	W
10./NJG3	Unknown	B-17		12.16	-	W
12./NJG3	Ofw Rudolf Mangelsdorf	B-17		12.20	3	W
11./NJG3	Lt Herbert Koch	B-17		12.45	6	W
12./NJG3	Lt Hermann Stock	B-17	SE Diepholz	12.59	5	W
St.III./NJG5	Hptm Leopold Fellerer	B-17		13.00	21	W
3./NJG5	Oblt Rolf Bokemeyer	B-17		13.11	5	W
3./NJG5	Oblt Rolf Bokemeyer	B-17		13.12	6	W
11./NJG1	Oblt Martin Drewes	B-17	1km W Almelo	13.15	9	W
IV./NJG1	Hptm Ernst Metz	B-17	NW Enschede	13.15	1	W
11./NJG3	Fw Hugo Fuetscher	B-17	4km NW Plantlünne	13.16	3	W
11./NJG1	Oblt Martin Drewes	B-17	NW Rotterdam	13.45	10	W

12-13 January 1944

St.I./NJG100	Hptm Alois Lechner	R-5	Russia	19.35	43	O

13-14 January 1944

8./NJG1	Oblt Dietrich Schmidt	Mosquito	Nr. Kleve	05.36	11	W

14-15 January 1944

11./NJG1	Oblt Martin Drewes	Lancaster	Klazienaveen	18.38	11	W
Stab./NJG3	Unknown	E/A		18.47	-	W
4./NJG2	Lt Wendelin Breukel	Lancaster		18.47	9	W
Stab./NJG3	Maj Helmut Lent	Lancaster		18.49	78	W
4./NJG2	Lt Wendelin Breukel	Lancaster		19.00	10	W
6./NJG2	Uffz Villforth	Lancaster		19.02	1	W
St.III./NJG3	Hptm Walter Barte	Lancaster	W Bennebostel	19.05	17	W
Stab./NJG3	Maj Helmut Lent	Lancaster		19.05	79	W
5./NJG3	Oblt Rudolf Szardenings	Lancaster	W Berlin	19.08	6	W
4./NJG2	Lt Wendelin Breukel	Lancaster		19.10	11	W
Stab./NJG3	Maj Helmut Lent	Lancaster		19.10	80	W
5./NJG2	Oblt Walter Riedlberger	Lancaster		19.10	4	W
9./NJG3	Hptm Erhard Peters	Lancaster		19.12	8	W
4./NJG2	Lt Wendelin Breukel	Lancaster		19.15	12	W
3./NJG5	Lt Robert Wolf	Lancaster		19.15	3	W
St.III./NJG5	Hptm Leopold Fellerer	Lancaster		19.20	22	W
9./NJG3	Hptm Erhard Peters	Lancaster		19.22	9	W
3./NJG5	Ofw A. Klaiber	Lancaster		19.23	2	W
3./NJG5	Lt Robert Wolf	Lancaster		19.23	4	W
6./NJG2	Uffz Villforth	Lancaster		19.25	2	W
9./NJG3	Uffz Koenig	Lancaster		19.27	4	W
9./NJG3	Hptm Erhard Peters	Lancaster		19.28	11	W
Stab./NJG5	Obstlt Gunther Radusch	Lancaster		19.29	41	W
8./NJG1	Fw Horst Becker	Lancaster		19.30	2	W
5./NJG1	Lt Otto Fries	Halifax	3km SE Schalenburg	19.30	6	W
5./NJG5	Lt Peter Spoden	Lancaster	W Braunschweig	19.30	4	W
9./NJG3	Hptm Erhard Peters	Lancaster		19.33	11	W

Peter Spoden was another of the 'unsung heroes' of the nightfighter force. Flying first with 5./NJG 5 then Stab II./NJG 6 he achieved 23 night victories in the defence of central Germany, plus a lone B-17 by day. He survived the war and is a well-known commentator on the night air war in various television documentaries. (Peter Spoden)

5./NJG1	Lt Otto Fries	Lancaster	5km E Bad Munder	19.35	7	W
3./NJG6	Hptm H-M Hadeball	Lancaster	Nr. Lamspinge	19.35	27	W
5./NJG5	Lt Peter Spoden	Lancaster	W Braunschweig	19.35	5	W
9./NJG3	Uffz Koenig	Lancaster		19.38	5	W
4./NJG2	Lt Wendelin Breukel	Lancaster		19.40	13	W
1./NJG6	Oblt Walter Walter	Lancaster	3km N Hornburg	19.40	7	W
4./NJG2	Lt Wendelin Breukel	Lancaster		19.45	14	W
St.III./NJG5	Hptm Leopold Fellerer	E/A		19.45	23	W
6./NJG2	Oblt Günther Köberich	Lancaster		19.56	6	W
10./NJG1	Oblt Hermann Greiner	Lancaster	Nr. St.Dizier	20.00	14	W
Stab./NJG2	Maj H. Sayn-Wittgenstein	Lancaster		20.00	71	W
4./NJG5	Oblt Josef Kraft	Lancaster	Berlin	20.20	11	W
12./NJG1	Lt Kurt Matzak	Lancaster	60km NW Terschelling	20.33	2	W
11./NJG1	Ofw Heinz Vinke	Halifax		20.33	39	W
8./NJG1	Oblt Dietrich Schmidt	Lancaster	Kolhorn	20.35	12	W
III./NJG2	Ofw Schulte-Heinrich	Lancaster		20.57	1	W
St.III./NJG2	Hptm Berthold Ney	Lancaster		21.19	1	W

19-20 January 1944

11./NJG1	Ofw Heinz Vinke	Lancaster			40	W
8./NJG2	Unknown	E/A		02.17	-	W

20-21 January 1944

4./NJG2	Ofw Hans Berschwinger	Halifax		18.56	8	W
4./NJG2	Ofw Hans Berschwinger	Halifax		19.10	9	W
Stab./NJG2	Maj H. Sayn-Wittgenstein	Halifax	Nr. Berlin	19.10	72	W
Stab./NJG2	Maj H. Sayn-Wittgenstein	Halifax	Nr. Berlin	19.15	73	W
4./NJG5	Uffz Josef Handing	Lancaster		19.19	1	W
St.III./NJG5	Hptm Werner Hoffmann	Lancaster	20km NNE Rathenow	19.20	20	W
1./NJG4	Hptm Ludwig Meister	Halifax	2km SW Liebenwalds	19.20	28	W
1./JG302	Hptm Heinrich Wurzer	Halifax	W Berlin	19.24	1	W
Stab./NJG2	Maj H. Sayn-Wittgenstein	Halifax	Nr. Berlin	19.25	74	W
6./NJG4	Oblt Hans Autenrieth	Halifax	8km SE Dannenberg	19.30	16	W
Stab./NJG5	Obstlt Günther Radusch	Lancaster		19.30	42	W
8./NJG3	Oblt Paul Zorner	Lancaster	N Berlin	19.31	24	W
4./NJG2	Ofw Hans Berschwinger	Halifax		19.32	10	W
3./JG300	Lt Dieterle	Halifax		19.32	1	W

5./NJG5	Oblt Wilhelm Johnen	Lancaster	NNW Berlin	19.36	12	W
1./NJG4	Hptm Ludwig Meister	Halifax	25km NE Neuruppin	19.37	29	W
6./NJG5	Oblt Manfred Tischtau	Lancaster		19.37	5	W
Stab./NJG5	Obstlt Günther Radusch	Lancaster		19.39	43	W
St.III./NJG5	Hptm Leopold Fellerer	Halifax		19.40	24	W
6./NJG4	Oblt Hans Autenrieth	Lancaster	7km NE Berlin	19.45	17	W
5./NJG5	Fw Peter Ingwersen	Lancaster		19.45	1	W
4./NJG5	Oblt Josef Kraft	Lancaster	Berlin	19.45	12	W
4./NJG5	Uffz Franz Zawadka	Halifax		19.45	1	W
8./NJG3	Oblt Paul Zorner	Lancaster	N Berlin	19.45	25	W
St.III./NJG5	Unknown	E/A		19.47	-	W
7./JG301	Hptm Helmut Suhr	Lancaster	Berlin	19.47	2	W
3./JG300	Lt Dieterle	Halifax		19.48	2	W
St.IV./NJG5	Unknown	E/A		19.50	-	W
St.III./NJG5	Hptm Leopold Fellerer	Halifax		19.50	25	W
11./NJG5	Oblt Helmuth Schulte	Halifax	Berlin - Falkenberg	19.51	2	W
8./NJG5	Oblt Gustav Tham	Halifax	Berlin	19.57	7	W
St.III./NJG5	Hptm Leopold Fellerer	Halifax		19.58	26	W
11./NJG5	Unknown	E/A		20.03	-	W
St.III./NJG5	Hptm Leopold Fellerer	Halifax		20.05	27	W
8./NJG5	Oblt Peter Ehrhardt	Lancaster		20.08	8	W
St.III./NJG5	Hptm Leopold Fellerer	Lancaster		20.11	28	W
2./NJG5	Lt Bernhard Jungmann	Halifax		20.23	1	W
11./NJG1	Ofw Heinz Vinke	Lancaster		21.07	41	W
1,.JG301	Fw Fritz Yung	Halifax	Approaching Berlin		3*	W
Stab./NJG1	Obstlt Werner Streib	Lancaster		02.55	68	W

21-22 January 1944

St.III./NJG2	Hptm Hoffmann	Halifax			2	W
III./NJG2	Uffz Rannefeld	Halifax			2	W
1./NJG2	Hptm Albert Schulz	Halifax			14	W
9./NJG3	Unknown	E/A		21.38	-	W
Stab./NJG2	Maj H. Sayn-Wittgenstein	Lancaster	Nr. Magdeburg	22.00	75	W
8./NJG2	Oblt Friedrich Tober	Lancaster	25km WSW Cuxhaven	22.09	2	W
Stab./NJG2	Maj H. Sayn-Wittgenstein	Lancaster	Nr. Magdeburg	22.10	81	W
2./NJG5	Oblt Hans Leickhardt	Halifax		22.13	6	W
2./NJG5	Ofw Fritz Nischwitz	Halifax		22.15	1	W
III./NJG2	Uffz Emil Nonnemacher	Halifax	S Magdeburg	22.16	1	W
Stab./NJG3	Maj Helmut Lent	Lancaster		22.28	76	W
5./NJG1	Lt Otto Fries	Halifax	10k NW Scheveningen	22.30	8	W
3./NJG2	Hptm Eduard Jacob	Halifax		22.30	3	W
Stab./NJG2	Maj H. Sayn-Wittgenstein	Lancaster	Nr. Magdeburg	22.30	82	W
2./NJG5	Ofw Fritz Nischwitz	Halifax		22.35	2	W
9./NJG3	Uffz Walter Röhlfing	Halifax	6km SE Uelzen	22.39	6	W
4./NJG5	Hptm Julius Schurbel	Halifax		22.42	1	W
Stab./NJG2	Maj H. Sayn-Wittgenstein	Halifax	Nr. Magdeburg	22.45	78	W
5./NJG2	Hptm H-H Hissbach	Lancaster		22.52	2	W
6./NJG2	Uffz Villforth	Lancaster	50km N Magdeburg	22.52	3	W
3./NJG5	Lt Robert Wolf	Halifax		22.54	5	W
9./NJG3	Hptm Erhard Peters	Lancaster		22.55	12	W
Stab./NJG3	Maj Helmut Lent	Lancaster		22.58	77	W

The nightfighter force suffered a double blow on the night of 21st January with the loss of two highly successful crews. Hauptmann Manfred Meurer, Kommandeur of I./NJG 1, (photo left, with Ritterkreuz) collided with wreckage from a shot-down bomber, his 65th victory. Major Heinrich, Prinz zu Sayn Wittgenstein (right), Kommodore of NJG 2, claimed five bombers, bringing his score to 83, before being shot down and killed near Magdeburg.

6./NJG3	Fw Rudolf Frank	Lancaster	Magdeburg	22.59	26	W
5./NJG2	Hptm H-H Hissbach	Lancaster		23.00	3	W
3./NJG5	Fw Ernst Reitmeyer	Lancaster			5*	W
9./NJG5	Unknown	E/A		23.01	-	W
4./NJG5	Ofw Heinz Säwert	Halifax		23.03	11	W
6./NJG3	Fw Rudolf Frank	Halifax	Magdeburg	23.04	27	W
5./JG300	Unknown	E/A		23.05	-	W
St.I./NJG1	Unknown	E/A		23.10	-	W
2./NJG2	Lt Edgar Honold	Halifax		23.10	2	W
5./NJG2	Uffz Wolfram Möckel	Halifax		23.13	1	W
9./NJG5	Uffz Baader	Lancaster		23.15	2	W
6./JG302	Fwndreas Hartl	Four-eng	Approaching Berlin	23.15	8*	W
4./NJG5	Oblt Josef Kraft	Halifax	Magdeburg	23.15	13	W
9./NJG3	Hptm Erhard Peters	Lancaster		23.15	13	W
Stab./NJG5	Obstlt Günther Radusch	Halifax		23.15	44	W
Stab./NJG2	Maj H. Sayn-Wittgenstein	Halifax	Nr. Magdeburg	23.15	79	W
3./NJG5	Lt Robert Wolf	Lancaster	Magdeburg	23.15	6	W
11./NJG5	Unknown	E/A		23.16	-	W
1./NJG6	Lt Wilhelm Engel	Halifax	Nr. Eikendorf	23.20	9	W
5./NJG5	Fw Peter Ingwersen	Halifax		23.20	2	W
4./NJG5	Oblt Josef Kraft	Halifax	Magdeburg	23.20	14	W
4./NJG5	Uffz Franz Zawadka	Lancaster		23.20	2	W
3./JG300	Lt Dieterle	Lancaster		23.24	3	W
12./NJG5	Oblt Otto Karl Klemenz	Halifax		23.29	2	W
1./NJG3	Uffz Alfred Thiele	Halifax		23.34	1	W
4./NJG5	Uffz Josef Handing	Lancaster		23.44	2	W
St.I./NJG1	Unknown	E/A		23.50	-	W
3./NJG6	Hptm H-M Hadeball	Lancaster	Rossdorf, near Bruchkobel		28*	W
11./NJG3	Lt Herbert Koch	Lancaster			7*	W
III./NJG1	Oblt Witzleb	Halifax		00.32	3	W
7./NJG1	Oblt H-H Augenstein	Halifax	4km ESE Oldenzaal	00.35	26	W
11./NJG1	Oblt Martin Drewes	Lancaster	20km W Sneek	00.46	12	W
11./NJG1	Ofw Heinz Vinke	Lancaster		01.12	42	W

22-23 January 1944

I./NJG1	Hptm Manfred Meurer	Halifax	Magdeburg	23.10	63	W
I./NJG1	Hptm Manfred Meurer	Lancaster	20km SW Magdeburg	23.50	64	W

27-28 January 1944

2./JG302	Ofw Eduard Ries	Four-eng			2	W
12./NJG1	Lt Kurt Matzak	Lancaster	Quakenbrück	19.35	3	W
12./NJG1	Oblt H-W Schnaufer	Lancaster	4km NW Quakenbrück	19.45	43	W
7./NJG2	Unknown	E/A		19.50	-	W
4./NJG5	Ofw Heinz Säwert	Lancaster		20.14	12	W
5./NJG5	Lt Peter Spoden	Lancaster	30km W Berlin	20.27	6	W
12./NJG3	Hptm Eduard Schröder	Lancaster	Sautour	20.30	7	W
3./NJG5	Lt Robert Wolf	Lancaster		20.35	7	W
5./NJG5	Oblt Wilhelm Johnen	Lancaster	SSE Berlin	20.36	13	W
6./NJG3	Oblt Horst-Helmut Orpky	Lancaster		20.36	3	W
5./NJG5	Lt Peter Spoden	Lancaster	20km W Berlin	20.36	7	W
4./NJG3	Fw Bruno Rupp	Lancaster			2*	W
8./NJG5	Oblt Gustav Tham	Lancaster	Berlin	20.38	8	W
6./NJG5	Ofw Walter Kammerer	Lancaster		20.46	2	W
7./NJG5	Ofw Alfred Ohm	Lancaster	SE Berlin	20.50	1	W
6./NJG5	Ofw Walter Kammerer	Lancaster		20.54	3	W
5./NJG5	Oblt Wilhelm Johnen	Lancaster	SSW Berlin	20.57	14	W
9./NJG3	Uffz Walter Rohlfing	Lancaster	9km SE Chemnitz	21.13	7	W
5./NJG5	Oblt Wilhelm Johnen	Lancaster	SW Berlin	21.18	15	W
III./NJG2	Uffz Emil Nonnemacher	Lancaster	SW Münster	21.35	2	W
9./NJG2	Uffz Heinz Misch	Lancaster	2km N Northeim	21.40	2	W
9./NJG2	Uffz Heinz Misch	Lancaster	4km N Northeim	21.45	3	W
1./NJG6	Lt Wilhelm Engel	Lancaster	S Wurges	22.16	10	W
1./NJG6	Oblt Walter Walter	Lancaster	1km W Katzenelnbogen	22.40	8	W
3./NJG1	Oblt Werner Baake	Lancaster	16km SW Aachen	22.50	24	W
I./NJG4	Maj Wilhelm Herget	Lancaster	Sautour	23.10	48	W

28-29 January 1944

5./JG301	Ofw Kurt Welter	Halifax	Berlin		11	W
6./JG302	Fw Andreas Hartl	Halifax	Berlin area		9*	W
3./JG300	Fw Hermann Wischnewski	Halifax			6*	W
3./JG300	Fw Hermann Wischnewski	Halifax			7*	W
11./NJG2	Hptm Paul Semrau	Lancaster			16	W
4./NJG3	Oblt Gerhard Raht	Halifax		02.34	17	W
4./NJG3	Oblt Gerhard Raht	Lancaster		02.37	18	W
3./NJG3	Unknown	E/A		02.46	-	W
5./NJG3	Uffz Adam Baumann	Halifax		02.58	1	W
4./NJG2	Ofw Hans Berschwinger	Lancaster		02.59	11	W
St.II./NJG5	Oblt Leopold Fellerer	Halifax		03.01	29	W
3./NJG3	Oblt Heinz Ferger	Halifax		03.02	5	W
9./NJG3	Hptm Erhard Peters	Lancaster		03.02	14	W
3./NJG5	Lt Robert Wolf	Halifax		03.07	8	W
4./NJG3	Fw Bruno Rupp	Halifax		03.08	3	W
St.I./NJG5	Hptm Werner Hoffmann	Halifax	5km N Neuruppin	03.12	21	W
Stab./NJG5	Obstlt Günther Radusch	Halifax		03.12	45	W
4./NJG3	Fw Bruno Rupp	Halifax		03.14	4	W
St.II./NJG5	Oblt Leopold Fellerer	Halifax		03.15	30	W
St.II./NJG5	Unknown	E/A		03.17	-	W
4./NJG2	Ofw Hans Berschwinger	Halifax		03.17	12	W

The crew of this crash-landed Ju88G of II./NJG 3 seem to have escaped, judging from its condition. Note the Schrage Musik upward firing 20mm cannon in the fuselage.

8./NJG2	Oblt Friedrich Tober	Halifax		03.17	3	W
5./NJG2	Hptm H-H Hissbach	Lancaster		03.20	4	W
8./JG301	Ofw Herbert Seifert	Lancaster	Berlin	03.20	1	W
4./NJG2	Oblt Johann Werth	Halifax	Greifwald	03.20	3	W
7./JG301	Oblt Kucharsowsky	Four-eng	Berlin	03.21	1	W
4./NJG5	Hptm Julius Schurbel	Lancaster		03.23	2	W
3./NJG5	Lt Robert Wolf	Halifax		03.23	9	W
Stab./NJG3	Maj Helmut Lent	Four-eng		03.25	83	W
10./NJG5	Ofw Kaschig	Halifax	N Berlin	03.26	1	W
St.I./NJG5	Unknown	E/A		03.27	-	W
St.I./NJG5	Hptm Werner Hoffmann	Halifax	25km NE Berlin	03.27	22	W
4./NJG5	Oblt Josef Kraft	Lancaster	Berlin	03.28	15	W
6./NJG2	Uffz Villforth	Lancaster	E Berlin	03.28	4	W
1./NJG5	Lt Joachim Hanss	Halifax		03.30	2	W
5./NJG5	Uffz Josef Sadzenica	Lancaster		03.31	1	W
5./NJG2	Hptm H-H Hissbach	Lancaster		03.35	5	W
6./NJG2	Uffz Villforth	Lancaster	E Berlin	03.35	5	W
2./NJG5	Oblt Hans Leickhardt	Lancaster		03.44	7	W
3./NJG5	Fw Ernst Reitmeyer	Lancaster		03.50	6	W
St.I./NJG5	Hptm Werner Hoffmann	Halifax	25km NE Berlin	03.55	23	W
11./NJG1	Ofw Heinz Vinke	Lancaster		05.10	43	W
11./NJG1	Ofw Heinz Vinke	Lancaster		05.55	44	W
11./NJG1	Ofw Heinz Vinke	Lancaster		06.13	45	W

29 January 1944

3./NJG102	Unknown	E/A		11.45	-	W
5./NJG6	Uffz Robert Krohn	P-47		11.50	2	W
5./NJG6	Uffz Alois Nitterl	P-38	Laudert	11.50	1	W
5./NJG6	Lt Karl-Heinz Grunert	B-17	S Seemoos	11.55	1	W
5./NJG6	Uffz Robert Krohn	P-47		11.55	3	W

5./NJG6	Uffz Alois Nitterl	P-38	Laudert	11.55	2	W
5./JG300	Lt Klaus Bretschneider	Four-eng			9*	W

29-30 January 1944

I./JG301	Lt Heinrich Alven.von	Four-eng			1	W
Stab./NJG2	Lt Bruno Heilig	Lancaster			3	W
St.II./NJG6	Unknown	E/A		00.39	-	W

30 January 1944

2./NJG3	Uffz Horst Kümmer	E/A		12.10	1	W
9./NJG2	Uffz Gunther Schmidt	P-47	NW Lingen	12.30	1	W
3./NJG2	Ofw Robert Lüddeke	B-17	SW Minden	12.56	4	W
3./NJG2	Ofw Robert Lüddeke	B-17		12.56	5	W
10./NJG1	Oblt Hermann Greiner	B-17		13.00	15	W
6./NJG2	Uffz Villforth	B-17	N Rheine	13.00	6	W
1./NJG2	Hptm Walter Schulz	B-17	SW Minden	13.30	15	W

30-31 January 1944

3./NJG5	Lt Robert Wolf	Lancaster		20.08	10	W
4./NJG5	Oblt Josef Kraft	Lancaster	Berlin	20.09	16	W
St.III./NJG3	Hptm Walter Barte	Lancaster		20.10	16	W
3./NJG6	Fw Gunther Bahr	Lancaster	Warwitz	20.14	11	W
5./NJG3	Uffz Otto Kutzner	Lancaster		20.14	5	W
St.I./NJG5	Hptm Werner Hoffmann	Lancaster	middle of Berlin	20.15	24	W
3./NJG5	Fw Ernst Reitmeyer	Lancaster		20.15	7	W
1./NJG5	Ofw Hans-Joachim Schmidt	Lancaster		20.15	2	W
4./NJG5	Oblt Josef Kraft	Lancaster	Berlin	20.16	17	W
9./NJG3	Hptm Erhard Peters	Lancaster		20.18	15	W
1./NJG5	Lt Joachim Hanss	Lancaster		20.20	3	W
3./NJG5	Lt Robert Wolf	Lancaster		20.20	11	W
10./NJG5	Ofw Kaschig	Lancaster	Wittstock	20.23	2	W
2./NJG5	Oblt Hans Leickhardt	Lancaster		20.23	8	W
Stab./NJG5	Obstlt Günther Radusch	Lancaster		20.24	46	W
3./NJG6	Fw Günther Bahr	Lancaster	10km W Oranienburg	20.25	12	W
7./NJG5	Fw Robert Schratter	Lancaster		20.25	1	W
Stab./NJG5	Obstlt Günther Radusch	Lancaster		20.27	47	W
4./NJG5	Oblt Josef Kraft	Lancaster	Berlin	20.28	18	W
5./NJG5	Lt Peter Spoden	Lancaster	SW Stettin	20.29	8	W
St.I./NJG6	Oblt Hansjorg Birkenstock	Lancaster	Nr. Spandau,Berlin	20.30	5	W
Stab./NJG5	Obstlt Günther Radusch	Lancaster		20.30	48	W
StII./NJG6	Oblt Richard	Lancaster	Krummendorf	20.30	1*	W
3./NJG3	Oblt Heinz Ferger	Lancaster		20.34	6	W
St.I./NJG3	Oblt Fritz Brinkmann	Lancaster	N Berlin	20.35	3	W
St.I./NJG5	Hptm Werner Hoffmann	Lancaster	middle of Berlin	20.35	25	W
St.III./NJG2	Hptm Berthold Ney	Lancaster		20.35	2	W
9./NJG3	Hptm Erhard Peters	Lancaster		20.35	16	W
3./NJG5	Lt Robert Wolf	Lancaster		20.38	12	W
7./NJG1	Oblt Hans-Heinz Augenstein	Lancaster		21.49	27	W
11./NJG1	Lt Guido Krupinski	Lancaster		22.00	2	W
7./NJG1	Oblt Hans-Heinz Augenstein	Lancaster		22.00	28	W
11./NJG1	Ofw Heinz Vinke	Lancaster	Kolhorn	22.10	46	W

Two Halifaxes destroyed on 15th February brought Oberleutnant Wilhelm Johnen's tally to seventeen. Within a short time this aircraft would become the focus of much controversy when he force-landed the 5./NJG 5 Messerschmitt in Switzerland. (S.Hall)

12./NJG1	Oblt H-W Schnaufer	Lancaster	10km W Amsterdam	22.15	44	W
11./NJG1	Ofw Heinz Vinke	Lancaster		22.25	47	W
St.II./NJG2	Hptm Paul Semrau	Lancaster			17	W
3./NJG6	Hptm H-M Hadeball	Lancaster			29	W

1-2 February 1944

6./NJG101	Oblt Hans Krause	Wellington	SE Maribor	20.45	5	W

5-6 February 1944

6./NJG4	Oblt Karl-Heinz Schaffer	Stirling	10km SE Avallon	09.30	1	W

7-8 February 1944

1./NJG100	Ofw Karl Strohecker	U-2	N Polosk	21.17	3	O

10 February 1944

10./NJG1	Oblt Hermann Greiner	B-17	Nr. Osnabruck	12.25	16	W

10-11 February 1944

4./NJG100	Lt Klaus Scheer	DB-3	20km E Kotka	00.50	1	O

11 February 1944

3./NJG102	Oblt Philipp	P-38	300m N Oberalgen	12.40	1	W

12-13 February 1944

7./NJG1	Lt Michael Bogensperger	Mosquito	W Mönchengladbach	21.05	1	W

13-14 February 1944

7./KG2	Unknown	Beaufighter	Thames estuary	20.58	-	W

15-16 February 1944

9./NJG3	Hptm Erhard Peters	Lancaster	N Flensburg	19.57	17	W
4./NJG3	Oblt Gerhard Raht	Lancaster	NE Hallersleben	20.03	19	W
9./NJG3	Hptm Erhard Peters	Lancaster	E Flensburg	20.04	18	W
8./NJG3	Oblt Paul Zorner	Lancaster	3km SE Ribuitz	20.22	26	W
10./NJG3	Uffz Hans Weingartner	Lancaster	DC	20.31	2	W
9./NJG3	Hptm Erhard Peters	Lancaster	Fehmarn Belt	20.32	19	W
8./NJG5	Oblt Peter Ehrhardt	Lancaster	NW Berlin	20.34	9	W
5./NJG5	Oblt Wilhelm Johnen	Halifax	SE Damgarten	20.34	16	W
4./NJG5	Hptm Ernst Zechlin	Lancaster	Belzig area	20.37	4	W
9./NJG3	Hptm Erhard Peters	Lancaster	SE Gedser Island	20.38	20	W

3./NJG3	Fw Rudolf Frank	Four-eng	NE Luneburg	20.40	28	W
5./NJG5	Oblt Wilhelm Johnen	Halifax	Kummenerer See	20.45	17	W
St.III./NJG5	Hptm Leopold Fellerer	Halifax	Schwerinsee	20.46	31	W
4./NJG5	Oblt Josef Kraft	Lancaster	SW Rostock	20.46	19	W
St.I./NJG5	Hptm Werner Hoffmann	Halifax	40km NW Rostock	20.48	26	W
9./NJG3	Hptm Erhard Peters	Halifax	NE Rostock	20.52	21	W
St.III./NJG5	Oblt Helmuth Schulte	Lancaster	30km NE Treptow	20.53	3	W
4./NJG5	Hptm Ernst Zechlin	Lancaster	Brandenburg area	20.57	5	W
St.III./NJG2	Maj Paul Semrau	Lancaster	Nr. Berlin	20.58	18	W
3./NJG3	Fw Rudolf Frank	Four-eng	Nr. Luneburg	20.59	29	W
St.III./NJG5	Oblt Helmuth Schulte	Lancaster	45km N Berlin	21.07	4	W
Stab./NJG10	Hptm Rudolf Schoenert	Halifax	NNW Berlin	21.08	59	W
1./NJG5	Fw Ernst Reitmeyer	Lancaster	BQ-2	21.09	8	W
St.IV./NJG3	Lt Achim Woeste	Four-eng	Nr. Neumünster	21.10	3	W
8./NJG3	Oblt Paul Zorner	Lancaster	NW Berlin	21.11	27	W
4./NJG5	Oblt Josef Kraft	Halifax	Neuruppin	21.13	20	W
4./NJG3	Uffz Lorenz Gerstmayer	Lancaster	30km from Parchim	21.15	4	W
5./NJG5	Oblt Wilhelm Johnen	Halifax	NNW Berlin	21.20	18	W
St.III./NJG5	Oblt Helmuth Schulte	Lancaster	20km NNW Berlin	21.29	5	W
11./NJG1	Ofw Heinz Vinke	Lancaster	Ijsselmeer	22.38	48	W
7./NJG1	Oblt H-H Augenstein	Halifax	Emmerich	22.48	29	W
12./NJG1	Lt Kurt Matzak	Lancaster	Nr. Texel	22.50	4	W
12./NJG1	Oblt H-W Schnaufer	Lancaster	150km NW Den Helder	22.58	45	W
10./NJG1	Oblt Hermann Greiner	Lancaster	Beacon "Leopold"	22.59	17	W
8./NJG2	Uffz Emil Nonnenmacher	Lancaster	10km SSW Stuttgart	23.10	5*	W
12./NJG1	Oblt H-W Schnaufer	Lancaster	Nr. Hoorn	23.19	46	W
12./NJG1	Oblt H-W Schnaufer	Lancaster	In Waddensee	23.33	47	W
6./NJG5	Oblt Manfred Tischtau	Halifax			6*	W

18-19 February 1944

St.I./NJG100	Hptm Alois Lechner	PS-84	Russia	20.15	44	E

19-20 February 1944

11./NJG1	Ofw Heinz Vinke	Lancaster	BM-77	01.46	49	W
11./NJG1	Lt Guido Krupinski	Lancaster	DL-55	01.56	3	W
11./NJG1	Ofw Heinz Vinke	Lancaster	BM-78	01.57	50	W
12./NJG1	Lt Friedrich Potthast	Lancaster	CN-56	02.08	5	W
11./NJG1	Ofw Heinz Vinke	Halifax	CK-62	02.13	51	W
7./NJG1	Oblt H-H Augenstein	Lancaster	ES-5	02.30	30	W
9./NJG3	Hptm Erhard Peters	Four-eng		02.33	22*	W
3./NJG3	Ofw Rudolf Frank	Halifax	NW Delmenhorst	02.38	30	W
5./NJG1	Lt Otto Fries	Halifax	SW Eisleben	02.40	9	W
8./JG302	Lt Fries	Lancaster	Leipzig	02.40	1	W
9./NJG3	Lt Hans Raum	Four-eng	50km SW Hanover	02.41	2	W
3./NJG3	Oblt Heinz Ferger	Four-eng	W Steinhuder Meer	02.45	7	W
9./NJG3	Hptm Erhard Peters	Four-eng		02.45	23*	W
1./NJG4	Hptm Ludwig Meister	Lancaster	13km NE Celle	02.48	30	W
3./NJG3	Ofw Rudolf Frank	Four-eng	S Bremen	02.50	31	W
7./NJG2	Lt Franz Brinkhaus	Lancaster	E Bremen	02.51	4	W
St.I./NJG5	Hptm Werner Hoffmann	Lancaster	25km E Braunschweig	02.51	27	W
St.III./NJG3	Lt Hermann Leube	Four-eng	60km N Braunschweig	02.55	7	W

On the night of 15th February 1944, Oberleutnant Heinz-Wolfgang Schnaufer of 12./NJG 1 destroyed two Lancasters to bring his score to 47 confirmed victories. During the course of the next twelve months the victory bars on his Bf110G would become legendary.

2./NJG5	Oblt Hans Leickhardt	Lancaster	NE Gifhorn	02.57	9	W
St.IV./NJG5	Oblt Werner Hopf	Lancaster	Stendal	02.58	3	W
2./NJG6	Oblt Martin Becker	Halifax	NW Hanover	02.59	7	W
3./NJG3	Oblt Heinz Ferger	Four-eng	10km from Hanover	02.59	8	W
3./NJG3	Oblt Heinz Ferger	Four-eng	Nr. Hanover	03.00	9	W
7./NJG2	Lt Franz Brinkhaus	Lancaster	E Bremen	03.01	5	W
2./NJG6	Oblt Martin Becker	Lancaster	NE Hanover	03.02	8	W
4./NJG5	Hptm Julius Schürbel	Lancaster	Stendal	03.02	3	W
3./NJG3	Ofw Rudolf Frank	Lancaster	E Celle	03.04	32	W
9./NJG3	Hptm Erhard Peters	Four-eng		03.04	24*	W
8./NJG3	Oblt Paul Zorner	Lancaster	Wesendorf	03.04	28	W
St.IV./NJG5	Oblt Werner Hopf	Lancaster	Tangerhütte	03.12	4	W
7./NJG5	Lt Otto Keller	Halifax	Gardelegen	03.12	13	W
10./NJG5	Fw Josef Handing	Lancaster	SW Stendal	03.15	3	W
St.I./NJG5	Hptm Werner Hoffmann	Halifax	Nr. Stendal	03.17	28	W
3./NJG5	Oblt Josef Kraft	Four-eng	Nr. Brandenburg	03.17	21	W
8./NJG3	Oblt Paul Zorner	Lancaster	Nr. Wesendorf	03.17	29	W
1./NJG4	Hptm Ludwig Meister	Lancaster	Wesendorf	03.19	31*	W
3./NJG6	Fw Günther Bahr	Lancaster	15-20km SSW Stendal	03.20	13	W
2./NJG6	Oblt Martin Becker	Halifax	SE Hanover	03.20	9	W
3./NJG3	Oblt Heinz Ferger	Four-eng	Beacon "Raum"	03.20	10	W
9./NJG3	Hptm Erhard Peters	Four-eng		03.20	25*	W
2./NJG5	Lt Robert Wolf	Halifax	"Probable"	03.20	13	W
10./NJG3	Lt Walter Briegleb	Four-eng	W Berlin	03.22	1	W
3./NJG4	Hptm Fritz Söthe	Lancaster		03.22	5	W
St.IV./NJG5	Oblt Werner Hopf	Lancaster	SW Frankfurt-am-Oder	03.26	5	W
5./NJG5	Ofw Walter Kammerer	Lancaster	SW Berlin	03.26	4	W
8./NJG3	Oblt Paul Zorner	Lancaster	Gardelegen	03.26	30	W
5./JG302	Ofw Walter	Lancaster	S.E. Krahne	03.27	1	W
5./JG301	Ofw Kurt Welter	Lancaster	Berlin	03.27	12	W
5./NJG3	Uffz Adam Baumann	Four-eng	N Zerbst	03.30	2	W
4./NJG5	Ofw Heinz Säwert	Four-eng		03.30	13	W
I./NJG5	Oblt Helmut Jankowski	Four-eng	Belzig-Wittenburg	03.33	1	W
4./NJG5	Hptm Julius Schürbel	Four-eng	SW Berlin	03.33	4	W
7./NJG5	Lt Otto Keller	Halifax		03.34	2	W
1./NJG6	Oblt Albert Walter	Lancaster		03.34	9	W

1./NJG5	Lt Joachim Hanss	Lancaster	Stendal area	03.36	4	W
3./NJG6	Lt Wilhelm Engel	Lancaster	SE Brandenburg	03.38	11	W
9./NJG3	Lt Hans Raum	Four-eng	Nr. Briest	03.39	3	W
St.III./NJG5	Oblt Helmuth Schulte	Four-eng	S Brandenburg	03.40	6	W
8./NJG3	Oblt Paul Zorner	Lancaster	S Briest	03.41	31	W
St.IV./NJG3	Lt Nikolas Völker	Four-eng	Belzig area, nr Jüterbog	03.42	1	W
2./NJG5	Lt Robert Wolf	Lancaster	SW Frankfurt-am-Oder	03.42	14	W
7./NJG5	Lt Otto Keller	Halifax		03.43	3	W
St.I./NJG6	Hptm Heinz Reschke	Lancaster	N Torgau	03.55	2	W
4./NJG5	Hptm Ernst Zechlin	Lancaster	NW Nauen	03.58	6	W
6./JG300	Uffz Harald Löwenich	Halifax	S Leipzig	04.12	1	W
1./NJG10	Hptm Friedrich-Karl Müller	Lancaster	Leipzig	04.14	20	W
5./JG302	Ofw Walter	Lancaster	Leipzig	04.16	2	W
5./JG301	Ofw Kurt Welter	Lancaster	Leripzig	04.16	13	W
2./NJG6	Oblt Martin Becker	Lancaster	Halle - Leipzig	04.18	10	W
2./NJG5	Ofw Fritz Nischwitz	Lancaster	SW Leipzig	04.18	3	W
2./NJG5	Oblt Manfred Tischtau	Lancaster	SW Leipzig	04.18	7	W
8./JG301	Lt Glaas	Lancaster	Leipzig	04.20	1	W
5./JG300	Lt Norbert Graziadei	Lancaster		04.20	2	W
5./JG300	Lt Paul-Georg Schaffrath	Lancaster	Lingen	04.40	1	W
4./NJG5	Ofw Heinz Säwert	Four-eng		04.22	14	W
St.III./NJG5	Oblt Helmuth Schulte	Four-eng	25km S Leipzig	04.26	7	W
3./NJG3	Ofw Rudolf Frank	Lancaster	W Vechta	04.33	33	W
Stab./NJG10	Hptm Rudolf Schoenert	Halifax	6-10km SW Leipzig	04.33	60	W
9./NJG3	Hptm Erhard Peters	Four-eng		04.50	26*	W
11./NJG1	Ofw Heinz Vinke	Lancaster	HK 3-1	05.08	52	W
3./NJG3	Ofw Rudolf Frank	Lancaster	N Meppen	05.22	34	W
11./NJG1	Ofw Heinz Vinke	Lancaster	15km NW Appeldoorn	05.34	53	W
2./NJG1	Lt Heinz Oloff	Lancaster	9km N Halle	05.38	8	W
Stab./NJG2	Maj Günther Radusch	Four-eng	KH-9	05.39	49	W
6./NJG1	Hptm E-W Bonin.von	Lancaster	12km N Dessau	05.40	32	W
12./NJG1	Lt Friedrich Potthast	Lancaster	EL 8-4	06.36	6	W

20 February 1944

4./NJG6	Lt Funder	B-17	80km SE Dessau	13.43	2	W
3./NJG6	Lt Wilhelm Engel	B-17	W Leipzig	14.02	12	W
St.III./NJG5	Lt "Zorkenhafer"	B-17		14.44	1	W

20-21 February 1944

4./JG300	Ofw Hans Wegener	Halifax	S Stuttgart	22.34	1	W
I./JG300	Hptm Iro Ilk	Lancaster	30km NNW Stuttgart	22.41	1	W
1./NJG4	Hptm Ludwig Meister	Lancaster	Rastatt area	03.17	32	W
5./NJG2	Hptm H-H Hissbach	Four-eng	Stuttgart	04.09	6	W
5./NJG3	Uffz Otto Kutzner	Halifax	Nr. Stuttgart	04.10	6	W
4./NJG3	Uffz Lorenz Gerstmayer	Four-eng	10-15km W Stuttgart	04.12	5	W
5./NJG3	Oblt Rudolf Szardenings	Lancaster	Nr. Stuttgart	04.16	7	W
3./NJG1	Oblt Josef Nabrich	Halifax	Doullens area	05.43	7	W

22 February 1944

10./NJG3	Oblt Hans-Hermann Müller	B-17	Thistedt	15.07	7	W
10./NJG3	Oblt Hans-Hermann Müller	B-17	Thistedt	15.09	8	W

On 18th August 1943, Oberfeldwebel Heinz Vinke of II./NJG 1 had made a lucky escape after being shot down into the sea by Wing Commander J.R.D.Braham of 141 Squadron. On 26th February 1944, after winning the Ritterkreuz and being credited with 54 night victories, his luck ran out. He was caught by 198 Squadron Typhoons near Dunkirk and was shot down by Pilot Officer R.A.Lallemant (left) and Flying Officer G.E.A.Hardy. Vinke was awarded the Eichenlaub posthumously.

24 February 1944

8./NJG5	Oblt Peter Ehrhardt	B-17		13.05	10	W
8./NJG5	Lt Johannes Hager	B-17		13.15	23	W
10./NJG3	Uffz Heinz Koppe	B-17		13.30	4	W
4./NJG5	Willi Lauterbach	E/A		13.40	1	W
St.III./NJG6	Maj Rolf Leuchs	B-17	15km NW Warnenmünde	13.44	6	W
4./NJG6	Lt Wolfgang Henning	B-17	N Werneuchen	13.47	1	W
2./NJG102	Oblt	B-17		13.56	-	W
10./NJG3	Lt Lippert	B-17	W Vejle	14.20	1	W
10./NJG3	Oblt Hans-Hermann Muller	Lancaster		14.55	9	W
4./NJG6	Lt Funder	B-17	Nr. Rostok	15.34	3	W
5./NJG5	Lt Peter Spoden	B-17			9*	W

24-25 February 1944

6./NJG1	Hptm E-W Bonin.von	Four-eng	Brieg	21.36	33	W
Stab./NJG3	Maj Helmut Lent	Four-eng	Aero Island	21.37	84	W
8./NJG4	Ofw Reinhard Kollak	Lancaster	SE Metz	21.40	33	W
10./NJG3	Uffz Hans Weingartner	B-17	NE Siegen	21.40	3	W
7./NJG2	Hptm Billig	Four-eng	80km from beacon 'Puls'	21.46	1	W
Stab./NJG3	Maj Helmut Lent	Four-eng	Langeland Island	21.47	85	W
II./NJG1	Hptm E-W Bonin.von	Four-eng	Marling	21.48	34	W
III./NJG4	Ofw Reinhard Kollak	Lancaster	Metz	21.57	34	W
II./NJG3	Lt Hermann Leube	Four-eng	SW Frankfurt-a-Main	21.59	8	W
II./NJG3	Hptm Gerhard Raht	Four-eng	Metz - Baden Baden	22.02	20	W
III./NJG4	Ofw Reinhard Kollak	Lancaster	SE Metz	22.10	35	W
III./NJG3	Oblt Paul Zorner	Lancaster	W Stuttgart	22.15	32	W

II./NJG1	Ofw Fritz Schellwat	Lancaster	SE Saverne	22.22	17	W
III./NJG3	Oblt Paul Zorner	Lancaster	WSW Stuttgart	22.22	33	W
III./NJG3	Oblt Paul Zorner	Lancaster	SW Stuttgart	22.30	34	W
1./NJG6	Oblt Walter Walter	Lancaster	3 km. W. Bösingen	22.38	10	W
7./NJG4	Oblt Richard Delakowitz	Four-eng	Stuttgart-Tübingen	22.40	1	W
2./NJG2	Oblt Heinz Rökker	Stirling	S Heilbronn	22.43	10	W
5./NJG101	Oblt Meyer	Lancaster	Twalmen	22.50	1	W
2./NJG3	Fw Hans Meissner	Lancaster	NW Heilbronn	22.52	13	W
LBSt 7	Ofw Karsten	Halifax	30km NNW Stuttgart	22.56	1	W
5./NJG102	Hptm Wagner	Halifax	20km S Schweinfurt	23.15	1	W
2./NJG1	Lt Heinz Oloff	Lancaster			9*	W
8./NJG3	Oblt Paul Zorner	Lancaster	W Stuttgart	00.25	35	W
8./NJG3	Oblt Paul Zorner	Lancaster	N Stuttgart	00.51	36	W
9./NJG3	Lt Hans Raum	Four-eng	Nr. Schweinfurt	01.15	4	W
9./NJG3	Lt Hans Raum	Four-eng	Nr. Würzburg	01.32	5	W
3./NJG102	Lt Kehrberg	Lancaster	4km NNW Gräfendorf	01.37	1	W
9./NJG1	Fw Rauer	Halifax	Tunburg	01.38	1	W
7./NJG1	Oblt Hans-Heinz Augenstein	Halifax	S Augsburg	02.30	31*	W

25-26 February 1944

4./NJG2	Oblt Johann Werth	Lancaster	S Schirmbeck	20.00	4*	W
4./NJG100	Uffz Kollrep	PS-84	Russia	20.04	1	E
4./NJG100	Hptm Theodor Bellinghausen	DB-3	Russia	20.10	1	E
1./NJG4	Hptm Ludwig Meister	Lancaster	25km W Rethel	20.55	33	W
1./NJG4	Hptm Ludwig Meister	Lancaster	13km SE Rethel	20.59	34	W
1./NJG4	Hptm Ludwig Meister	Lancaster	2km S Vouziers	21.14	35	W
St.III./NJG2	Hptm Berthold Ney	Lancaster	30km SW Metz	21.15	3	W
St.III./NJG6	Oblt Hansjorg Birkenstock	Lancaster	60km SW Fuf Kuli	21.25	6	W
3./NJG6	Fw Günther Bahr	Lancaster	SE Neufchateau	21.28	14	W
2./NJG6	Oblt Martin Becker	Lancaster	12km E Luneville	21.30	11	W
2./NJG2	Oblt Heinz Rökker	Lancaster	W Hagenau	21.43	11	W
2./NJG6	Oblt Martin Becker	Lancaster	Nr. Roggenbach	21.49	12	W
5./NJG2	Hptm H-H Hissbach	Four-eng	Schönbach area	21.51	7	W
4./NJG3	Hptm Gerhard Raht	Four-eng	S Augsburg	22.43	21	W
Stab./NJG6	Maj Heinrich Wohlers	Lancaster	30km W Augsburg	22.45	29	W
10./NJG3	Lt Lippert	Halifax	OS 6-8	23.43	2	W
Stab./NJG2	Maj Günther Radusch	Halifax	Schnee-Eifel	24.00	50	W
St.III./NJG6	Hptm Heinz Reschke	Lancaster	Stettin	00.27	3	W
8./NJG4	Ofw Reinhard Kollak	Four-eng	18km E Vervins	00.28	36	W
Stab./NJG2	Maj Günther Radusch	Halifax	S Rottweil	00.30	51	W
9./NJG1	Fw Rauer	Halifax	40km from Augsburg	01.33	2	W
1./NJG6	Hptm Gerhard Friedrich	Lancaster	Seeburg	01.33	10	W
5./NJG3	Oblt Rudolf Szardenings	Four-eng	SE Stuttgart	01.36	8	W
7./NJG1	Oblt H-H Augenstein	Lancaster		02.30	32*	W

26-27 February 1944

4./NJG100	Hptm Theodor Bellinghausen	E/A		19.22	2	E
I./JG302	Ofw Naumeier	Il-2	W Helsinki	20.00	1	W
3./JG302	Ofw Egbert Jaacks	TB-7	Helsinki	21.01	2	W
3./JG302	Fw Dieter Rusche	TB-7	Helsinki	21.11	1	W
1./JG302	Uffz Werner Dienst	DC-3	Helsinki	23.35	4	W

(left) Ju88G flown in daylight by Leutnant Horst Rudiger Blume, who flew with 8./NJG 2 and 11./NJG 3. During the early months of 1944 the nightfighter units were frequently called upon to oppose the large American daylight incursions and were sometimes savagely handled by the fighter escorts. On 30th January NJG 2 lost seven Ju88s to P-47s and other units lost five more, with many others badly damaged. Blume (right) claimed three victories in 1943 and 1944, but lost his life in a crash on 14th November. (S.Hall)

1-2 March 1944

St.III./NJG2	Hptm Hoffmann	Lancaster	Nr. Stuttgart	03.08	3	W
8./NJG2	Uffz Emil Nonnenmacher	Lancaster	Over Stuttgart	03.10	3*	W
8./NJG2	Uffz Emil Nonnenmacher	Lancaster	Over Stuttgart	03.10	4*	W
7./NJG2	Lt Heinz Reuter	Lancaster	Nr. Stuttgart	03.15	1	W
St.III./NJG6	Hptm Heinz Reschke	Lancaster	Stettin	03.30	4	W
1./NJG6	Hptm Gerhard Friedrich	Halifax	60km SW Strasbourg	04.00	11	W

3 March 1944

II./NJG3	Lt Manfred Scheunpflug	B-17		12.10	1	W

4 March 1944

5./NJG4	Ofw Kunker	Stirling	20km N Dijon	00.58	1	W

4-5 March 1944

III./NJG100	Lt G E Francsi	DB-3	Russia	18.14	1	E

5-6 March 1944

4./NJG100	Uffz Opitz	DB-3	Russia	04.10	1	E

6 March 1944

12./NJG1	Lt Friedrich Potthast	B-17			7	W
11./NJG1	Oblt Hermann Greiner	B-17	11km SW Quakenbrück	12.46	18	W
2./NJG5	Ofw Herger	B-17		13.23	1	W
6./NJG5	Uffz Philipp	B-24	20km NW Berlin	13.45	1	W
11./NJG1	Oblt Hermann Greiner	B-24	Münsterbrück	13.53	19	W
6./NJG5	Ofw Fritz Rathkamp	B-24	5km SE Münster	14.19	1	W

6-7 March 1944

4./NJG100	Lt Klaus Scheer	PS-84	Russia	23.24	2	E
4./NJG100	Uffz Gutschmidt	DB-3	Russia	23.40	1	E
4./NJG100	Lt Klaus Scheer	PS-84	Russia	23.52	3	E

7 March 1944

4./NJG100	Lt Klaus Scheer	PS-84	Russia	00.24	4	E
4./NJG100	Lt Klaus Scheer	PS-84	Russia	00.43	5	E
4./NJG100	Lt Klaus Scheer	PS-84	Russia	00.46	6	E

7-8 March 1944

2./NJG1	Hptm E-W Modrow	Mosquito	2km S Venlo airport	22.30	1	W

8 March 1944

6./NJG5	Lt Rudolf Thun	B-24		14.46	1	W

9-10 March 1944

4./NJG100	Lt Klaus Scheer	DB-3	Russia	18.12	7	E
4./NJG100	Lt Klaus Scheer	DB-3	Russia	18.24	8	E
4./NJG100	Uffz Kollrep	Twin-eng	Russia	18.55	2	E
4./NJG100	Uffz Kollrep	DB-3	Russia	19.04	3	E
4./NJG100	Uffz Kollrep	Twin-eng	Russia	19.07	4	E
4./NJG100	Hptm Theodor Bellinghausen	PS-84	Russia	19.08	3	E
4./NJG100	Uffz Kollrep	Twin-eng	Russia	19.21	5	E
4./NJG100	Lt Klaus Scheer	DB-3	Russia	19.26	9	E
4./NJG100	Hptm Theodor Bellinghausen	PS-84	Russia	19.32	4	E
4./NJG100	Hptm Theodor Bellinghausen	PS-84	Russia	19.45	5	E
4./NJG100	Hptm Theodor Bellinghausen	PS-84	Russia	20.01	6	E
4./NJG100	Uffz Leunig	B-25	Russia	20.18	1	E
4./NJG100	Hptm Theodor Bellinghausen	PS-84	Russia	20.29	7	E
3./NJG100	Ofw Karl Strohecker	SB-2	SSW Polosk	21.00	4	E
3./NJG100	Ofw Karl Strohecker	SB-2	SSE Polosk	21.20	5	E
4./NJG100	Uffz Opitz	DB-3	Russia	01.39	2	E
4./NJG100	Uffz Opitz	DB-3	Russia	01.58	3	E
4./NJG100	Lt Klaus Scheer	DB-3	Russia	02.14	10	E
4./NJG100	Uffz Haase	DB-3	Russia	02.43	1	E

10-11 March 1944

4./NJG4	Ofw Manfred Meier	Stirling	1km NE Brazey	23.23	1	W

11-12 March 1944

12./NJG1	Uffz Fischer	Lancaster	Off Vlieland	21.35	1	W

15-16 March 1944

6./NJG1	Hptm E-W Bonin.von	Four-eng	SE St.Dizier	22.10	35	W
2./NJG2	Oblt Heinz Rökker	Lancaster	50km SW Strassbourg	22.26	12	W
St.III./NJG6	Hptm Heinz Reschke	Four-eng	SW Strasbourg	22.32	8	W
12./NJG1	Lt Kurt Matzak	Lancaster		22.34	5	W
6./NJG1	Uffz Deh	Four-eng	NE beacon "Christa"	22.35	1	W
St.III./NJG6	Maj Rolf Leuchs	Lancaster	25km SW St.Die	22.35	7	W
2./NJG2	Oblt Heinz Rokker	Lancaster	W Hagenau	22.35	13	W

While the night skies over Germany were filled with noise, fire and death, the night war in the est went largely unnoticed. Here, an unidentified crew from 4./NJG 100 pose in front of a Ju88C-6, probably at Orscha. Note the absence of radar antennae.
(S.Hall)

11./NJG 3	Lt Herbert Koch	Lancaster	SE Selestat	22.35	8*	W
6./NJG1	Hptm E-W Bonin.von	Four-eng	NE Lure	22.41	36	W
2./NJG2	Oblt Heinz Rökker	Halifax	W Stuttgart	22.55	14	W
LBSt.7	Ofw Kellermann	Lancaster	10km W Uberlinger See	22.58	1	W
2./NJG1	Fw Rauer	Four-eng	BR-CR	22.58	3	W
4./NJG3	Hptm Gerhard Raht	Four-eng	SW Stuttgart	23.12	22	W
6./NJG2	Oblt Gunther Köberich	Lancaster	Bodensee - Ulm	23.15	7	W
4./NJG6	Fw Helmut Bunje	Halifax	Wurmlingen	23.19	1	W
4./NJG1	Oblt Fritz Lau	Lancaster	S Stuttgart	23.19	4	W
11./NJG 3	Lt Herbert Koch	Lancaster	SW Freiburg	23.21	9*	W
I./JG301	Uffz "Hans" Müller	Lancaster	N Stuttgart	23.23	1	W
I./JG301	Uffz Gerhard Koch	Lancaster	Stuttgart	23.25	1	W
5./NJG2	Hptm H-H Hissbach	Four-eng	70km S Stuttgart	23.26	8	W
I./JG301	Fw F. Laubenheimer	Lancaster	N Stuttgart	23.28	2	W
St.III./NJG2	Hptm Berthold Ney	Four-eng	Stuttgart	23.28	4	W
4./NJG3	Uffz Bruno Rupp	Four-eng	SW Stuttgart	23.28	5	W
8./NJG2	Uffz Oppermann	Four-eng	N Stuttgart	23.32	1	W
10./NJG5	Oblt Rolf Bussmann	Four-eng	E Offenburg	23.32	20	W
6./NJG2	Lt Ternieden	Halifax	2km E Mossingen	23.32	4	W
I./JG301	Uffz Gerhard Koch	Lancaster	SW Stuttgart	23.35	2	W
7./NJG1	Uffz Heinz Amsberg	Lancaster	Nr. Stuttgart	23.36	4	W
9./JG300	Lt Willi Trabert	Four-eng	S Stuttgart	23.36	1	W
III./NJG102	Hptm Karl Floitgraf	Halifax	Bernhausen	23.37	3	W
4./NJG3	Uffz Lorenz Gerstmayer	Four-eng	SE Stuttgart	23.37	6	W
9./JG300	Lt Kurt Gabler	Four-eng	S Stuttgart	23.39	1	W
10./NJG5	Oblt Rolf Bussmann	Four-eng	Schlettert-Strasbourg	23.45	21	W
10./NJG5	Oblt Rolf Bussmann	Four-eng	Nr. St.Die	23.56	22	W
8./NJG1	Oblt Dietrich Schmidt	Lancaster	1km SE Langweiler	00.56	13	W
1./NJG4	Hptm Ludwig Meister	Halifax	S Aachen	01.18	36	W
1./NJG4	Hptm Ludwig Meister	Lancaster	2km. S Berine	01.30	37	W

18-19 March 1944

5./JG301	Ofw Kurt Welter	Lancaster			14	W
5./JG301	Ofw Kurt Welter	Lancaster			15	W
1./NJG100	Lt August Kayser	PS-84	Russia	19.59	1	E
5./NJG2	Uffz "Pollner"	Four-eng	Nr. Mönchengladbach	21.43	1	W
6./NJG2	Lt Franz Brinkhaus	Lancaster	48km NW Limburg	21.49	6	W
3./NJG6	Hptm H-M Hadeball	Halifax	10km N Montabaur	21.55	30	W
4./NJG3	Oblt Ruprecht Panzer	Four-eng	NW Frankfurt-am-Main	22.05	3	W
3./NJG6	Lt Lomberg	Halifax	Griesheim	22.12	1	W
9./NJG5	Fw Robert Schratter	Four-eng	NNW Frankfurt-am-Main	22.13	2	W
5./NJG6	Uffz Pata	Lancaster	Frankfurt-am-Main	22.15	1	W
V./NJG5	Lt Johannes Jockenhofer	Four-eng	S Mainz	22.18	1	W
9./JG300	Fw Knoll	Four-eng	N Frankfurt-a-M	22.20	3	W
6./JG300	Uffz Willy Paulus	Four-eng	20km WSW Frankfurt-a-M	22.20	1	W
5./NJG2	Uffz Villforth	Four-eng	20-70km NE Saarbrücken	22.25	7	W
1./NJG6	Hptm Gerhard Friedrich	Halifax	Preist	22.32	12	W
8./NJG1	Oblt Dietrich Schmidt	Halifax	Kaiserslautern	22.35	14	W
12./NJG1	Ofw Karl-Heinz Scherfling	Halifax	RP-2	22.44	29	W
St.III./NJG6	Oblt Hansjorg Birkenstock	Halifax	Nr. Baumholder	23.05	7	W
5./JG300	Lt Klaus Bretschneider	Four-eng			10*	W
4./NJG100	Uffz Haase	DB-3	Russia	00.43	2	E

19-20 March 1944

1./NJG100	Fw Heinz De Fries	PS-84	Russia	18.58	1	E
3./NJG100	Lt Truber	PS-84	Russia	21.30	1	E

21-22 March 1944

1./KG54	Ogfr Grossmann	Mosquito	London area	01.06	1	E

22-23 March 1944

3./NJG100	Fw Helmut Konter	U-2	12km E Polosk	20.54	2	E
3./NJG3	Ofw Rudolf Frank	Four-eng	GR-2	21.18	35	W
Stab./NJG3	Obstlt Helmut Lent	Lancaster	Nr. Bielefeld	21.26	86	W
4./NJG3	Hptm Gerhard Raht	Lancaster	Nr. Hanover	21.28	23	W
St.III./NJG3	Oblt Fritz Brinkmann	Lancaster	Nr. Quakerbrück	21.30	4	W
5./NJG2	Ofw Willi Glitz	Lancaster	Minden	21.30	2	W
5./NJG3	Ofw Konrad Hardorfer	Four-eng	Bremen-Hamburg	21.30	1	W
2./NJG2	Oblt Heinz Rökker	Lancaster	S Aurich	21.30	15	W
4./NJG3	Uffz Lorenz Gerstmayer	Four-eng	NW Hanover	21.33	7	W
Stab./NJG3	Obstlt Helmut Lent	Four-eng	S Bielefeld	21.35	87	W
7./NJG4	Hptm	Four-eng	Minden-Osnabrück	21.37	-	W
3./NJG3	Ofw Rudolf Frank	Four-eng	Gütersloh	21.40	36	W
St.III./NJG2	Maj Paul Semrau	Four-eng	Nr. Minden	21.40	19	W
2./NJG6	Oblt Martin Becker	Halifax	10km SE Hamm	21.42	13	W
8./NJG3	Hptm Paul Zorner	Four-eng	Nr. Giessen	21.43	37	W
2./NJG2	Uffz Alfred Walther	Four-eng	5km N Frankfurt-a-M	21.45	1	W
4./NJG3	Hptm Gerhard Raht	Lancaster	NW Frankfurt-am-Main	21.48	24	W
1./NJG4	Hptm Ludwig Meister	Four-eng	Gütersloh	21.51	38	W
2./NJG6	Oblt Martin Becker	Lancaster	Meyrich	21.53	14	W
4./NJG3	Hptm Gerhard Raht	Lancaster	N Frankfurt-am-Main	21.54	25	W
7./JG300	Lt Arnold Döring	Four-eng	NE Frankfurt-am-Main	21.57	1	W

Leutnant Klaus Bretschneider of 5./JG 300, who would become the Staffel leader in July, gets ready for take-off in his FW190 'Rauhbautz II' (Tough Guy II). He would include fourteen night victories in his tally of 34 before being killed in combat with American fighters on 24th December, a month after receiving the Ritterkreuz. He claimed a four-engined bomber destroyed near Krefeld on 25th March.

9./JG300	Lt Kurt Gabler	Four-eng	NE Frankfurt-am-Main	21.57	2	W
7./NJG3	Uffz Hans Werthner	Lancaster	15km NE Frankfurt-a-M	21.58	2	W
12./NJG3	Lt Hermann Stock	Four-eng	150km NW Frankfurt-a-M	22.05	7	W
6./NJG1	Uffz Deh	Four-eng		22.10	2*	W
2./NJG6	Oblt Martin Becker	Halifax	SE Frankfurt-am-Main	22.10	15	W
5./NJG3	Ofw Konrad Hardorfer	Four-eng	SE Frankfurt-am-Main	22.11	2	W
4./NJG3	Hptm Gerhard Raht	Lancaster	WNW Frankfurt-am-Main	22.14	26	W
8./NJG3	Hptm Paul Zorner	Four-eng	Nr. Giessen	22.18	38	W
4./NJG3	Uffz Lorenz Gerstmayer	Four-eng	SW Frankfurt-a-M	22.22	8	W
2./NJG2	Oblt Heinz Rökker	Lancaster	Koblenz - Limburg	22.27	16	W
2./NJG6	Oblt Martin Becker	Lancaster	NW Frankfurt-a-M	22.29	16	W
2./NJG2	Oblt Heinz Rökker	Halifax	N Koblenz	22.35	17	W
8./NJG1	Oblt Dietrich Schmidt	Halifax	NNW Paderborn-Minden	22.36	15	W
2./NJG6	Oblt Martin Becker	Halifax	NW Frankfurt-a-M	22.38	17	W
I./NJG6	Oblt Martin Becker	Halifax	NW Frankfurt-a-M	22.39	18	W
6./NJG1	Uffz Deh	Four-eng		22.40	3*	W
12./NJG3	Hptm Eduard Schröder	Four-eng	Koblenz	22.40	8	W
12./NJG1	Oblt H-W Schnaufer	Lancaster	Halle	23.04	48	W
2./NJG4	Uffz Altmann	Four-eng		23.05	1	W
8./JG300	Hptm Gerhard Stamp	Lancaster			3*	W

24-25 March 1944

5./JG301	Ofw Kurt Welter	Lancaster	Nr Berlin		16	W
12./NJG3	Hptm Eduard Schröder	Four-eng	Berlin area	21.??	9	W
2./NJG3	Fw Hans Meissner	Lancaster	Heligoland	21.18	14	W
Stab./NJG3	Obstlt Helmut Lent	Four-eng	Berlin area	21.30	88	W
Stab./NJG5	Lt	Four-eng	Near Berlin	21.55	-	W
4./NJG3	Lt Gerhard Schmalen	Four-eng	Berlin	21.56	1	W
Stab./NJG3	Obstlt Helmut Lent	Four-eng	Lubeck	22.00	89	W
St.III./NJG2	Hptm Berthold Ney	Four-eng	Werneuchen	22.01	5	W
St.III./NJG3	Lt Hermann Leube	Four-eng	Near Borkheide	22.18	9	W
5./NJG2	Hptm H-H Hissbach	Four-eng	20-80km from Kassel	22.22	9	W
2./NJG5	Lt Robert Wolf	Four-eng	EF-5	22.22	15	W
1./NJG5	Oblt Ernst-Georg Drünkler	Four-eng	Nr Langen	22.23	6*	W
4./NJG5	Oblt Josef Kraft	Four-eng	5km S Berlin	22.23	22	W
1./NJG5	Fw Ernst Reitmeyer	Four-eng		22.23	9	W

7./NJG2	Lt Heinz Reuter	Four-eng	50km S Berlin	22.25	2	W
1./NJG5	Lt Friedhelm Wustenhagen	Four-eng	Parchim	22.27	2	W
2./NJG5	Lt Heinz Bergel	Four-eng	50km SW Berlin	22.28	1	W
4./NJG3	Hptm Gerhard Raht	Four-eng	Nr. Berlin	22.28	27	W
7./JG302	Lt German Merz	Halifax	Berlin	22.32	1	W
St.III./NJG5	Lt Joachim Hanss	Four-eng	NE Dessau	22.34	5	W
2./NJG6	Oblt Martin Becker	Lancaster	Reuden	22.35	19	W
6./JG302	Fw Andreas Hartl	Four-eng	20km S Berlin	22.37	1	W
7./JG301	Fw Robert Siegfahrt	Lancaster	Giessen	22.37	3	W
I./JG302	Fw Anton Benning	Lancaster	Berlin	22.40	3	W
5./NJG2	Ofw Willi Glitz	Four-eng	Berlin area	22.42	3	W
4./JG302	Ofw Eberhard Kroker	Four-eng	S Berlin	22.42	1	W
2./NJG5	Lt Robert Wolf	Lancaster	GG-8	22.43	16	W
8./NJG2	Uffz Emil Nonnenmacher	Halifax	Jüterbog	22.44	6*	W
5./JG301	Ofw Kurt Welter	Lancaster	Nr. Berlin	22.44	17	W
V./NJG5	Lt Johannes Jockenhofer	Four-eng	30km SW Berlin	22.45	2	W
9./NJG5	Lt Otto Keller	Four-eng	50 km fron Berlin beacon	22.45	4	W
St.III./NJG5	Oblt Wolfgang Knieling	Four-eng	5-10 km 220° from Berlin	22.45	2	W
9./JG300	Lt Willi Trabert	Four-eng	WSW Berlin	22.46	2	W
9./JG302	Lt Glass	Lancaster	Trebbin	22.50	3	W
5./NJG2	Hptm H-H Hissbach	Four-eng	20-50km from Berlin	22.50	10	W
2./NJG2	Oblt Heinz Rökker	Lancaster	Berlin - Leipzig	22.50	18	W
III./JG300	Hptm Iro Ilk	Four-eng	Leipzig	22.53	2	W
5./JG302	Oblt Karl-Heinz Seeler	Four-eng	10km S Jüterbog	22.57	3	W
8./NJG3	Hptm Paul Zorner	Four-eng	E Leipzig	22.57	39	W
8./NJG5	Oblt Peter Ehrhardt	Lancaster	20km from Erfurt	22.58	11	W
4./NJG5	Oblt Josef Kraft	Four-eng	Wertheim	22.58	23	W
10./NJG5	Oblt Heinrich Schneider	Four-eng	W Halle	22.59	1	W
10./NJG3	Lt Walter Briegleb	Halifax	5km NE Bitterfeld	23.01	2	W
10./NJG5	Oblt Heinrich Schneider	Four-eng	W Halle	23.03	2	W
8./NJG1	Oblt Dietrich Schmidt	Lancaster	Erfurt	23.04	16	W
8./NJG1	Oblt Dietrich Schmidt	Lancaster	Gotha	23.04	17	W
1./NJG5	Oblt E-G Drünkler	Four-eng	Nr. Langen	23.10	7	W
7./JG300	Fw Karl Kraft	Four-eng	Nr. Berlin	23.14	3	W
2./NJG2	Oblt Heinz Rökker	Halifax	SW Bernburg	23.20	19	W
St.III./NJG1	Oblt Martin Drewes	Lancaster	N Lippstadt	23.26	13	W
2./NJG3	Oblt Gunter Rogge	Four-eng	Langensalza	23.30	3	W
8./NJG5	Oblt Peter Ehrhardt	Lancaster	Kassel	23.34	12	W
5./NJG2	Uffz Villforth	Halifax	Nr. Kassel	23.38	8	W
1./NJG5	Oblt Ernst-Georg Drünkler	Four-eng	Nr Langensalza	23.40	8*	W
9./JG301	Fw Hans Todt	Lancaster	SW Berlin	23.41	1	W
2./NJG2	Oblt Heinz Rökker	Lancaster	E Kassel	23.48	20	W
3./NJG3	Uffz Gottfried Schneider	Lancaster	Berlin centre	23.58	1	W
8./NJG3	Fw Alfred Thiele	Lancaster		23.58	2	W
4./NJG2	Oblt Johann Werth	Four-eng			5*	W
3./JG300	Fw Hermann Wischnewski	Four-eng			8*	W
3./JG300	Fw Hermann Wischnewski	Four-eng			9*	W
8./NJG5	Oblt Peter Ehrhardt	Lancaster	20-30km W Rheine	00.05	13	W
2./NJG3	Oblt Berger	Four-eng	Near Twente	00.09	2	W
St.III./NJG1	Oblt Martin Drewes	Lancaster	50km W Rheine	00.10	14	W
12./NJG1	Oblt H-W Schnaufer	Lancaster	3km N Neuwarendorf	00.12	49	W

(Left) Oberleutnant Martin Drewes on the far left with his crew of III./NJG 1. On the night of 24th March they claimed their 13th - 15th night victories. (right) Oberfeldwebel Rudolf Frank of 3./NJG 3. He would win the Ritterkreuz in April 1944, but would be dead within the month, colliding with his 45th shot-down opponent. He was awarded the Eichenlaub posthumously.

3./NJG10	Uffz Benno Gramlich	Four-eng	NW Dortmund	00.17	3	W
8./NJG5	Oblt Peter Ehrhardt	Lancaster	20km N Rheine	00.18	14	W
7./NJG2	Lt Heinz Reuter	Four-eng	25km NE Berlin	00.18	3	W
3./NJG3	Ofw Rudolf Frank	Four-eng	S Meppen	00.19	37	W
12./NJG1	Oblt H-W Schnaufer	Lancaster	Neuwarendorf E Münster	00.21	50	W
3./NJG3	Ofw Rudolf Frank	Four-eng	15km from Meppen	00.24	38	W
6./JG300	Lt Klaus Bretschneider	Four-eng	N Krefeld	00.27	11	W
3./NJG1	Oblt Josef Nabrich	Lancaster	Eindhoven	00.30	8	W
3./NJG1	Oblt Heinz Strüning	Four-eng	230° from Dortmund	00.30	42	W
St.III./NJG1	Oblt Martin Drewes	Lancaster	40km W Twente	00.35	15	W
St.III./NJG3	Lt Achim Woeste	Four-eng		00.36	4	W
7./NJG4	Uffz Leopold Weinberger	Halifax	Leiden-Utrecht	00.37	1	W
St.III./NJG3	Hptm Werner Husemann	Four-eng	W Münster	00.39	15	W
6./JG300	Lt Klaus Bretschneider	Four-eng	N Krefeld	00.41	12	W
12./NJG1	Oblt H-W Schnaufer	Lancaster	Nr. Varsseveld	00.41	51	W
3./NJG3	Ofw Rudolf Frank	Four-eng	10km N Almelo	00.42	39	W
5./NJG2	Oblt Walter Riedlberger	Four-eng	S Krefeld	00.42	5	W
11./NJG1	Lt Friedrich Potthast	Lancaster	10km N Venlo	00.45	8	W

25-26 March 1944

3./NJG100	Fw Helmut Konter	U-2	21km SSW Polosk	20.57	3	E
3./NJG100	Fw Helmut Konter	U-2	28km SSE Polosk	21.08	4	E
3./NJG100	Fw Helmut Konter	U-2	26km SSE Polosk	21.28	5	E
3./NJG100	Fw Helmut Konter	U-2	22km SSE Polosk	23.06	6	E

26-27 March 1944

4./NJG100	Hptm Theodor Bellinghausen	DB-3	Russia	19.50	8	E
4./NJG100	Lt Klaus Scheer	DB-3	Russia	21.30	11	E

8./NJG3	Hptm Paul Zorner	Four-eng	French frontier	22.00	40	W
4./NJG3	Uffz Lorenz Gerstmayer	Four-eng	Beacon "Philipp"	22.27	9	W
7./NJG2	Uffz Erich Killmann	Four-eng	Near Beacon "Ida"	22.34	1	W
11./NJG5	Lt Wilhelm Seuss	Lancaster	Mönchengladbach area	22.36	2	W
I./NJG4	Maj Wilhelm Herget	Four-eng	Dinant	22.55	49	W
2./NJG2	Uffz Walter Reichenberger	Lancaster	Malmedy	22.55	1	W
2./NJG2	Oblt Heinz Rökker	Four-eng	Monchengladbach	22.55	21	W
I./NJG4	Maj Wilhelm Herget	Four-eng	N Philippeville	23.00	50	W
8./NJG3	Hptm Paul Zorner	Four-eng	W St.Trond	23.04	41	W
7./NJG2	Uffz Erich Killmann	Four-eng	Twente	23.25	2	W
4./NJG2	Oblt Johann Werth	Halifax	Mönchengladbach		6*	W

29-30 March 1944

4./NJG100	Uffz Haase	DB-3	Russia	20.20	3	E
NJG Finn'd	Hptm Hunschans	Pe-2	NW Petsamo	22.08	1	E

30-31 March 1944

4./NJG100	Uffz Opitz	TB-7	Russia	20.08	4	E
St.III./NJG1	Oblt Martin Drewes	Lancaster	NE Spa	00.05	16	W
St.III./NJG1	Oblt Witzleb	Lancaster	2km E Gmund	00.12	4	W
7./NJG4	Oblt Richard Delakowitz	Four-eng	Daun	00.13	2	W
8./NJG5	Ofw Herbert Altner	Four-eng		00.15	6	W
St.III./NJG3	Lt Hermann Leube	Four-eng	Nr. Bonn	00.15	10	W
StIII./JG302	Hptm Helmut Suhr	Four-eng		00.15	3*	W
7./NJG2	Uffz Hans Werthner	Lancaster	10km S Eitorf	00.15	3*	W
7./NJG2	Lt Hans Schäfer	Lancaster	10km W Montabaur	00.17	3	W
2./NJG6	Oblt Martin Becker	Halifax	SW Cologne	00.20	20	W
9./NJG3	Lt Hans Raum	Four-eng	20km SW Bonn	00.20	6	W
3./NJG6	Lt Lomberg	Lancaster	Klofeld-Bockeroth	00.21	2	W
2./NJG6	Oblt Martin Becker	Halifax	SE Rosbach	00.23	21	W
St.IV./NJG1	Lt Georg Fengler	Lancaster	Prüm area	00.24	1	W
3./NJG3	Ofw Rudolf Frank	Four-eng	Alsfeld	00.24	40	W
St.III./NJG3	Lt Hermann Leube	Four-eng	20-30km SE Bonn	00.24	11	W
7./NJG2	Lt Hans Schafer	Lancaster	W Geissen	00.24	4	W
6./NJG2	Fw Wolfram Mockel	Lancaster	W Bonn	00.25	2	W
9./NJG3	Lt Hans Raum	Four-eng	30km S Bonn	00.25	7	W
4./NJG3	Hptm Gerhard Raht	Four-eng	SW Bonn nr beacon "Ida"	00.26	28	W
9./NJG3	Uffz Walter Rohlfing	Four-eng	Nr. Koblenz	00.26	8	W
3./NJG10	Uffz Benno Gramlich	Four-eng	W Nuremburg	00.28	4	W
2./NJG4	Hptm Josef Krahforst	Four-eng	nr Schweinfurt	00.28	1	W
St.III./NJG3	Lt Achim Woeste	Four-eng	nr Koblenz	00.28	5	W
3./NJG6	Lt Wilhelm Engel	Lancaster	8km NE Kastellaun	00.30	13	W
8./NJG3	Uffz Werner Rominger	Four-eng	Rhein-Main area	00.30	2	W
5./NJG3	Lt Fehre	Four-eng	80-150km E beacon "Ida"	00.31	2	W
2./NJG3	Fw Hans Meissner	Lancaster	NW Frankfurt-am-Main	00.31	15	W
8./NJG1	Oblt Walter Prues	Lancaster	N Limburg	00.31	1	W
St.III./NJG3	Lt Hermann Leube	Four-eng	40km ESE Bonn	00.32	12	W
3./NJG4	Hptm Hubert Rauh	Four-eng	Kehlen-Giessen	00.32	10	W
8./NJG5	Lt Gerhard Wagner	Four-eng	40-60km NNW Frankfurt	00.32	3	W
2./NJG6	Oblt Martin Becker	Halifax	5km NE Bad Honningen	00.33	22	W
8./NJG5	Ofw Herbert Altner	Four-eng		00.34	7	W

8./NJG2	Uffz Kleibauer	Halifax	between Cologne & Bonn	00.34	1	W
3./NJG6	Lt Lomberg	Halifax	Herborn	00.34	3	W
9./NJG3	Lt Hans Raum	Halifax	Dorf Roth	00.34	8	W
9./NJG3	Lt Hans "Flach"	Four-eng	50-100km SW Kassel	00.35	1	W
2./NJG6	Oblt Martin Becker	Halifax	5km NE Bendorf	00.35	23	W
6./JG300	Lt Klaus Bretschneider	Four-eng	Stuttgart area	00.35	13	W
3./NJG3	Ofw Rudolf Frank	Four-eng	Vogelsberg	00.35	41	W
5./NJG3	Uffz Otto Kutzner	Four-eng	NW Frankfurt-am-Main	00.35	7	W
9./NJG3	Lt Hans Raum	Four-eng	50km SE Kassel	00.35	9	W
11./NJG5	Hptm Gustav Tham	Lancaster	Vogelsberg	00.35	9	W
2./NJG6	Lt Hans Zettel	Stirling	Montabaur	00.35	1	W
Stab./NJG2	Obst Günther Radusch	Four-eng	Adenau	00.36	52	W
4./NJG3	Uffz Bruno Rupp	Lancaster	Nr. Giessen	00.36	5	W
2./NJG3	Fw Hans Meissner	Lancaster	NW Frankfurt-am-Main	00.37	16	W
4./NJG3	Uffz Lorenz Gerstmayer	Four-eng	SE Bonn	00.38	10	W
1./NJG5	Fw Ernst Reitmeyer	Lancaster	Eckelshausen	00.38	10	W
St.III./NJG3	Lt Achim Woeste	Four-eng	Bützbach area	00.38	6	W
St.III./NJG2	Hptm Berthold Ney	Lancaster	Westerburg, nr Becken	00.39	6	W
2./NJG6	Oblt Martin Becker	Halifax	12km WNW Wetzlar	00.40	24	W
St.III./NJG2	Lt Ernst Hövermann	Four-eng	120km ENE beacon "Ida"	00.40	1	W
5./NJG3	Uffz Otto Kutzner	Four-eng	N Frankfurt-am-Main	00.40	8	W
Stab./NJG2	Obst Günther Radusch	Four-eng	SW Koblenz	00.41	53	W
Stab./NJG2	Oblt Johann Werth	Lancaster	OR-PR	00.41	7	W
St.III./NJG2	Maj Paul Semrau	Four-eng	Wüzburg-Nürnberg	00.42	20	W
4./NJG1	Oblt Fritz Lau	Halifax	S Hamm	00.43	5	W
5./NJG2	Ofw Robert Luddeke	Halifax	N Frankfurt-a-M	00.44	6	W
12./NJG3	Fw Klaus Möller	Four-eng	Fulda area	00.45	3	W
4./NJG3	Oblt Ruprecht Panzer	Four-eng	N beacon "Otto"	00.45	4	W
12./NJG1	Ofw Karl-Heinz Scherfling	Lancaster	NE Venlo	00.45	30	W
4./NJG5	Oblt Helmuth Schulte	Halifax	NE Frankfurt-am-Main	00.48	8	W
5./NJG2	Hptm Heinz-Horst Hissbach	Four-eng	Rheine area	00.49	11	W
8./NJG1	Oblt Dietrich Schmidt	Halifax	N Würzburg	00.49	18	W
2./NJG2	Uffz Alfred Walther	Four-eng	30km NW Nürnberg	00.49	2	W
2./NJG6	Oblt Martin Becker	Halifax	10km NW Alsfeld	00.50	25	W
7./NJG4	Oblt Richard Delakowitz	Four-eng	S Vogelsberg	00.50	3	W
St.III./NJG1	Oblt Martin Drewes	Lancaster	Wetzlar	00.50	17	W
5./NJG2	Hptm H-H Hissbach	Four-eng	SW Buer	00.50	12	W
11./NJG5	Lt Wilhelm Seuss	Lancaster	N Fulda	00.52	3	W
3./NJG3	Lt Wolf Gütter	Four-eng	50-60km from Würzburg	00.53	2	W
2./NJG5	Oblt Hans Leickhardt	Halifax	NE Frankfurt-am-Main	00.53	10	W
3./NJG3	Ofw Rudolf Frank	Four-eng		00.54	42	W
7./NJG2	Lt Hans Schäfer	Halifax	Vogelsberg area	00.54	5	W
1./NJG5	Oblt E-G Drünkler	Halifax	NE Frankfurt-am-Main	00.55	9	W
9./NJG5	Lt Robert Wolf	Lancaster	Nr. Fulda	00.55	17	W
6./NJG6	Hptm Fritz Radusch	Halifax	NE Fulda	00.56	1	W
6./NJG6	Hptm Heinz Reschke	Halifax		00.56	6	W
11./NJG5	Lt Wilhelm Seuss	Lancaster	Nr Bamburg	00.56	4	W
11./NJG5	Uffz Herbert Ter.Steegen	Lancaster	S Vogelsberg	00.56	5	W
1./NJG6	Hptm Gerhard Friedrich	Halifax	NW Nürnberg	00.57	13	W
4./NJG3	Uffz Lorenz Gerstmayer	Four-eng	E Frankfurt-am-Main	00.58	11	W

2./NJG5	Ofw Maier	Four-eng	SE Nürnberg	01.00	1	W
11./NJG5	Hptm Gustav Tham	Lancaster	N Schweinfurt	01.02	10	W
III./JG302	Hptm Gerlach	Four-eng		01.03	1	W
11./NJG5	Lt Wilhelm Seuss	Lancaster	SE Schweinfurt	01.04	5	W
3./NJG10	Uffz Brandt	Four-eng	Giessen	01.05	1	W
1./NJG5	Oblt Ernst-Georg Drünkler	Halifax	2km E Oberndorf	01.06	10	W
St.III./NJG5	Lt Joachim Hanss	Lancaster	Erfurt area	01.07	6	W
6./NJG6	Hptm Heinz Reschke	Halifax		01.07	7	W
3./NJG4	Lt Klaus Helm	Four-eng	Westerwald	01.08	1	W
4./NJG3	Uffz Bruno Rupp	Lancaster	N Bamberg	01.08	7	W
12./NJG3	Fw Klaus Möller	Four-eng	Kalenbrück-Erlangen	01.09	4	W
11./NJG5	Lt Wilhelm Seuss	Lancaster	Nr. Neustadt	01.12	6	W
6./NJG6	Lt Otto Drahne	Lancaster	30-50km NW Nürnberg	01.13	1	W
4./NJG5	Oblt Helmuth Schulte	Lancaster	30-50km N Nürnberg	01.13	9	W
3./NJG3	Uffz Heinz Krause	Four-eng	Nr. Nürnberg	01.14	2	W
1./NJG5	Fw Ernst Reitmeyer	Lancaster	NW Nürnberg	01.15	11	W
St.III./NJG3	Lt Achim Woeste	Four-eng	French-Swiss frontier	01.15	7	W
3./NJG3	Uffz Heinz Krause	Four-eng	Nr. Nürnberg	01.18	3	W
5./NJG2	Ofw Robert Luddeke	Halifax	NW Nürnberg	01.18	7	W
St.III./NJG1	Oblt Martin Drewes	Lancaster	Nr. Limburg	01.20	18	W
Stab./NJG3	Obstlt Helmut Lent	Four-eng	20-40km N Nürnberg	01.21	90	W
1./NJG5	Fw Ernst Reitmeyer	Lancaster	NW Nürnberg	01.21	12	W
8./NJG2	Uffz Gunther Schmidt	Four-eng	NE Nürnberg	01.24	2	W
9./JG300	Lt Kurt Gabler	Four-eng	Nr. Nürnberg	01.25	3	W
4./NJG5	Oblt Helmuth Schulte	Halifax	30-50km NW Nürnberg	01.26	10	W
1./NJG5	Oblt E-G Drünkler	Halifax	W Erlangen	01.30	11	W
4./NJG5	Oblt Helmuth Schulte	Halifax	Augsburg-Stuttgart	01.30	11	W
5./NJG3	Uffz Otto Kutzner	Four-eng	S Nürnberg	01.33	9	W
7./NJG2	Uffz Emil Nonnenmacher	Lancaster	10km NW Nürnberg	01.41	7	W
8./NJG2	Uffz Emil Nonnenmacher	Lancaster	Poxdorf	01.41	8	W
5./NJG2	Hptm H-H Hissbach	Four-eng	SW Thuringerwald	01.52	13	W
3./NJG6	Lt Wilhelm Engel	Halifax	Limmwerder-Cochem	02.38	14	W
9./NJG4	Lt Hromadnik	Four-eng	S Villers-deux-Eglise	02.52	1	W
2./NJG6	Oblt Martin Becker	Halifax	NW Kettenburg	03.15	26	W
7./NJG1	Fw Hans Zeidler	Lancaster	5km SW Waterloo	03.30	1	W
4./NJG1	Unknown	Four-eng		03.30	-	W
9./NJG4	Lt Hromadnik	Four-eng	NW Charleville	03.36	2	W
2./NJG1	Hptm E-W Modrow	Halifax	NNW. Abbeville	04.13	2	W
9./NJG4	Lt Hromadnik	Lancaster	3½km N Namur	04.16	3	W
2./NJG1	Hptm E-W Modrow	Halifax	S Abbeville	04.30	3	W
II./KG51	Maj Wolf Dietrich Meister	B-17	E Cambridge	04.33	2	W
12./NJG1	Lt Kurt Matzak	Lancaster		22.34	6	W

30 March-1 April 1944

III./NJG100	Lt G E Francsi	U-2	Russia	00.07	2	E

Chapter Nine

Change of Direction

1st April - 5th June 1944

So the deep-penetration winter battles over Germany came to an end with greatest defeat for Bomber Command that it was ever to experience. Had the *Luftwaffe* nightfighter force been able to repeat this event with any regularity, it is doubtful that the strategic air offensive could have continued. However, in April, things changed dramatically. While the now heavily escorted USAAF 8th Bomber Command stormed into Germany by day, it was decided, much against the wishes of Harris, that the night bombing objectives be altered to that of a tactical bombing force under *Operation Point-Blank*. With the impending Allied assault upon France, it was now necessary to strike at such targets as transport, bridges, marshalling yards, barracks, armoured concentrations and ammunition dumps in order to facilitate the landings when they came.

By the spring of 1944, the *Nachtjagd* had reached almost it's full size and was disposed as follows:

Stab NJG 1	*Obstlt* Jabs	Bönninghardt	Bf110/He219
I./NJG 1	*Maj* Förster	Venlo	Bf110/He219
II./NJG 1	*Maj* v.Bonin	St. Dizier	Bf110
III./NJG 1	*Maj* Drewes	Laon-Athies	Bf110
IV./NJG 1	*Maj* Schnaufer	St. Trond	Bf110
Stab NJG 2	*Obstlt* Radusch	Deelen	Ju88C
I./NJG 2	*Hptm* Zechlin	Langensalza	Ju88C/R
II./NJG 2	*Maj* Semrau	Quackenbrück	Ju88C
III./NJG 2	*Maj* Ney	Langendiebach/Twente	Ju88C/G
Stab NJG 3	*Obstlt* Lent	Stade	Ju88C
I./NJG 3	*Maj* Husemann	Vechta	Bf110/Ju88C
II./NJG 3	*Hptm* Havenstein	Langendiebach/Vechta	Ju88C
III./NJG 3	*Maj* Barte	Stade	Bf110
IV./NJG 3	*Hptm* Buschmann	Grove	Ju88C
Stab NJG 4	*Maj* Thimmig	Chenay	Bf110/Ju88C
I./NJG 4	*Maj* Herget	Florennes	Bf110/Do217Ju88
II./NJG 4	*Hptm* Autenrieth	Dijon	Bf110
III./NJG 4	*Hptm* Kamp	Juvincourt	Bf110/Do217/Ju88
Stab NJG 5	*Obstlt* Borchers	Deelen	Bf110
I./NJG 5	*Maj* Hoffmann	Stendal	Bf110
II./NJG 5	*Hptm* Fellerer	Parchim	Bf110/Do217
III./NJG 5	*Maj* Zorner	Brandis	Bf110
IV./NJG 5	*Hptm* Altendorf	Brandis	Bf110
V./NJG 5	*Hptm* Peters	Insterburg	Bf110
Stab NJG 6	*Maj* v.Reeken	Schleissheim	Bf110
I./NJG 6	*Maj* Reschke	Illesheim	Bf110
II./NJG 6	*Maj* Leuchs	Echterdingen	Bf110
IV./NJG 6	*Maj* Lütje	Otopeni, Zilistea	Bf110
I./NJG 7	*Maj* Bengsch	Handorf	Ju88G
Stab NJG 100	*Hptm* Bonow	Bryansk	Ju88
I./NJG 100	Unknown	Various (Russia)	Ju88

Stab NJG 101	*Maj* Sewing	Ingolstadt-Manching	Bf110/Ju88/Do217
I./NJG 101	*Maj* Herber	Ingolstadt-Manching	Bf110/Do217/Ju88
II./NJG 101	*Hptm* Schwab	Riem	Bf110/217
III./NJG 101	*Maj* Jung	Kitzingen	Bf110/Do217
Stab./NJG 102	*Hptm* Evers	Kitzingen	Bf110
I./NJG 102	*Maj* Fehling	Kitzingen	Bf110
II./NJG 102	*Hptm* Floitgraf	Echterdingen	Bf110
NJG 200	Unknown	Various (Russia)	Bf110/Ju88/FW190
NJGr.10	*Maj* Schoenert	WerneuchenHangelar	FW190A
Stab JG 300	*Obstlt* Kettner	Deelen	FW190A
I./JG 300	*Hptm* Stamp	Bonn-Hangelar	Bf109G
II./JG 300	*Hptm* Peters	Rheine	FW190A
III./JG 300	*Maj* Ilk	Erbenheim	Bf109G
Stab/JG 301	Maj Mössinger	Schleissheim	Bf109G/FW190A
I./JG 301	Unknown	Neubiburg	Bf109G
II./JG 301	*Hptm* Fulda	Vienna-Seyring	Bf109G
III./JG 301	*Maj* Wegner	Vienna-Seyring	Bf109G
Stab./JG 302	*Maj* Peters	Doberitz	Bf109G
I./JG 302	*Hptm* Lewens	Jüterbog-Waldlager	Bf109G
II./JG 302	Unknown	Ludwigslust	Bf109G/FW190A
III./JG 302	Unknown	Oldenburg	Bf109G

1-2 April 1944

1./NJG100	Lt August Kayser	U-2	18km SE Polotsk	21.14	2	E
III./NJG100	Lt G E Francsi	U-2	Russia	22.05	3	E
III./NJG100	Lt G E Francsi	U-2	Russia	22.18	4	E

3-4 April 1944

6./NJG101	Oblt Hans Krause	Wellington	N Plattensee	23.46	6	W
6./NJG101	Ofw Hunderz	Wellington	20km S Budapest	00.08	3	W
6./NJG101	Ofw Hunderz	Wellington	20km S Budapest	00.16	4	W

4 April 1944

10./NJG6	Ofw Peter Tiedemann	B-24		13.56	1	E
11./NJG6	Oblt Martin "Bauer"	B-24		14.08	1	E
St.IV./NJG6	Oblt Hans Meyer	B-24		14.08	1	E

4-5 April 1944

St.I./NJG100	Lt G E Francsi	U-2	Russia	01.54	5	E

5 April 1944

12./NJG6	Karl Fleischmann	B-17		15.27	1	E

5-6 April 1944

I./NJG100	Lt G E Francsi	U-2	Russia	21.09	6	E
I./NJG100	Fw Regnini	U-2	Russia	21.10	1	E
I./NJG100	Lt G E Francsi	R-5	Russia	21.57	7	E
3./NJG100	Oblt Hans Schmid	U-2	Russia	23.30	1	E
1./NJG100	Fw Rudi Düding	R-5	Russia	23.34	3	E

5-6 April 1944

1./NJG100	Fw Heinz De Fries	U-2	Russia	00.22	2	E

Leutnant Gustav-Eduard Francsi of Stab I./NJG 100 (far right) joined the Nachtjagd in late 1941. It was not until he went to the east in late 1943 that he achieved success. On 5th April 1944 he claimed his 5th - 7th victories and would end the war as the most successful Ostfront nightfighter Experte with 56 credited kills. By comparison, Leutnant Hermann Muth of III./NJG 5 had no chance to prove himself. In a daylight interception of American bombers near Berlin on 6th April, he was shot down and killed by the escort fighters. (S.Hall)

6-7 April 1944

1./NJG100	Hptm Godbersen	R-5	Russia	20.50	2	E

7-8 April 1944

4./NJG100	Lt Richter	PS-84	Russia	20.40	2	E
4./NJG100	Lt Richter	PS-84	Russia	21.00	3	E
3./NJG100	Uffz "Hahn"	R-5	Russia	21.25	1	E
4./NJG100	Uffz Kick	PS-84	Russia	21.38	1	E
4./NJG100	Hptm Günther Köberich	PS-84	Russia	21.38	8	E
4./NJG100	Lt Klaus Scheer	PS-84	Russia	21.50	12	E

8-9 April 1944

3./NJG100	Lt Truber	U-2	Russia	22.19	2	E

9 April 1944

St.III./NJG3	Hptm Walter Barte	B-17		15.30	17	W

9-10 April 1944

12./NJG3	Ofw Rudolf Mangelsdorf	Four-eng	UT-34	23.48	4	W
12./NJG3	Hptm Eduard Schröder	Four-eng	NW Horsbaek	00.02	10	W
9./NJG4	Hptm Kurt Fladrich	Stirling	Somme Estuary	03.13	9	W
2./NJG3	Fw Hans Meissner	Four-eng	S Aarhus	03.34	17	W
3./NJG3	Oblt Heinz Ferger	Four-eng	Horsens	03.38	11	W
St.I./NJG3	Hptm Werner Husemann	Four-eng	Hersfeld	03.40	16	W
3./NJG3	Lt Wolf Gütter	Four-eng	Marburg-Giessen	03.45	3	W
I./NJG3	Hptm Werner Husemann	Four-eng	Hersfeld area	03.45	17	W
2./NJG3	Oblt Günter Rogge	Four-eng	30-40 km. S Grove	03.50	4	W
St.III./NJG3	Lt Achim Woeste	Four-eng	Nr. Grove	03.50	8	W
12./NJG3	Uffz Carlhans Hurth	Four-eng	Gesing N Esbjerg	03.54	1	W
2./NJG3	Oblt Berger	Four-eng	Grove area	03.57	3	W
2./NJG3	Uffz Erich Scheding	Four-eng	NW Vejle	03.58	1	W
1./NJG2	Ofw Rudolf Winn	Lancaster	30-50km W Westerland	03.58	2	W

2./NJG3	Oblt Berger	Four-eng	Grove area	04.13	4	W
7./NJG3	Lt Heinz Reuter	Four-eng	SW Westerland	04.27	4	W
2./NJG3	Oblt Berger	Four-eng	Grove area	04.30	5	W

10-11 April 1944

9./NJG4	Hptm Kurt Fladrich	Halifax	1km N Lechelle	00.01	10	W
7./NJG4	Uffz Leopold Weinberger	Halifax	La Fere	00.02	2	W
1./NJG6	Hptm Gerhard Friedrich	Lancaster	Montdidier	00.03	14	W
7./NJG4	Uffz Leopold Weinberger	Halifax	Ercheu	00.07	3	W
7./NJG1	Fw Hans Zeidler	Halifax	Rouvroy	00.07	2	W
7./NJG4	Uffz Leopold Weinberger	Halifax	Rouvroy	00.08	4	W
1./NJG6	Hptm Gerhard Friedrich	Halifax	Nr. Montdidier	00.13	15	W
7./NJG4	Uffz Leopold Weinberger	Halifax	Lavarières	00.16	5	W
9./NJG4	Oblt Hans Külow	Halifax	Rosières	00.17	2	W
9./NJG4	Hptm Kurt Fladrich	Halifax	Ercheu	00.27	11	W
8./NJG4	Lt Josef Springer	Halifax	NE Dieppe	00.28	2	W
8./NJG4	Hptm Helmut Bergmann	Lancaster	Vieux Mereuil	02.20	21	W
8./NJG4	Hptm Helmut Bergmann	Lancaster	6km N Salarnes	02.30	22	W
8./NJG4	Hptm Helmut Bergmann	Lancaster	1km SE Sailly	02.35	23	W
8./NJG4	Hptm Helmut Bergmann	Lancaster	NNW Achiet-le-Petit	02.43	24	W
8./NJG4	Hptm Helmut Bergmann	Lancaster	Beauquesne	02.50	25	W
8./NJG4	Hptm Helmut Bergmann	Lancaster	Near Vignacourt	02.54	26	W
8./NJG4	Hptm Helmut Bergmann	Lancaster	N Guignemicourt	03.06	27	W

11 April 1944

10./NJG3	Ofw Fritz Wilkenning	B-17		12.00	1	W

11-12 April 1944

3./NJG1	Fw Rauer	Lancaster	42km E Antwerp	23.02	4	W
4./NJG3	Fw Lorenz Gerstmayer	Four-eng	NW Aachen	23.03	12	W
4./NJG3	Fw Lorenz Gerstmayer	Four-eng	NW Aachen	23.07	13	W
12./NJG1	Uffz Fischer	Lancaster	10km W Roermond	23.15	2	W
St.IV./NJG1	Oblt H-W Schnaufer	Lancaster	Nr. Beese	23.15	52	W
4./NJG3	Fw Lorenz Gerstmayer	Four-eng	NW Aachen	23.19	14	W
St.IV./NJG1	Oblt H-W Schnaufer	Lancaster	2km N Saintlenaerts	23.25	53	W
2./NJG1	Oblt Werner Baake	Lancaster	100km W Haarlem	23.37	25	W
St.III./KG51	Lt Wolfgang Wenning	B-17	N Ipswich	00.42	1	W
5./KG51	Hptm Dietrich Puttfarken	Spitfire	England	00.50	1	W
5./KG51	Hptm Dietrich Puttfarken	Halifax	Engl;and	01.25	2	W

14-15 April 1944

St.I./NJG100	Lt G E Francsi	R-5	Russia	21.27	8	E

15=16 April 1944

3./NJG100	Uffz Mayer	U-2	Russia	23.23	1	E

16-17 April 1944

6./NJG101	Oblt Hans Krause	Wellington	N Breca	23.58	7	W

18-19 April 1944

2./NJG4	Hptm Josef Krahforst	Halifax	Montdidier area	23.46	2	W

The Luftwaffe had discontinued intruder operations in late 1941 because Hitler reasoned that the German peoples' morale would be better served by seeing British bombers falling over Germany. However, in the spring of 1944, the Me410-equipped II./KG 51 began a new wave of such missions. On 22nd April, a large number of Me410s stalked a returning American raid and shot down at least nine Liberators and one RAF Albemarle for the loss of two crews. One loss was that of the Staffelkapitan of 5./KG 51, Hauptmann Dietrich Puttfarken. He had won the Ritterkreuz as a bomber pilot and, at his death, was credited with five victories over England, including two on the night of his loss.

6./NJG1	Oblt Johannes Hager	Halifax	16km NE Montdidier	23.52	24	W
St.III./NJG1	Oblt Martin Drewes	Halifax	30km NE Rouen	00.01	19	W
St.III./NJG1	Oblt Martin Drewes	Halifax	SW Dieppe	00.11	20	W
IV./NJG3	Lt Bruno Heilig	Stirling	W Haderslev	00.23	4	W
IV./NJG3	Lt Bruno Heilig	Stirling	W Rømø Island	00.46	5	W
5./KG51	Oblt Claus Breissner	Lancaster	Cambridge	02.25	1	W
5./KG51	Hptm Dietrich Puttfarken	Lancaster	England	02.25	3	W
St.III./KG51	Lt Wolfgang Wenning	Halifax	NW Hull	03.47	2	W

19-20 April 1944

4./NJG100	Lt Richter	DB-3	Russia	22.51	4	E

20-21 April 1944

3./NJG100	Lt Schreiber	DB-3F	Russia	20.58	1	E
St.I./NJG100	Lt G E Francsi	R-5	Russia	22.24	9	E
1./NJG100	Lt G E Francsi	U-2	Russia	23.27	10	E
St.I./NJG100	Lt G E Francsi	U-2	Russia	23.42	11	E
I./NJG100	Lt Jansen	U-2	Russia	23.43	2	E
3./NJG100	Hptm August Fischer	DB-3	Russia	23.58	1	E
7./NJG1	Oblt Witzleb	Lancaster	60-80km NW Lille	00.06	5	W
Stab./NJG4	Maj Wolfgang Thimmig	Four-eng	25km S Paris	00.38	20	W
St.III./NJG1	Oblt Martin Drewes	Lancaster	30km SSW Paris	00.40	21	W
3./NJG4	Hptm Fritz Söthe	Four-eng	N Paris	00.50	6	W
3./NJG4	Hptm Fritz Söthe	Four-eng	50-100km N Paris	00.53	7	W
St.III./NJG5	Hptm Paul Zorner	Mosquito	20km SE Brussel	03.30	42	W
6./KG51	Fw Gustav Delp	P-47	N Cambridge	05.10	1	W

21-22 April 1944

4./NJG100	Uffz Haase	PS-84	Russia	21.38	4	E

22-23 April 1944

6./KG51	Fw Johann Trenke	B-17	England	22.04	1	W
St.III./KG51	Lt Wolfgang Wenning	Halifax	England	22.05	3	W
6./KG51	Uffz Baier	Halifax	England	22.10	1	W
6./KG51	Fw Johann Trenke	Halifax	England	22.10	2	W
4./KG51	Hptm Werner Duro	E/A	England	22.14	1	W
6./KG51	Fw Gustav Delp	Halifax	S Norwich	22.15	2	W
6./KG51	Fw Johann Trenke	Halifax	England	22.16	3	W
6./KG51	Uffz Baier	Halifax	England	22.22	2	W
6./KG51	Uffz Baier	Halifax	Orfordness	22.26	3	W
5./KG51	Lt Uwe Laas	Halifax	England	22.26	1	W
5./NJG1	Lt Otto Fries	Stirling	19km SW Soissons	23.31	10	W
1./NJG4	Uffz Konrad Beyer	Four-eng	near Maizy/Aisne	23.32	1	W
3./NJG4	Oblt Kadleik	Four-eng	NW Rheims	23.43	1	W
I./NJG4	Maj Wilhelm Herget	Four-eng	Soissons	23.50	51	W
5./KG 51	Hptm Dietrich Puttfarken	Halifax	England		4*	W
5./KG 51	Hptm Dietrich Puttfarken	Halifax	England		5*	W
6./NJG1	Oblt Johannes Hager	Lancaster	SSW Laon	00.16	25	W
8./NJG1	Oblt Dietrich Schmidt	Lancaster	S Laon	00.21	19	W
4./NJG1	Uffz Heinz Filipzig	Lancaster	NW Laon	00.26	1	W
8./NJG1	Oblt Dietrich Schmidt	Lancaster	SW Laon	00.26	20	W
St.III./NJG1	Hptm Martin Drewes	Lancaster	10-20km SE Rosieres	00.27	22	W
12./NJG3	Oblt Herbert Koch	Stirling	DW Svendborg	00.34	10*	W
3./NJG4	Lt Klaus Helm	Halifax	18km SSE Namur	00.52	2	W
I./NJG4	Maj Wilhelm Herget	Lancaster	9km SSE Namur	00.57	52	W
St.III./NJG5	Hptm Paul Zorner	Halifax	Nr. Aachen	01.03	43	W
8./NJG5	Oblt Peter Ehrhardt	Lancaster	100km from Düsseldorf	01.07	15	W
3./NJG4	Lt Klaus Helm	Lancaster	Eschweiler	01.08	3	W
2./NJG1	Hptm E-W Modrow	Lancaster	Düsseldorf	01.10	4	W
.III./JG300	Hptm Iro Ilk	Lancaster	E Düsseldorf	01.15	3	W
8./JG302	Lt Otto Schwamb	Lancaster	Düsseldorf	01.15	1	W
1./JG300	Ofw Hermann Wischnewski	Halifax	5-8km N Düsseldorf	01.22	10	W
4./NJG3	Hptm Gerhard Raht	Halifax	SSW Düsseldorf	01.23	29	W
8./NJG5	Oblt Peter Ehrhardt	Four-eng	Düsseldorf	01.25	16	W
III./JG300	Hptm Iro Ilk	E/A	Düsseldorf	01.25	4	W
11./NJG5	Uffz Herbert Ter.Steegen	Halifax	5-10km from Düsseldorf	01.25	6	W
8./JG302	Lt Otto Schwamb	Four-eng	Düsseldorf	01.26	2	W
I./JG300	Fw Janssen	Lancaster	5-10km from Düsseldorf	01.30	1	W
6./JG300	Fw Kohler	Lancaster	Düsseldorf	01.30	1	W
11./NJG5	Uffz Herbert Ter.Steegen	E/A	Düsseldorf	01.30	7	W
8./JG300	Uffz Franz Bollerey	Lancaster	Köln	01.35	1	W
III./JG300	Hptm Iro Ilk	Lancaster	Düsseldorf	01.35	5	W
2./NJG6	Lt Hans Zettel	Lancaster	Düsseldorf	01.38	2	W
11./NJG1	Oblt Hermann Greiner	Lancaster	10km NW Düsseldorf	01.39	20	W
2./NJG1	Maj Hans Karlowski	Lancaster	10km S Xanten	01.39	1	W
St.I./NJG5	Lt Joachim Hanss	Halifax	Holzminden	01.40	7	W
Stab./NJG2	Oblt Johann Werth	Lancaster	3-5km from Düsseldorf	01.40	8	W

Hauptmann Iro Ilk, Gruppenkommandeur of III./JG 300 (left) was credited with six night victories in his total of nine. Three of his victories came on the night of 24th April. He eventually fell to a Canadian Spitfire pilot on 25th September. In early 1943 Oberleutnant Johannes Hager (right) joined 6./NJG 1 and fought right through to the armistice, like Ilk, winning the Ritterkreuz. In 99 missions he achieved 48 victories, his 25th on 23rd April.

11./NJG5	Uffz Herbert Ter.Steegen	Halifax	Nr. Düsseldorf	01.42	8	W
2./NJG1	Uffz Karl-Heinz Wildhagen	Lancaster	50-70km from Düsseldorf	01.43	1	W
4./NJG3	Hptm Gerhard Raht	Lancaster	WNW Düsseldorf	01.44	30	W
St.I./NJG2	Hptm Ernst Zechlin	Halifax	Düsseldorf	01.50	7	W
3./NJG3	Ofw Rudolf Frank	Lancaster	10-30km NW Venlo	01.55	43	W
2./NJG1	Hptm E-W Modrow	Lancaster	30km SE Gilze	01.55	5	W
8./NJG3	Ofw "Watten"	Four-eng	N Düsseldorf	01.57	1	W
2./NJG2	Oblt Heinz Rökker	Lancaster	50km NW Düsseldorf	01.58	22	W
St.I./NJG2	Hptm Ernst Zechlin	Stirling	Düsseldorf	02.03	8	W
2./NJG1	Hptm E-W Modrow	Lancaster	25km SE Gilze Rijn	02.04	6	W
12./NJG3	Hptm Eduard Schroder	Lancaster	S Wunstorf	02.25	11	W
2./NJG2	Uffz Alfred Walther	Four-eng	N Düsseldorf	02.55	3	W
4./KG51	Uffz Bongert	Halifax	England	03.35	1	W

23 April 1944

St.III./NJG3	Hptm Walter Barte	Four-eng	SW Düsseldorf	14.08	18	W

23-24 April 1944

2./NJG3	Oblt Berger	Four-eng	N Schleswig	23.17	6	W
12./NJG3	Lt Herbert Koch	Four-eng	Faborg	23.18	11	W
3./NJG3	Ofw Rudolf Frank	Four-eng	Laaland Island	23.22	44	W
5./NJG1	Lt Otto Fries	Halifax	Beacon "Chameleon"	23.38	11	W
10./NJG3	Lt Walter Briegleb	Halifax	15km S Agerso	23.42	3	W
8./NJG3	Oblt Fritz Brinkmann	Halifax	Falster Island area	23.43	5	W
St.1./NJG3	Hptm Werner Husemann	Four-eng	Laaland Island	23.56	18	W

24-25 April 1944

3./NJG100	Fw Herbert Simross	R-5	Russia	23.52	4	E
3./NJG4	Hptm Fritz Söthe	Four-eng	St Dizier area	00.04	8	W
2./NJG1	Hptm E-W Modrow	Lancaster	18km NE Lüttich	00.05	7	W
6./NJG4	Oblt Karl-Heinz Schaffer	Lancaster	La Chapelle-Theile	00.08	2	W
2./NJG1	Oblt Werner Baake	Lancaster	N Tilburg	00.26	26	W
4./NJG4	Oblt Jakob Schaus	Four-eng	Nancy-Karlsruhe	00.29	5	W
St.III./NJG1	Hptm Martin Drewes	Lancaster	30-50km W Mannheim	00.37	23	W

III./JG301	Oblt Otto-Ernst Hocker	Lancaster	Mannheim	00.43	1	W
2./NJG6	Oblt Martin Becker	Halifax	SW Kahlsruhe	00.58	27	W
4./NJG4	Oblt Jakob Schaus	Four-eng	20km SW Karlsruhe	00.59	6	W
7./NJG4	Uffz Leopold Weinberger	Four-eng	SW Kahlsruhe	01.00	6	W
3./NJG101	Oblt Ernst Schuren	Halifax	Mannheim	01.20	1	W
3./NJG101	Oblt Ernst Schuren	Lancaster	30km NE Mannheim	01.28	2	W
5./NJG2	Lt Edgar Honold	Four-eng	15km NW Munich	02.03	3	W
St.IV./NJG1	Oblt H-W Schnaufer	Lancaster	9km SE St.Trond	02.03	54	W
IV./NJG1	Ofw Gerhard Graefe	Halifax	10km NW Namur	02.05	1	W
7./JG300	Uffz Emil Kriegbaum	Four-eng	15km N Munich	02.05	1	W
2./NJG2	Oblt Heinz Rökker	Lancaster	30km NE Munich	02.05	23	W
FührKurSt	Oblt Lembeke	Lancaster	Nr Dachau	02.09	1	W
I./NJG4	Oblt Friedrich Thorl	Four-eng	Chimay	02.13	1	W
2./NJG1	Oblt Werner Baake	Halifax	W Gorinchem	02.18	27	W
11./NJG1	Oblt Hermann Greiner	Halifax	Near Breda	02.18	21	W
St.IV./NJG1	Oblt H-W Schnaufer	Lancaster	3km N Mechelen	02.28	55	W
St.IV./NJG1	Oblt H-W Schnaufer	Halifax	1km W Haasdonk	02.30	56	W
St.IV./NJG1	Oblt H-W Schnaufer	Lancaster	In the Sea	02.40	57	W
4./NJG3	Hptm Gerhard Raht	Four-eng	30-40km S Echterdingen	02.43	31	W
2./NJG1	Hptm E-W Modrow	Lancaster	25km SE Gilze Rijn	03.34	8	W
6./KG51	Uffz Baier	Halifax	England	03.58	4	W
St.III./KG51	Lt Wolfgang Wenning	Halifax	20km NE Cambridge	04.22	4	W

26-27 April 1944

3./NJG100	Fw Herbert Simross	R-5	Russia	23.53	5	E
3./NJG100	Fw Herbert Simross	R-5	Russia	00.02	6	E
8./NJG1	Lt Peter Ehrhardt	Lancaster	S St.Dizier	00.24	17	W
St.III./NJG1	Hptm Martin Drewes	Lancaster	Biberach	00.35	24	W
8./NJG1	Oblt Dietrich Schmidt	Four-eng	Dijon	00.47	21	W
8./NJG4	Hptm Helmut Bergmann	Lancaster	15km SE St. Dizier	00.50	28	W
St.III./NJG1	Hptm Martin Drewes	Lancaster	N Chaumont	00.50	25	W
8./NJG4	Hptm Helmut Bergmann	Lancaster	30km E St. Dizier	00.57	29	W
8./NJG4	Hptm Helmut Bergmann	Lancaster	15km SE Toul	01.01	30	W
4./NJG5	Oblt Josef Kraft	Lancaster	10-20km S Strasbourg	01.24	24	W
4./NJG5	Oblt Helmuth Schulte	Lancaster	Strasbourg	01.28	12	W
2./NJG6	Oblt Martin Becker	Lancaster	Beacon "Christa"	01.30	28	W
3./NJG6	Uffz Benno Kratz	Lancaster	Near Reichental	01.34	3	W
3./NJG101	Oblt Christian	Lancaster	SW Stuttgart	01.34	1	W
2./NJG6	Oblt Martin Becker	Lancaster	W Stuttgart	01.36	29	W
5./JG300	Ofw Richard Heemsoth	Lancaster	SW Essen	01.36	4	W
4./NJG5	Oblt Josef Kraft	Lancaster	Strassbourg area	01.45	25	W
9./NJG5	Lt Otto Keller	Lancaster	AS-6	01.47	5	W
3./NJG6	Uffz Benno Kratz	Lancaster	Near Gansbach	01.48	4	W
3./NJG3	Ofw Rudolf Frank	Halifax	LM	01.55	45	W
St.IV./NJG1	Oblt H-W Schnaufer	Lancaster	1km SE Achtmaal	02.05	58	W
2./NJG1	Fw Josef Strohlein	Lancaster	3km NW St.Trond	02.05	1	W
9./NJG5	Lt Otto Keller	Lancaster	15km SE Heidelberg	02.12	6	W
4./NJG3	Hptm Gerhard Raht	Lancaster	Schweinfurt	02.15	32	W
St.IV./NJG1	Oblt H-W Schnaufer	Lancaster	In the Sea	02.18	59	W
FührKurSt	Hptm Boruschen	Lancaster		20.12	1	W
3./NJG6	Fw Gunther Bahr	Lancaster	15km NW Kirchenhall	02.45	15	W

On the night of 27th April 1944, Oberleutnant Wilhelm Johnen landed his 8./NJG 6 Bf110G at Dubendorff airfield in Switzerland. He had just claimed his 19th victory, a Halifax near Strasbourg, and had become lost. The German authorities, fearing that he had defected, immediately ordered the arrest of family members, but it had been a genuine error. He was quickly repatriated, a German agent was allowed to blow up the Messerschmitt to destroy the radar equipment, and the Swiss Government received a squadron of Bf109E fighters!

Johnen went on to receive the Ritterkreuz in October 1944 and increase his victory tally to 34. He survived the war as Gruppenkommander of III./NJG 6 and later wrote a book about his experiences.

7./JG302	Lt German Merz	Lancaster	Schweinfurt	02.59	2	W
9./NJG5	Lt Otto Keller	Lancaster	20km WSW Landau	03.03	7	W
6./KG51	Fw Johann Trenke	E/A	England	04.25	4	W

27-28 April 1944

3./NJG100	Fw Herbert Simross	U-2	Russia	20.12	7	E
3./NJG100	Lt Eberhard Kaross	U-2	SSE Polotsk	22.17	2	E
3./NJG100	Lt Eberhard Kaross	U-2	SSE Polotsk	01.14	3	E
2./NJG6	Oblt Martin Becker	Lancaster	W Stuttgart	01.19	30	W
LBSt.7	Hptm Ernst Wallner	Lancaster	Beacon 'Christa'	01.19	1	W
St.III./NJG5	Hptm Paul Zorner	Lancaster	SE Nancy	01.20	44	W
St.III./NJG5	Hptm Leopold Fellerer	Halifax	Mulhouse	01.30	32	W
11./NJG1	Lt Friedrich Potthast	Halifax	Maastricht	01.30	9	W
St.IV./NJG1	Oblt H-W Schnaufer	Halifax	15km SW Arnhem	01.30	60	W
8./NJG6	Oblt Wilhelm Johnen	Halifax	30-60km SW Strasbourg	01.31	19	W
3./NJG100	Lt Eberhard Kaross	U-2	SSE Polotsk	01.31	4	E
St.IV./NJG5	Unknown	E/A		01.32	-	W
12./NJG1	Oblt H-H Augenstein	Halifax	4km ESE Maastricht	01.33	33	W
6./NJG1	Oblt Johannes Hager	Halifax	N Arnhem	01.33	26	W
2./NJG6	Oblt Martin Becker	Halifax	Near beacon "Christa"	01.35	31	W
3./NJG6	Lt Lomberg	Halifax	W beacon "Christa"	01.35	4	W
3./NJG6	Fw Gunther Bahr	Halifax	30-60km NW Fried'hafen	01.37	16	W
6./NJG1	Oblt Johannes Hager	Halifax	W Arnhem	01.37	27	W
12./NJG5	Oblt Otto Karl Klemenz	Halifax	15km S Colmar	01.38	3	W
2./NJG2	Oblt Heinz Rökker	Halifax	Freiburg im Breisgau	01.39	24	W
St.IV./NJG1	Lt Georg Fengler	Halifax	2km W Witten	01.40	2	W
3./NJG6	Hptm H-M Hadeball	Lancaster	5km S beacon "Christa"	01.40	31	W
St.IV./NJG1	Oblt H-W Schnaufer	Lancaster	32km SW Arnhem	01.40	61	W
St.III./NJG1	Hptm Martin Drewes	Lancaster	30-50km SE Arnhem	01.48	26	W
4./NJG3	Fw Lorenz Gerstmayer	Halifax	SE beacon "Christa"	01.51	15	W
St.III./NJG5	Hptm Paul Zorner	Lancaster	SW Strassbourg	01.51	45	W
I./NJG4	Oblt Friedrich Thörl	Halifax	3km NE St.Trond	01.52	2	W

4./NJG1	Oblt Fritz Lau	Halifax	Dahren	01.53	6	W
2./NJG6	Oblt Martin Becker	Lancaster	SW Bremen	01.57	32	W
11./NJG1	Oblt Hermann Greiner	Lancaster	6km W Genk	01.58	22	W
1./NJG1	Oblt Wilhelm Henseler	Lancaster	St.Trond	01.58	6	W
11./NJG1	Oblt Hermann Greiner	Halifax	2km SE Oerle	02.00	23	W
6./NJG1	Oblt Johannes Hager	Lancaster	W St.Trond	02.00	28	W
St.III./NJG2	Lt Ernst Hovermann	Halifax	30km E Brussels	02.01	2	W
1./NJG1	Oblt Wilhelm Henseler	Halifax		02.02	7	W
3./NJG1	Lt Hittler	Halifax	30km NE Namur	02.05	1	W
1./NJG6	Hptm Gerhard Friedrich	Halifax	Switz. W Friedrichshafen	02.09	16	W
Stab./NJG1	Maj Hans-Joachim Jabs	Lancaster	2km SE Dienst	02.09	27	W
8./NJG5	Fw Robert Schratter	Lancaster	Friedrichshafen	02.10	3	W
St.III./NJG5	Hptm Paul Zorner	Lancaster	NW Friedrichshafen	02.10	46	W
3./NJG4	Hptm H-J Thomaschat	Halifax	6km NE Grimbergen	02.18	1	W
2./NJG6	Hptm Martin Becker	Halifax	30-50km W Fried'hafen	02.19	33	W
St.III./NJG5	Hptm Leopold Fellerer	Halifax	Borensen	02.20	33	W
4./NJG5	Oblt Josef Kraft	Halifax	Friedrichshafen	02.24	26	W
Stab./NJG1	Maj Hans-Joachim Jabs	Halifax	Rotterdam	02.25	47	W
1./NJG6	Hptm Gerhard Friedrich	Halifax	Schaffhausen	02.34	17	W
8./JG301	Lt Glaas	Four-eng	NE Friedrichshafen	02.38	2	W
I./NJG4	Maj Wilhelm Herget	Lancaster	NW Dunkirk	02.50	53	W
3./NJG6	Fw Gunther Bahr	Lancaster	50-80km NW Fried'hafen	02.51	17	W
5./NJG2	Hptm H-H Hissbach	Halifax	St. Dizier	03.06	14	W
1./NJG4	Uffz Konrad Beyer	Halifax	E Landrecies	03.10	2	W
6./NJG2	Lt Erich Jung	Lancaster	ESE Neufchateau	03.10	2	W
5./KG51	Uffz Döhlenz	E/A	North Sea	04.15	1	W

28-29 April 1944

I./NJG100	Lt G E Francsi	U-2	Russia	22.00	12	E
I./NJG100	Lt G E Francsi	U-2	Russia	22.30	13	E

29 April 1944

St.III./NJG3	Hptm Walter Barte	B-17	Javenitz, SW Stendal	13.28	19	W
Stab./NJG1	Maj Hans-Joachim Jabs	Spitfire	3km S Valburg	15.15	29	W
Stab./NJG1	Maj Hans-Joachim Jabs	Spitfire	6km S Apeldoorn	15.15	30	W

30 April-1 May 1944

St.I./NJG100	Lt G E Francsi	U-2	Russia	21.26	14	E
St.I./NJG100	Lt G E Francsi	R-5	Russia	21.46	15	E
St.I./NJG100	Lt G E Francsi	U-2	Russia	22.30	16	E

1-2 May 1944

St.I./NJG100	Hptm Kurt Bonow	Il-4	Russia	21.23	6	E
3./NJG100	Lt Eberhard Kaross	DB-3	Russia	23.17	5	E
3./NJG100	Lt Eberhard Kaross	DB-3	Russia	23.46	6	E
11./NJG1	Oblt Hermann Greiner	Halifax	15km S Brussels	00.15	24	W
2./NJG1	Hptm E-W Modrow	Halifax	30-50km NW Brussels	00.25	9	W
11./NJG1	Oblt Hermann Greiner	Halifax	2km NW Audenaarde	00.30	25	W
12./NJG1	Oblt H-H Augenstein	Lancaster	Nr. Ghent	00.32	34	W
3./NJG4	Lt Klaus Helm	Four-eng	Ghent	00.33	4	W
8./NJG4	Ofw Reinhard Kollak	Lancaster	3km S Beaumont	00.35	37	W

Oberstleutnant Hans-Joachim Jabs had flown with II./ZG 76 in the French campaign and the Battle of Britain. When he joined the nightfighter force he had fourteen confirmed victories credited and had won the Ritterkreuz. He claimed his first night victory on the night of 25th June 1942 and, by late April 1944, was the Geschwader Kommodore of NJG 1, with the Eichenlaub and 42 credited kills, including a B-17. On 29th April, he flew a day mission and was intercepted on final approach to Deelen by Spitfire IXs of 132 Squadron, led by Squadron Leader A.G.Page. In the fight that followed he destroyed two Spitfires, but was then shot down by Page and crash-landed on Deelen airfield unhurt. He survived the war with 50 victories and later became firm friends with Geoffrey Page and also with John Caulton, one of the two shot-down Spitfire pilots.

4./NJG4	Oblt Jakob Schaus	Four-eng	NW Paris	00.40	7	W
8./NJG4	Ofw Reinhard Kollak	Lancaster	3km SW Châteaudun	00.41	38	W
St.III./NJG4	Hptm Hans-Karl Kamp	Lancaster	Saint-Arnoud	00.43	14	W
St.III./NJG1	Hptm Martin Drewes	Lancaster	60km NW Paris	00.45	27	W
7./NJG1	Oblt Walter Prues	Lancaster	18km SW Poix	00.47	2	W
4./NJG4	Hptm Fritz Söthe	Four-eng	SSW Paris	00.55	9	W
KdoKunkel	Lt Artur Ewert	Halifax	Toulouse	01.50	1	W

3-4 May 1944

7./NJG1	Oblt Walter Brieglеb	Four-eng	Maibach-Harz	23.48	4	W
St.III./NJG1	Oblt Scheer	Lancaster	S Laon/Athies	23.55	1	W
8./NJG1	Oblt Dietrich Schmidt	Lancaster	SW St.Dizier	00.03	22	W
8./NJG4	Hptm Helmut Bergmann	Lancaster	30 km SSE Champlain	00.10	31	W
1./NJG5	Ofw A. Klaiber	Lancaster	S Montier-en-Dur	00.13	3	W
8./NJG4	Hptm Helmut Bergmann	Lancaster	50-60km ESE Champlain	00.15	32	W
3./SKG10	Ofw Tirchow	Lancaster	12km S Mailly-le-Camp	00.15	1	W
4./NJG4	Oblt Jakob Schaus	Four-eng	N Paris	00.16	8	W
St.III./NJG1	Hptm Martin Drewes	Lancaster	S Troyes	00.18	28	W
8./NJG1	Oblt Dietrich Schmidt	Lancaster	W St.Dizier	00.19	23	W
2./NJG1	Oblt Werner Baake	B-17	50km W Venlo	00.20	28	W
1./NJG4	Oblt Schubert	Four-eng	Ste. Ménéhould	00.21	1	W
7./NJG4	Oblt Richard Delakowitz	Four-eng	SW Châlons-sur-Marne	00.21	4	W
8./NJG4	Hptm Helmut Bergmann	Lancaster	60-70km SE Champla1n	00.22	33	W
7./NJG4	Oblt Richard Delakowitz	Four-eng	SW Châlons-sur-Marne	00.22	5	W
3./SKG10	Fw Otto Heinrich	Four-eng	Near Mailly-le-Camp	00.25	1	W
8./NJG4	Hptm Helmut Bergmann	Lancaster	N Paris	00.26	34	W
St.III./NJG1	Hptm Martin Drewes	Lancaster	NE Troyes	00.26	29	W
9./NJG4	Ofw Reinhard Kollak	Lancaster	15km SW Châlons-sur-M	00.26	39	W
3./SKG10	Fw Otto Heinrich	Four-eng	Hirson-Rocroi area	00.27	2	W
St.III./NJG4	Lt Heinz Wilberg	Four-eng	N Romilly	00.27	3	W
2./NJG5	Uffz Wester	Mosquito	S. Bar-sur-Aube	00.30	1	W
1./SKG10	Lt Walter Starik	Lancaster	Roye/Amy	00.30	1	W
7./NJG4	Oblt Richard Delakowitz	Four-eng	SW Châlons-sur-Marne	00.31	6	W
1./NJG5	Uffz Hengsbach	Four-eng	Caumont L'Evente	00.32	1	W

St.III./NJG1	Hptm Martin Drewes	Lancaster	Romilly area	00.36	30	W
St.I./NJG5	Hptm Werner Hoffmann	Lancaster	15-20km S Mailly-le-Camp	00.36	29	W
St.I./NJG5	Hptm Werner Hoffmann	Lancaster	15-20km S Mailly-le-Camp	00.36	30	W
7./NJG1	Oblt Walter Prues	Lancaster	SW Romilly	00.37	3	W
8./NJG4	Hptm Helmut Bergmann	Lancaster	N Paris	00.39	35	W
9./NJG4	Oblt Wilhelm Breimer	Lancaster	Formérie	00.44	1	W
St.III./NJG1	Hptm Martin Drewes	Lancaster	Nogent-sur-Seine	00.44	31	W
8./NJG1	Lt Peter Ehrhardt	Lancaster	Montargis area	00.44	18	W
4./NJG4	Uffz Walter Krause	Four-eng	Near Romilly-sur-Seine	00.45	1	W
4./NJG4	Oblt Jakob Schaus	Four-eng	SE Paris	00.46	9	W
8./NJG1	Oblt Dietrich Schmidt	Lancaster	SSW Romilly-sur-Seine	00.46	24	W
St.III./NJG1	Hptm Martin Drewes	Lancaster	80-100km SE Paris	00.50	32	W
St.I./NJG6	Oblt Gaul	Lancaster	SW Avallon	00.52	1	W
St.I./NJG4	Lt Heinz Wilberg	Four-eng	Sézanne area	00.53	4	W
3./SKG10	Fw Otto Heinrich	Four-eng	Near Mailly-le-Camp	00.54	3	W
9./NJG4	Uffz Philipp Feilen	Four-eng	10km SE Montereaux	00.55	1	W
3./SKG10	Ofw Tirchow	Lancaster	Sézanne	00.55	2	W
Kdo.Kunkel	Hptm Artur Ewert	Stirling	S Paris	00.58	2	W
11./NJG1	Lt Wilhelm Seuss	Halifax	18km N Mielec	00.58	7	W
9./NJG4	Uffz Philipp Feilen	Four-eng	10km SE Monterean	01.00	1	W
1./NJG4	Oblt Walter Riedlberger	Four-eng	NE Alençon	01.07	6	W
St.III./NJG1	Oblt Scheer	Lancaster	NE Chartres	01.25	1	W

4-5 May 1944

11./NJG5	Lt Wilhelm Seuss	Halifax	18km N Mielec, Krakow	00.19	7	W
1./NJG6	Uffz Herbert Nahlik	Wellington	W edge Budapest	01.02	1	W

5-6 May 1944

6./NJG4	Oblt Karl-Heinz Schaffer	B-24		00.43	3	W
St.IV./NJG6	Maj Herbert Lütje	Wellington	NW Ploesti	01.05	33	W

6-7 May 1944

2./NJG1	Oblt Werner Baake	Mosquito	1km NE Herkendorf	23.35	29	W
2./NJG1	Oblt Werner Baake	B-17	W Venlo	00.09	30	W
2./NJG100	Hptm Ulrich Meien.von	Wellington		00.16	6	W
2./NJG100	Ofw Karl Maisch	Wellington	10km N Alexandrowo	00.30	5	W
1./NJG6	Oblt Bauer	B-24	SW Budapest	01.06	1	W
St.I./KG51	Oblt Czerwinski	Halifax	SSW. La Prisière	02.27	1	W
St.III./NJG4	Lt Heinz Wilberg	Four-eng	Brionne area	02.33	5	W
12./NJG3	Fw Hugo Fuetscher	Four-eng	N Ebsjerg	02.34	4	W
3./SKG10	Ofw Tirchow	Lancaster	SE Alencon	03.15	3	W

7-8 May 1944

St.IV./NJG6	Maj Herbert Lütje	Wellington	Marosheviz	23.54	34	W
10./NJG6	Uffz Josef Brunner	B-24	W Bucharest	00.05	5	W
St.IV./NJG6	Maj Herbert Lutje	Wellington	SW Blejesti	00.14	35	W
St.III./NJG4	Hptm Hans-Karl Kamp	Four-eng	Romorantin	00.15	15	W
10./NJG6	Oblt Franz Günther	Halifax	Alexandrowo	00.20	1	W
St.III./NJG4	Hptm Hans-Karl Kamp	Four-eng	E Bourges	00.20	16	W
2./NJG100	Hptm Ulrich Meien.von	Wellington	SW Blejesti	00.20	7	W
St.III./NJG4	Hptm Hans-Karl Kamp	Four-eng	SE Bourges	00.32	17	W

Hans Krause joined the nightfighter force in late 1941, joining NJG 3 in northern Germany. His first successes came in October 1942, but he transferred to 6./NJG 101, an independent Jagdstaffel, as its Staffel leader in mid-1943. Based near Vienna, the task of the unit was air defence against the Red Air Force and the night bombers of No. 205 Group, operating from Italy. In February 1945, having won the Ritterkreuz for 28 combat victories, he moved to command I./NJG 4, but did not add to his score and survived the war. He claimed his 8th victory, a Halifax near Budapest, on the night of 10th May 1944.

8./NJG4	Ofw Reinhard Kollak	Four-eng	15km NE Romorantin	00.36	40	W
1./SKG10	Lt Walter Starik	Lancaster	Nr. Hirson	00.40	2	W
2./NJG100	Hptm Ulrich Meien.von	Wellington	SE Craiova	00.46	8	W
5./NJG4	Uffz Georg Matt	Four-eng	30km N Bourges	00.47	1	W
9./NJG4	Lt Hromadnik	Lancaster	30-50km SW Orléans	00.48	4	W
6./NJG4	Oblt Karl-Heinz Schaffer	Stirling	Off Ushant	01.02	4	W
3./NJG1	Lt Hittler	Mosquito	Grevenbroich	01.30	2	W
12./NJG3	Fw Klaus Möller	Four-eng	60km N Thisted	02.19	5	W
10./NJG6	Oblt Franz Gunther	Wellington		02.45	2	W
1./KG51	Oblt Czerwinski	Halifax	Tours	03.02	2	W
7./NJG1	Oblt Walter Prues	Lancaster	Ushant	03.44	4	W

8-9 May 1944

4./NJG100	Oblt Günter Bertram	Boston	Russia	22.31	14	E
4./NJG100	Oblt Günter Bertram	PS-84	Russia	23.53	15	E
4./NJG100	Oblt Günter Bertram	PS-84	Russia	24.00	16	E
4./NJG100	Oblt Günter Bertram	PS-84	Russia	00.30	17	E
1./NJG4	Uffz Konrad Beyer	Four-eng	Maubeuge	03.25	3	W
1./NJG4	Uffz Konrad Beyer	Four-eng	Charleroi	03.31	4	W
1./NJG4	Uffz Konrad Beyer	Four-eng	Charleroi area	03.33	5	W
St,IV./NJG1	Oblt H-W Schnaufer	Halifax	30km SW Charleroi	03.34	62	W
1./NJG4	Uffz Konrad Beyer	Four-eng	Landrecies	03.37	6	W
St.IV./NJG1	Lt Georg Fengler	Halifax	17km NNE Courtrai	03.44	3	W
1./NJG4	Lt Wolfgang Marstaller	Four-eng	Landrecies	03.45	1	W
St.IV./NJG1	Hptm Adolf Breves	Halifax	6km SW Courtrai	03.53	1	W
11./NJG1	Oblt Hermann Greiner	Halifax	21km NNE Courtrai	03.57	26	W
11./NJG1	Lt Friedrich Potthast	Halifax	1km S Courtrai	03.59	10	W

9-10 May 1944

2./NJG100	Uffz Josef Czajica	R-5	15-20km S Saki	23.18	1	E
1./SKG10	Fw Walter Nikolaus	Lancaster	N Paris	00.43	1	W
1./NJG4	Oblt Heinrich Schulenburg	Four-eng	500m N Evreux airport	00.53	1	W
4./NJG4	Uffz Walter Krause	Four-eng	TB-7	00.58	2	W

6./NJG4	Hptm Hans Autenrieth	Lancaster	UB-3	01.00	18	W
7./NJG1	Oblt Walter Prues	Lancaster	26km SE Lisieux	01.02	5	W

10 May 1944

3./NJG3	StFw Alfred Kuhne	E/A		11.00	1	W

10-11 May 1944

2./NJG100	Uffz Emil Ohrisch	R-5	W Sevastopol	23.23	1	E
11./NJG1	Lt Friedrich Potthast	Lancaster	4km SE Lille	23.58	11	W
St.IV./NJG1	Hptm Adolf Breves	Lancaster	N Luxembourg	00.05	2	W
3./NJG1	Oblt Josef Nabrich	Lancaster	Sea W Ghent	00.12	9	W
4./NJG1	Lt Hans Schmitz	Lancaster	NW Lille	00.15	4	W
3./NJG1	Hptm Heinz Strüning	Lancaster	18km NE Brugge	00.15	43	W
6./NJG101	Oblt Hans Krause	Halifax	E Siegen	00.33	8	W

11-12 May 1944

2./NJG100	Oblt Georg Patzner	DB-3	Russia	22.48	1	E
4./NJG1	Oblt Fritz Lau	Lancaster	NW St.Trond	23.57	7	W
4./NJG1	Lt Hans Schmitz	Lancaster	WSW Rotterdam	00.15	5	W
2./NJG2	Oblt Heinz Rökker	Lancaster	20-50km NW Brussels	00.23	25	W
2./NJG1	Hptm E-W Modrow	Lancaster	15km SE Goes	00.26	10	W
11./NJG1	Oblt Hermann Greiner	Halifax	Leuven	00.27	27	W
4./NJG1	Uffz Hans Schwenke	Four-eng	NW Hasselt	00.27	1	W
2./NJG2	Oblt Heinz Rökker	Lancaster	off Zeebrugge	00.35	26	W
2./NJG1	Oblt Werner Baake	Lancaster	28km NE Huckhofen	00.42	31	W
2./NJG2	Ofw Vinzenz Giessubel	Lancaster	30km NE Brussels	00.42	2	W
1./NJG2	Fw Erwin Storck	Lancaster	Zuid Beveland	00.46	1	W
4./NJG3	Uffz Gerhard Fugmann	Four-eng	Brugge	00.48	1	W
6./NJG1	Lt Gottfried Hanneck	Four-eng	JJ-JH	00.48	1	W
2./NJG2	Oblt Heinz Rökker	Lancaster		00.49	u/c	W
2./NJG1	Hptm E-W Modrow	Lancaster	In sea	01.04	11	W
2./KG51	Uffz Walter Breugel	Four-eng	Bredstedt	02.38	1	W

12-13 May 1944

3./NJG100	Ofw Karl Strohecker	DB-3	3km S Strakeny-Wibinsk	22.35	6	E
3./NJG100	Ofw Karl Strohecker	Il-4	20km S Polotsk	23.10	7	E
3./NJG100	Hptm August Fischer	Il-4	Russia	23.51	2	E
2./NJG1	Hptm E-W Modrow	Halifax	15km NE Bergen	00.02	12	W
3./NJG1	Oblt Josef Nabrich	Lancaster	4km N Larum	00.20	10	W
12./NJG1	Oblt H-H Augenstein	Halifax	8km N Mechelen	00.25	35	W
11./NJG1	Oblt Hermann Greiner	Lancaster	Lieren, nr Appeldoorn	00.27	28	W
St.IV./NJG1	Lt Georg Fengler	Lancaster	NW Lieren	00.36	4	W
8./NJG2	Oblt Friedrich Tober	Halifax	30km NW St.Trond	00.37	4	W
St.III./NJG1	Hptm Martin Drewes	Lancaster	NW St.Trond	00.44	33	W
St.IV./NJG1	Oblt H-W Schnaufer	Halifax	6km NNE Baelen, Liège	00.44	64	W
2./NJG2	Uffz Martin Siegel	Lancaster	Dunkirk	00.45	1	W
St.IV./NJG1	Oblt H-W Schnaufer	Halifax	5km ENE Hasselt	00.46	63	W
St.IV./NJG1	Oblt H-W Schnaufer	Halifax	Hoogstraaten	00.48	65	W
3./NJG1	Hptm Heinz Strüning	Halifax	15km SSE Brussels	00.48	44	W
St.III./NJG1	Hptm Martin Drewes	Lancaster	NNW St Trond	00.49	34	W
2./NJG1	Hptm E-W Modrow	E/A	In sea	01.00	13	W
St.III./NJG1	Hptm Martin Drewes	Lancaster	S Gent	01.09	35	W

Leutnant Otto Fries of I./NJG 1 made history on the night of 19th May 1944. With his Bordfunker Feldwebel Arthur Staffa near Florennes they attacked by a 605 Squadron Mosquito. The cannon fire from Wing Commander N.V.Starr's fighter ripped into the He219, which rapidly became uncontrollable. The He219 was the first aircraft to be fitted with ejection seats and both Fries and Staffa ejected successfully, becoming the first aircrew to do so under combat conditions. The pair already had eleven victories to their credit and would add six more by the armistice. They were also to use their ejector seats again!

(S.Hall)

13-14 May 1944

3./NJG100	Lt Truber	Il-5	30km NE Polotsk	23.45	3	E
4./NJG100	Lt Richter	PS-84	Russia	00.10	5	E
4./NJG100	Lt Klaus Scheer	B-25	Russia	00.20	13	E

14-15 May 1944

5./NJG100	Fw Schleuchen	Il-4	Duchowno	22.29	1	E

15-16 May 1944

1./NJG100	Lt Grosskopf	R-5	Russia	21.46	1	E
IV./NJG3	Hptm Franz Buschmann	Four-eng	Russia	00.45	6	W
12./NJG3	Fw Hugo Fuetscher	Four-eng	Russia	00.47	5	W
10./NJG3	Uffz Heinz Koppe	Four-eng	Russia	00.50	5	W
IV./NJG3	Hptm Franz Buschmann	Four-eng	Russia	01.53	7	W

16-17 May 1944

1./NJG100	Oblt Waldemar Rathke	Il-4	Russia	21.25	14	E
1./NJG100	Fw Heinze	Il-4	Russia	21.30	2	E
3./NJG100	Fw Helmut Konter	Il-4	SSW Orscha	00.08	7	E

19-20 May 1944

4./NJG4	Oblt Jakob Schaus	Four-eng	Near Vimoutiers	00.25	10	W

21-22 May 1944

10./NJG3	Lt Walter Briegleb	Lancaster	25km W Odense	00.44	5	W
St.III./NJG1	Hptm Martin Drewes	Lancaster	Zwolle	00.50	36	W
St.III./NJG1	Hptm Martin Drewes	Lancaster	Deventer	00.58	37	W
3./NJG3	Uffz Hans Schadowski	Four-eng	Mönchengladbach	01.28	1	W
3./NJG1	Hptm Heinz Strüning	Lancaster	N Fessenhout	01.32	45	W
4./NJG3	Lt Hermann Leube	Four-eng	100km WNW Duisberg	01.34	13	W
St.IV./NJG1	Oblt H-W Schnaufer	Lancaster	3km S Mol	01.34	66	W
St.III./NJG2	Oblt Alfons Köster	Lancaster		01.36	18*	W
1./NJG1	Oblt Wilhelm Henseler	Lancaster	10km NE Dordrecht	01.37	8	W

St.III./NJG1	Hptm Martin Drewes	Lancaster	Eindhoven	01.38	38	W
3./NJG3	Uffz G Schneider	Four-eng	40-60km from Duisberg	01.38	2	W
IV./NJG3	Hptm Franz Buschmann	Four-eng	In sea PS-4	01.39	8	W
2./NJG1	Hptm E-W Modrow	Lancaster	20km SSW Finthorn	01.41	14	W
7./NJG2	Lt Hans Schafer	Lancaster	Eindhoven	01.43	6	W
7./NJG2	Obstlt . Günther Radusch	Four-eng	LK-LL-KL-KK	01.44	54	W
St.III./NJG1	Hptm Martin Drewes	Lancaster	Nr. Gouda	01.50	39	W
St.IV./NJG1	Oblt H-W Schnaufer	Lancaster	10km S Herentals	01.51	67	W
8./NJG2	Oblt Friedrich Tober	Lancaster	Rotterdam	01.52	5	W
8./NJG2	Obstlt .Günther Radusch	Four-eng	LK-LI-KI-KK	01.56	55	W
7./NJG2	Uffz Rannefeld	Four-eng	Limburg	01.58	1	W
10./NJG3	Uffz Heinz Koppe	Four-eng	NT1-6	02.00	6	W
7./NJG2	Obstlt Günther Radusch	Four-eng	IK-KK-KI-II	02.00	56	W
8./NJG1	Oblt Dietrich Schmidt	Lancaster	S Amsterdam	02.01	25	W
St.III./NJG1	Hptm Martin Drewes	Lancaster	Nr. Breda	02.04	40	W
2./NJG1	Hptm Paul Förster	Lancaster	12km NW Antwerpen	02.06	4	W
2./NJG1	Maj Hans Karlowski	Halifax	Off Hook of Holland	02.12	2	W
4./NJG1	Oblt Fritz Lau	Lancaster	NNW Rotterdam	02.20	10	W
6./KG51	Uffz Baier	B-24	England	02.51	5	W
6./KG51	Uffz Baier	B-24	England	02.56	6	W
6./KG51	Fw Johann Trenke	Lancaster	England	03.05	5	W
6./KG51	Fw Johann Trenke	Lancaster	England	03.14	6	W
6./KG51	Fw Johann Trenke	Lancaster	England	03.22	7	W

22-23 May 1944

7./NJG1	Uffz Fensk	Lancaster	S Den Helder	00.16	1	W
8./NJG1	Uffz Langer	Lancaster	5-10km NW Gröningen	00.16	1	W
3./NJG3	Oblt Heinz Ferger	Four-eng	320° from Gröningen	00.17	12	W
Stab./NJG3	Obstlt Helmut Lent	Four-eng	Osnabrück	00.21	91	W
St.I./NJG3	Hptm Werner Husemann	Four-eng	Assen-Emmen	00.25	19	W
I./NJG2	Hptm Eduard Jacob	Four-eng		00.25	4	W
St.III./NJG3	Lt Achim Woeste	Four-eng	N Haselunne	00.28	9	W
3./NJG3	Uffz Heinz Krause	Four-eng	Osnabrück	00.29	1	W
3./NJG3	Oblt Heinz Ferger	Four-eng	SE Quakenbrück	00.30	13	W
2./NJG3	Fw Hans Meissner	Four-eng	S Zwolle	00.36	18	W
1./NJG1	Oblt Wilhelm Henseler	Lancaster	S Rheine	00.39	9	W
Stab./NJG3	Obstlt Helmut Lent	Four-eng	Nr. Osnabrück	00.40	92	W
8./NJG1	Oblt Dietrich Schmidt	Lancaster	5km N Ibbenbüren	00.42	26	W
2./NJG1	Maj Hans Karlowski	Lancaster	NW Dortmund	00.44	3	W
I./JG300	Ofw Heinrich Lonnecker	Lancaster	Dortmund	00.46	3	W
5./NJG5	Oblt Werner Rapp	Lancaster	270° from Lingen	00.47	10	W
St.III./JG300	Hptm Kurt Peters	Four-eng	Dortmund	00.52	1	W
2./JG300	Fw Kurt Landl	Four-eng	Dortmund	00.53	1	W
I./JG300	Ofw Hermann Wischnewski	Four-eng	Dortmund	00.53	11	W
1./NJG10	Hptm Friedrich-Karl Müller	Halifax	10-15km N Dortmund	00.54	21	W
4./JG300	Oblt Werner Kucharsowsky	Four-eng	S Dortmund	00.55	1	W
2./NJG1	Oblt Werner Baake	Lancaster	SW Neerpelt	01.14	32	W
3./NJG1	Hptm Heinz Strüning	Lancaster	Near Giessen	01.14	46	W
1./NJG10	Oblt Jung	Four-eng	NW Düsseldorf	01.16	1	W
1./NJG10	Hptm Friedrich-Karl Müller	Lancaster	N Bonn	01.16	22	W
5./NJG2	Hptm H-H Hissbach	Lancaster		01.18	15	W

Heinz-Wolfgang Schnaufer (left) and Hans-Joachim Jabs (centre) in conversation beside a shot-down USAAF Thunderbolt.

12./NJG1	Oblt H-H Augenstein	Lancaster	Eindhoven	01.21	36	W
St.IV./NJG1	Oblt H-W Schnaufer	Lancaster	Nr. Neerpelt	01.23	68	W
2./NJG1	Hptm E-W Modrow	Lancaster	2½km S Assen	01.25	15	W
5./JG300	Fw Hans Kugler	Four-eng	Ruhr	01.32	5	W
12./NJG1	Oblt H-H Augenstein	Lancaster	Hoogstraten area	01.34	37	W
St.IV./NJG1	Oblt H-W Schnaufer	Lancaster	Nr. Brecht	01.36	69	W
9./NJG3	Uffz Walter Rohlfing	Four-eng	Westrem	02.12	9	W
5./JG302	Oblt Karl-Heinz Seeler	Four-eng	Dortmund area	02.16	4	W
5./JG302	Oblt Karl-Heinz Seeler	Four-eng	N Hamm	02.27	5	W
4./NJG4	Oblt Jakob Schaus	Four-eng	Salzwedel	02.35	11	W
8./NJG1	Oblt Dietrich Schmidt	Lancaster	North Sea	03.13	27	W
8./NJG1	Oblt Dietrich Schmidt	Four-eng	North Sea	03.48	28	W

23-24 May 1944

1./NJG100	Ofw Helmut Dahms	R-5	Russia	22.45	6	E

24-25 May 1944

11./NJG1	Oblt Hermann Greiner	Lancaster	10km S Aachen	00.29	29	W
4./NJG1	Oblt Fritz Lau	Halifax	SE Rotterdam	00.35	9	W
7./NJG2	Fw Erich Killmann	Four-eng	E Antwerp	00.41	3	W
7./NJG2	Oblt Johann Werth	Four-eng	LL-LM/ML-MM	00.43	9	W
2./NJG1	Maj Hans Karlowski	Halifax	30km N Aachen	00.45	4	W
3./NJG1	Hptm Heinz Strüning	Halifax	Near Leopoldsburg	00.47	47	W
12./NJG1	Lt Heinz Rolland	Lancaster	10km E Turnhout	00.51	10	W
1./NJG2	Oblt Adolf Kaiser	Halifax	E Turnhout	00.52	8	W
7./NJG2	Lt Hans Schäfer	Lancaster	NW Aachen	00.53	7	W
7./NJG5	Maj Walter Borchers	Four-eng	15km W Aachen	00.54	11	W
4./NJG3	Lt Hermann Leube	Four-eng	Walcheren	00.57	14	W
1./NJG10	Hptm Friedrich-Karl Müller	Four-eng	Roermond	01.06	23	W
St.IV./NJG1	Oblt H-W Schnaufer	Four-eng	3km NW Eindhoven	01.15	70	W
3./NJG1	Hptm Heinz Struning	Four-eng	Off Ostend	01.15	48	W
12./NJG1	Lt Heinz Rolland	Lancaster	N Antwerp	01.18	11	W
St.IV./NJG1	Oblt H-W Schnaufer	Four-eng	2km NNW Tilburg	01.18	71	W
St.IV./NJG1	Oblt H-W Schnaufer	Four-eng	1.5km W Goirle	01.22	72	W
12./NJG1	Ofw Karl-Heinz Scherfling	Lancaster	20km NNE Breda	01.23	31	W
St.IV./NJG1	Oblt H-W Schnaufer	Four-eng	Dongen - Tilburg	01.25	73	W
St.IV./NJG1	Oblt H-W Schnaufer	Lancaster	7km SW Tilburg	01.29	74	W
12./NJG1	Ofw Karl-Heinz Scherfling	Lancaster	15km N Breda	01.30	32	W
6./NJG6	Lt Karl-Heinz Grunert	Halifax	S Aachen	01.32	2	W

St.III./NJG2	Oblt Alfons Köster	Four-eng		01.36	19	W
III./NJG5	Oblt Peter Ehrhardt	Halifax	Off Ouddorp	01.45	19	W
7./NJG6	Oblt Josef Kraft	Lancaster	Listig	02.18	27	W
7./NJG6	Oblt Helmuth Schulte	Lancaster	20km E Aachen	02.19	13	W
St.I./NJG6	Oblt Gaul	Halifax	Eifel ON-4	02.20	2	W
7./NJG6	Oblt Josef Kraft	Halifax	Eifel PN-3	02.24	28	W
7./NJG6	Oblt Helmuth Schulte	Lancaster	20km SE Aachen	02.30	14	W
3./NJG1	Oblt Josef Nabrich	Lancaster	40km from Jülich	02.41	11	W
1./NJG1	Oblt Wilhelm Henseler	Four-eng	30km NW Eindhoven	02.48	10	W
3./NJG1	Oblt Josef Nabrich	Four-eng	N Tilburg	02.51	12	W

26-27 May 1944

3./NJG1	Fw Rauer	Mosquito	Werl	01.21	5	W
3./NJG1	Fw Wilhelm Morlock	Four-eng	20km N "Hamster"	01.27	1	W

27-28 May 1944

3./NJG100	Fw Helmut Konter	U-2	40km SSE Witebsk	22.29	8	E
3./NJG1	Fw Wilhelm Morlock	Four-eng	Knokke	01.44	2	W
11./NJG1	Uffz Grosser	Lancaster	Leeuwarden	02.04	1	W
12./NJG1	Lt Kurt Matzak	Halifax	S-E Leuven	02.04	7	W
11./NJG1	FhjFw Hans Gienanth.von	Halifax	N Brussels	02.05	2	W
Stab./NJG2	Obstlt Günther Radusch	E/A	Meerle	02.07	57	W
2./NJG2	Oblt Heinz Rökker	Halifax	20-40km NW Eindhoven	02.08	27	W
12./NJG1	Oblt H-H Augenstein	Halifax		02.13	38	W
12./NJG1	Oblt H-H Augenstein	Halifax		02.15	39	W
12./NJG1	Oblt H-H Augenstein	Halifax		02.17	40	W
11./NJG1	FhjFw Hans Gienanth.von	Halifax		02.17	3	W
2./NJG2	Uffz Walter Reichenberger	Four-eng	Heerlen, SW Geilenkirchen	02.22	2	W
I./NJG4	Oblt Friedrich Thorl	E/A	Brussels area	02.24	3	W
2./NJG1	Hptm E-W Modrow	Lancaster	Bisbar	02.25	16	W
6./NJG2	Lt Erich Jung	Lancaster	Nr. Brussels	02.27	3	W
11./NJG1	Oblt Hermann Greiner	Lancaster		02.29	30	W
2./NJG1	Hptm E-W Modrow	Lancaster	Bisbar	02.35	17	W
3./NJG1	Fw Rauer	Mosquito	Off Den Helder	02.44	6	W
11./NJG1	Oblt Hermann Greiner	Lancaster		02.48	31	W
St.III./NJG4	Hptm Hans-Karl Kamp	Lancaster	Zottegem	02.56	18	W
St.III./NJG4	Hptm Hans-Karl Kamp	Lancaster	W Brussels	03.01	19	W
St.III./NJG4	Hptm Hans-Karl Kamp	Lancaster	NW Brussels	03.05	20	W
9./NJG5	Fw Robert Schratter	Lancaster	NW Frankfurt-am-Main	03.08	4	W
9./NJG5	Lt Otto Keller	Lancaster	20-40km S Ostend	03.10	8	W
9./NJG5	Fw Robert Schratter	Lancaster	NW beacon "Goldammer"	03.11	5	W
6./NJG2	Fw Gajewski	Lancaster		03.13	2	W
9./NJG5	Fw Robert Schratter	Halifax	NW beacon "Goldammer"	03.16	6	W
I./NJG4	Maj Wilhelm Herget	E/A	Ostend-Bruges	03.17	54	W
9./NJG5	Lt Otto Keller	Four-eng	30-50km NW Ostmunde	03.26	9	W
8./NJG5	Lt Gerhard Wagner	Halifax	40km NE Dijon	03.26	4	W
2./NJG1	Hptm E-W Modrow	Lancaster	Durbuy	03.28	18	W

28-29 May 1944

3./NJG100	Uffz Dambler	R-5	40km NW Orscha	22.22	1	E
3./NJG100	Fw Helmut Konter	R-5	45km SE Witebsk	23.40	9	E

1./NJG100	Ofw Helmut Dahms	R-5	Russia	00.15	7	E
3./NJG100	Fw Helmut Konter	R-5	28km SSE Witebsk	00.22	10	E
2./NJG4	Hptm Josef Krahforst	Four-eng	1km SW Aaigem	02.10	3	W
6./KG51	Unknown	Stirling	Norfolk	02.40	-	W

29-30 May 1944

1./NJG100	Uffz Dahlmann	U-2	Russia	00.54	1	E
6./NJG101	Ofw Ludwig	Wellington	Sankt-Pölten	00.55	4	W
2./NJG4	Lt Karl Kern	Four-eng	SE Charleroi	01.25	1	W

30 May-1 June 1944

6./NJG101	Oblt Hans Krause	Halifax	30km S Subotica	22.40	9	W
3./NJG1	Fw Wilhelm Morlock	Hudson	SE Tholen Island	01.10	3	W
3./NJG1	Fw Wilhelm Morlock	Halifax	Off Walcheren	01.40	4	W
4./NJG4	Hptm Fritz Söthe	Four-eng	N Paris	01.50	10	W
4./NJG4	Hptm Fritz Söthe	Four-eng	Méru	01.54	11	W
1./NJG5	Oblt E-G Drunkler	Lancaster	5km SSW Rambouillet	02.05	12	W
St.III./NJG4	Hptm Hubert Rauh	Lancaster	7km W Rambouillet	02.05	11	W
St.III./NJG5	Hptm Paul Zorner	Lancaster	30km WNW Tergnier	02.35	47	W
9./NJG5	Lt Otto Keller	Lancaster	N Aumale	02.52	10	W

2-3 June 1944

3./NJG1	Lt Ernst-Ewald Hittler	B-24	Düsleben	23.56	3	W
3./NJG1	Hptm Heinz Strüning	Halifax	Schouwen Island	00.36	49	W
4./NJG4	Lt Frithjof Fensch	Four-eng	SW Paris	00.55	1	W
6./NJG4	Hptm Hans Autenrieth	Four-eng	Trappes	01.00	19	W
4./NJG4	Lt Frithjof Fensch	Four-eng	Trappes	01.00	2	W
4./NJG4	Oblt Jakob Schaus	Four-eng	Mantes-Rambouillet	01.00	12	W
St.III./NJG4	Hptm Hubert Rauh	Four-eng	Mantes-Rambouillet	01.01	12	W
1./NJG5	Oblt E-G Drunkler	Four-eng	Trappes	01.04	13	W
4./NJG4	Lt Frithjof Fensch	Four-eng	Dreux	01.06	3	W
4./NJG4	Oblt Jakob Schaus	Four-eng	Dreux area	01.06	13	W
St.II./NJG4	Hptm Hubert Rauh	Four-eng	Dreux area	01.10	13	W
4./NJG4	Oblt Jakob Schaus	Four-eng	Dreux area	01.12	14	W
4./NJG4	Hptm Fritz Söthe	Four-eng	Evreux-Louviers	01.19	12	W
St.III./NJG5	Hptm Paul Zorner	Lancaster	30km WNW Evreux	01.22	48	W
6./NJG4	Hptm Hans Autenrieth	Four-eng	Le Neubourg	01.23	20	W
St.I./NJG5	Hptm Werner Hoffmann	Four-eng	S Elbeouf	01.24	31	W
1./NJG5	Ofw A. Klaiber	Four-eng	40km NW Evreux	01.24	4	W
2./NJG5	Oblt Bernhard Jungmann	Four-eng	Yvetot	01.26	2	W

3-4 June 1944

4./NJG6	Fw Helmut Bunje	Wellington	SW Rome	01.03	2	M

5-6 June 1944

3./NJG100	Oblt Josef Pützkuhl	U-2	Russia	23.35	7	E
3./NJG100	Oblt Josef Pützkuhl	U-2	Russia	23.47	8	E
3./NJG100	Oblt Josef Pützkuhl	U-2	Russia	00.00	9	E
11./NJG1	Lt Otto Teschner	B-24	12km from Brussels	01.40	1	W
3./NJG1	Hptm Heinz Strüning	Mosquito	Noord Brabant	02.30	50	W
3./SKG10	Oblt Helmut Eberspächer	Lancaster	Isigny	05.01	1	W

3./SKG10	Fw Eisele	Lancaster	IIsigny-Lessay-Vire	05.02	1	W
3./SKG10	Oblt Helmut Eberspächer	Lancaster	Carentan	05.03	2	W
3./SKG10	Oblt Helmut Eberspächer	Lancaster	Carentan	05.04	3	W

Chapter Ten

Normandy and Retreat

7th June - 31st August 1944

The Allied assault against the Normandy coast on *D-Day*, 6th June 1944, had come as a bitter blow, but the Germans, never slow to respond, 'rolled with the punch' and held on. Nightfighter units already in the area to counter the tactical operations of *Operation Point-Blank* were quickly supplemented by others drawn out of Germany, for the night skies over Normandy had quickly become a target-rich environment. Not only Bomber Command raids were to be found, but the tactical bombers - Mitchells, Bostons, A-20s and Marauders of the British 2nd and American 9th Tactical Air Forces, were also present in great numbers.

In addition to air defence and interdiction missions over the beachheads, the *Luftwaffe* nightfighter crews found themselves performing a task for which none, except perhaps those in the east - had been trained; night ground attack. Thus the incidence of nightfighter crews being reported missing in action rose sharply. Also, with so many *Luftwaffe* nightfighter units now within easy range of the British intruders of 100 Group, 2nd Tactical Air Force and Air Defence of Great Britain, plus the two squadrons of P-61 Black Widows of the American 9th Air Force, losses to this cause were also to rise. For the *Nachtjager*, the 'happy time' was over.

In early June 1944, the nightfighter situation was as follows:

Stab NJG 1	*Obstlt* Jabs	Bonninghardt	Bf110/He219
I./NJG 1	*Maj* Förster	Venlo	Bf110/He219
II./NJG 1	*Maj* v.Bonin	St. Trond	Bf110
III./NJG 1	*Maj* Drewes	Twente/Leeuwarden	Bf110/Ju88G
IV./NJG 1	*Maj* Schnaufer	St. Trond	Bf110/He219
Stab NJG 2	*Obstlt* Radusch	Coulommieres	Ju88C
I./NJG 2	*Hptm* Zechlin	Eindhoven	Ju88C/G
II./NJG 2	*Maj* Semrau	Coulommieres	Ju88C
III./NJG 2	*Hptm* Merker	Langendiebach/Twente	Ju88C/G
Stab NJG 3	*Obstlt* Lent	Stade	Ju88C
I./NJG 3	*Maj* Husemann	Vechta	Bf110/Ju88C
II./NJG 3	*Hptm* Havenstein	Plantlunne	Ju88G
III./NJG 3	*Maj* Barte	Stade	Bf110
IV./NJG 3	*Hptm* Buschmann	Westerland	Ju88C?G
Stab NJG 4	*Maj* Thimmig	Chenay	Bf110/Ju88C
I./NJG 4	*Maj* Herget	Florennes	Bf110/Do217Ju88
II./NJG 4	*Hptm* Raht	Dijon	Bf110
III./NJG 4	*Hptm* Kamp	Juvincourt	Bf110/Do217/Ju88
Stab NJG 5	*Obstlt* Borchers	Laon-Athies	Bf110
I./NJG 5	*Maj* Hoffmann	St. Dizier	Bf110
II./NJG 5	*Hptm* Hans Leickhardt	Gutersloh	Bf110
III./NJG 5	*Maj* Zorner	Laon-Athies	Bf110
IV./NJG 5	*Hptm* Altendorf	Mainz-Finthen	Bf110
Stab NJG 6	*Maj* Griese	Schleissheim	Bf110
I./NJG 6	*Hptm* Hadeball	Neubiburg	Bf110/Ju88G
II./NJG 6	*Maj* Leuchs	Echterdingen	Bf110
III./NJG 6	*Hptm* Fellerer	Hagenau	Bf110
IV./NJG 6	*Maj* Lütje	Otopeni, Zilistea	Bf110

I./NJG 7	*Maj* Bengsch	Handorf	Ju88G
NJG 100	*Hptm* Bellinghausen	various (Russia)	Ju88
Stab NJG 101	*Maj* Sewing	Ingolstadt-Manching	Bf110/Ju88/Do217
I./NJG 101	*Maj* Herber	Ingolstadt-Manching	Bf110/Do217/Ju88
II./NJG 101	*Hptm* Schwab	Pandorf	Bf110/217
III./NJG 101	*Maj* Jung	Kitzingen	Bf110/Do217
Stab NJG 102	*Obstlt* Hülshoff	Kitzingen	Bf110
I./NJG 102	*Maj* Fehling	Powunden	Bf110
II./NJG 102	*Hptm* Floitgraf	Echterdingen	Bf110
NJG 200	Unknown	Various (Russia)	Bf110/Ju88/FW190
NJGr.10	*Maj* Schoernert	WerneuchenHangelar	FW190A

Note: The Wilde Sau Jagdgeschwadern, JG 300, 301 and 302, were by this time being employed mainly as day fighter units and will no longer be recorded a night fighter units in this work, although any night claims will be recorded.

Two further units would be formed within a fairly short time. *III./NJG 102* would be established at Kitzingen on 26th June, equipped with Bf110s. The *Gruppenkommandeur* is unknown. On 26th August, *Major* Friedrich-Karl Müller was to form *I./NJG 11*, at Lippstadt and Kothen. This unit, using Bonn-Hangelar as a forward base, would become a *Wilde Sau* unit equipped with Bf109Gs and FW190As.

As can be seen, several units had been brought into France, but within weeks the great retreat would begin, as the Allies moved further into the occupied territories.

6-7 June 1944

St.II./NJG4	Hptm Hubert Rauh	Four-eng	St. Lô area	00.20	14	W
St.II./NJG4	Hptm Hubert Rauh	Four-eng	St. Lô area	00.31	15	W
6./NJG2	Fw Gajewski	Lancaster	S Worthing	02.07	3	W
2./NJG5	Lt Joachim Hanss	Four-eng	Paris	02.27	8	W
2./NJG4	Uffz Altmann	Four-eng	TA-TB-UA	02.35	2	W
2./NJG5	Lt Joachim Hanss	Four-eng	W Creil	02.37	9	W
2./NJG2	Oblt Heinz Rökker	Lancaster	10-50km SW Caen	02.42	28	W
2./NJG4	Uffz "Melmann"	Four-eng	TA-TB-UA	02.45	3	W
2./NJG2	Oblt Heinz Rökker	Lancaster	10-50km SW Caen	02.48	29	W
2./NJG2	Oblt Heinz Rökker	Lancaster	SW Caen	02.51	30	W
2./NJG4	Uffz Altmann	Four-eng	TA-TB-UA-UB	03.01	4	W
2./NJG2	Oblt Heinz Rökker	Lancaster	W Caen	03.01	31	W
2./NJG2	Oblt Heinz Rökker	Lancaster	W Caen	03.08	32	W

7-8 June 1944

3./NJG100	Oblt Josef Putzkuhl	U-2	Russia	00.02	10	E
6./NJG2	Oblt Franz Brinkhaus	Four-eng	S Dieppe	01.04	7	W
4./NJG4	Oblt Jakob Schaus	Four-eng	SW Paris	01.15	15	W
4./NJG4	Oblt Jakob Schaus	Four-eng	SW Paris	01.16	16	W
6./NJG2	Oblt Franz Brinkhaus	Four-eng	NW Paris	01.19	8	W
6./NJG2	Oblt Franz Brinkhaus	Four-eng	W Beauvais	01.25	9	W
9./NJG4	Hptm Kurt Fladrich	Halifax	Meulan-les-Mureaux	01.26	12	W
6./NJG2	Oblt Franz Brinkhaus	Four-eng	S Abbeville	01.44	10	W
6./NJG2	Lt Martin Tschiersch	Four-eng	W Paris	01.45	1	W
8./NJG5	Lt Gerhard Wagner	Four-eng	Magny-en-Vexin	02.00	5	W
7./NJG2	Ofw Busch	Four-eng	NW Paris	02.01	1	W
6./NJG2	Lt Martin Tschiersch	Four-eng	NE Paris	02.03	2	W
8./NJG5	Lt Gerhard Wagner	Four-eng	Mantes area	02.03	6	W

Ju88G in flight. Note the excellent all-round vision and the cumbersome radar array. Two Abschussbalken are just visible in white on the fin.

7./NJG2	Ofw Walter Exner	Four-eng	NW Paris	02.06	1	W
8./NJG5	Lt Gerhard Wagner	Four-eng	Mantes	02.06	7	W
St.III./NJG2	Oblt Alfons Köster	Four-eng	TU1-5	02.07	20	W
7./NJG2	Ofw Walter Exner	Four-eng	NW Paris	02.08	2	W
6./NJG2	Lt Martin Tschiersch	Four-eng	NE Paris	02.08	3	W
8./NJG5	Uffz Hanke	Four-eng	15-25km NW Paris	02.09	1	W
8./NJG5	Lt Gerhard Wagner	Four-eng	30-40km SW Paris	02.11	8	W
4./NJG4	Uffz Kurt Stockbauer	Four-eng	Trappes	02.15	1	W
8./NJG2	Uffz Kleibauer	Halifax	W outskirts of Paris	02.20	2	W
Stab./NJG5	Maj Walter Borchers	Four-eng	30km W Paris	02.21	12	W
2./NJG5	Uffz Hengsbach	Four-eng	Trappes	02.21	2	W
II./NJG4	Hptm Hubert Rauh	Halifax	Rambouillet	02.21	16	W
4./NJG4	Hptm Fritz Söthe	Four-eng	Trappes	02.21	13	W
8./NJG5	FhjFw Herbert Altner	Four-eng	W Paris	02.22	8	W
9./NJG4	Lt Hromadnik	Four-eng	Mantes-Paris	02.24	5	W
Stab./NJG5	Maj Walter Borchers	Four-eng	35km W Paris	02.29	13	W
Stab./NJG5	Maj Walter Borchers	Four-eng	30km WSW Paris	02.31	14	W
6./NJG4	Ofw Richard Rother	Four-eng	E Paris	02.33	1	W
8./NJG5	FhjFw Herbert Altner	Four-eng	Nr. Paris	02.37	9	W
7./NJG2	Lt Heinz Reuter	Four-eng	60-80km SW Paris	02.40	5	W
2./NJG2	Uffz Martin Siegel	E/A	30-40km E Dreux	02.41	2	W
1./NJG2	Hptm Herbert Lorenz	Four-eng	10-20km NW Beauvais	02.55	1	W
I./NJG4	Maj Wilhelm Herget	Lancaster	NW Fécamp	03.05	55	W
1./NJG5	Lt Joachim Plass	Lancaster	SB-SC	03.12	1	W
I./NJG4	Maj Wilhelm Herget	Stirling	NW Fécamp	03.14	56	W

8-9 June 1944

1./NJG100	Unknown	R-5	Russia	23.41	-	E
1./NJG100	Unknown	R-5	Russia	00.19	-	E

9-10 June 1944

4./NJG4	Lt Frithjof Fensch	E/A	Elancourt	00.15	4	W
6./NJG4	Hptm Hans Autenrieth	Lancaster	5km E Dreux	00.30	21	W
8./NJG5	FhjFw Herbert Altner	Four-eng	SSW Paris	00.32	10	W
5./NJG2	Hptm H-H Hissbach	Lancaster	Étrepagny area	00.38	16	W

6./NJG2	Hptm Kamsties	Lancaster	Neufchateau	00.40	1	W
Stab./NJG2	Oblt Johann Werth	E/A	NW Paris	00.42	10	W
St.IV./NJG6	Maj Herbert Lütje	P-38	S Bucharest	08.29	36	W
St.IV./NJG6	Maj Herbert Lütje	P-38	S Bucharest	08.31	37	W

10-11 June 1944

6./NJG101	Oblt Hans Krause	B-25	SW Budapest	23.22	10	E
8./NJG5	FhjFw Herbert Altner	Four-eng	SW Paris	00.40	11	W
8./NJG5	FhjFw Herbert Altner	Four-eng	SW Paris	00.43	12	W
8./NJG5	Uffz Hanke	Lancaster	SW Paris	00.45	2	W
4./NJG4	Lt Frithjof Fensch	Four-eng	S Nogent	00.51	5	W
8./NJG5	Oblt Werner Hopf	Lancaster	SW Paris	00.53	6	W
3./NJG1	Oblt Josef Nabrich	Mosquito	N Münster	00.55	13	W
4./NJG4	Oblt Jakob Schaus	Four-eng	Dreux	00.58	17	W
II ./NJG2	Maj Paul Semrau	Four-eng	5-20km S Dreux	01.00	21	W
St.III./NJG5	Hptm Paul Zorner	Halifax	W Dreux	01.00	49	W
St.III./NJG5	Hptm Paul Zorner	Halifax	Verneuil-sur-Avre	01.01	50	W
St.III./NJG5	Hptm Paul Zorner	Halifax	Verneuil-sur-Avre	01.08	51	W
8./NJG2	Lt Josef Förster	Four-eng	Evreux-Melsbroek	01.12	1	W
10./NJG3	Lt Walter Briegleb	Lancaster	10km NW Paris	01.14	6	W
8./NJG2	Uffz Egon Engling	Lancaster	NW Paris	01.14	1	W
10./NJG3	Lt Walter Briegleb	Lancaster	Centre Paris	01.17	7	W
4./NJG4	Hptm Fritz Söthe	Four-eng	SW Evreux	01.18	14	W
II./NJG2	Maj Paul Semrau	Four-eng	15-30km NW Ranbouillet	01.20	22	W
10./NJG3	Lt Walter Briegleb	Lancaster	20km W Paris	01.24	8	W
11./NJG1	Oblt Hermann Greiner	Halifax		01.26	32	W
10./NJG3	Lt Walter Briegleb	Lancaster	25km SE Rouen	01.32	9	W
1./NJG2	Oblt Adolf Kaiser	Lancaster	Lisieux - Trouville	01.42	9	W
St.III./NJG5	Hptm Paul Zorner	Lancaster	10km SW Dreux	01.42	52	W
1./NJG1	Hptm E-W Modrow	Mosquito	8km S. Alkmaar	02.50	19	W
3./NJG100	Fw Helmut Konter	U-2	4km S Orscha	22.58	11	E

11-12 June 1944

3./NJG100	Fw Helmut Konter	U-2	20km SNOrscha	23.14	12	E
3./NJG100	Fw Helmut Konter	U-2	12km N Orscha	23.30	13	E
St.I./NJG2	Hptm Gerhard Raht	Four-eng	over French Coast	00.53	33	W
3./NJG1	Oblt Josef Nabrich	Mosquito	W Salzwebel	01.13	14	W
2./NJG100	Ofw Karl Maisch	Wellington	10km SW Som, Hungary	02.41	6	E

12-13 June 1944

I./NJG4	Oblt Friedrich Thörl	Four-eng	Amiens area	00.16	4	W
I./NJG4	Oblt Friedrich Thörl	Four-eng	Amiens area	00.21	5	W
8./NJG4	Ofw Reinhard Kollak	Four-eng	Formerie	00.22	41	W
7./NJG2	Uffz Hans Werthner	Four-eng	20-30km from Amiens	00.22	4	W
7./NJG2	Lt Hans Schäfer	Halifax	Amiens	00.23	8	W
St.IV./NJG1	Hptm H-W Schnaufer	Lancaster	Avennes les Auvert	00.27	75	W
8./NJG2	Uffz Egon Engling	Halifax	10km WNW Amiens	00.29	2	W
St.IV./NJG1	Hptm H-W Schnaufer	Lancaster	flugplatz Cambrai	00.31	76	W
12./NJG1	Oblt H-H Augenstein	Lancaster	Hazebrouck	00.33	41	W
5./NJG2	Ofw Robert Luddeke	Four-eng	N Compiègne	00.34	8	W
St.IV./NJG1	Hptm H-W Schnaufer	Lancaster	Tilloy	00.34	77	W

Hauptmann Gerhard Friedrich, a former transport pilot, joined the nightfighter force in 1942, going to III./NJG 1, gaining his first combat successes in the summer. Transferred to lead 10./NJG 4 in 1943 - the unit becoming 1./NJG 6 in August - he began to score regularly. A year later he became Gruppenkommandeur, credited with at least eighteen victories. The Ritterkreuz came on 15th March 1945. The and, next evening he took off for the last time. His 30th opponent, a Lancaster, exploded so violently that his Ju88G was torn apart, killing Friedrich and his crew. On the night of 12/13th June he brought down his eighteen victim, a Lancaster, probably LM158 of 90 Squadron, near Deelen

8./NJG4	Ofw Reinhard Kollak	Four-eng	N Rouen	00.35	42	W
6./NJG2	Lt Martin Tschiersch	Four-eng	N Beauvais	00.38	4	W
8./NJG4	Ofw Reinhard Kollak	Four-eng	N Rouen	00.40	43	W
1./NJG5	Oblt E-G Drünkler	Four-eng	3km N Arques	00.43	14	W
1./NJG5	Oblt E-G Drünkler	Four-eng	NW Dunkirk	00.56	15	W
4./NJG1	Oblt Fritz Lau	Lancaster	10km NW Eindhoven	00.58	10	W
6./NJG1	Uffz Gustav Sarzio	Lancaster	20km NW Venlo	01.03	1	W
St.I./NJG6	Hptm Joachim Bohner	Lancaster	20km S Deelen	01.22	1	W
8./NJG1	Oblt Dietrich Schmidt	Lancaster	N Venlo	01.24	29	W
1./NJG6	Hptm Gerhard Friedrich	Lancaster	Nr. Deelen	01.25	18	W
6./KG51	Uffz Ottren	E/A	England	01.25	1	W
11./NJG1	Lt Otto Teschner	Four-eng	Over Amiens	01.26	2	W
1./NJG1	Hptm E-W Modrow	Lancaster	Freiburg	01.27	20	W
3./NJG3	Ofw Gierscke	Four-eng	30-50km S Duisberg	01.28	1	W
3./NJG1	Lt Hittler	Lancaster	SW Ijsselmeer	01.30	4	W
St.IV./NJG1	Hptm Adolf Breves	Four-eng	St.Trond	01.31	3	W
1./NJG1	Hptm E-W Modrow	Lancaster	10km WNW Amiens	01.31	21	W
2./NJG3	Hptm Gerhard Keital	Four-eng	SE Suidbergen	01.32	1	W
11./NJG1	Fw Karl Pfeiffer	Four-eng	Raalte	01.34	10	W
7./NJG2	Lt Heinz Reuter	Four-eng	Nr. Abbeville	01.35	6	W
I./NJG4	Maj Wilhelm Herget	Lancaster	St Sauflieu/Amiens	01.36	57	W
2./NJG2	Ofhr Johannes Naskreut	Four-eng	Amiens	01.38	1	W
12./NJG1	Lt Heinz Rolland	Four-eng	Arnheim	01.38	12	W
8./NJG5	FhjFw Herbert Altner	Four-eng	Lens-Arras	01.39	13	W
8./NJG5	FhjFw Herbert Altner	Four-eng	Arras area	01.41	14	W
6./NJG1	Lt Gottfried Hanneck	Halifax	Ijsselmeer	01.43	2	W
I./NJG4	Maj Wilhelm Herget	Lancaster	St Just	01.45	58	W
I./NJG4	Oblt Friedrich Thörl	Four-eng	St Pol	01.45	6	W
1./NJG1	Hptm E-W Modrow	Lancaster	Near Krefeld	01.46	22	W
2./NJG2	Ofhr Johannes Naskreut	Four-eng	Amiens	01.48	2	W
Stab./NJG2	Obstlt Günther Radusch	Four-eng	Amiens	01.48	58	W
St.IV./NJG1	Hptm Adolf Breves	Four-eng	32km from Amsterdam	01.50	4	W
7./NJG2	Hptm Walter Piwarz	Halifax	Amiens	01.54	2	W

1./NJG5	Ofw A. Klaiber	Four-eng	20km ENE Rouen	01.55	5	W
5./NJG2	Lt Arnold	Four-eng		02.04	1	W
10./NJG3	Lt Walter Briegleb	Four-eng	Nr. Abbeville	02.06	10	W
5./NJG2	Hptm Collet	Four-eng	E Abbeville	02.06	1	W
5./KG51	Fw Visintin	Halifax	England	03.12	1	W

13-14 June 1944

1./NJG100	Unknown	Il-4	Russia	22.46	-	E
1./NJG100	Unknown	Il-4	Russia	22.53	-	E
1./NJG100	Unknown	Il-4	Russia	23.05	-	E
1./NJG100	Ofw Helmut Dahms	U-2	Russia	23.42	8	E
3./NJG100	Uffz Hausschild	U-2	N Bobruisk	00.02	1	E
1./NJG100	Unknown	Il-4	Russia	00.13	-	E
LBSt.7	Hptm Ernst Wallner	Wellington	Munich-Innsbruck	00.20	2	W
3./NJG100	Ofw Karl Strohecker	Li-2	Russia	00.28	8	E
3./NJG100	Ofw Karl Strohecker	Il-4	Russia	00.47	9	E
St.II./NJG6	Maj Rolf Leuchs	Four-eng	Kochelsee	01.00	8	W
3./NJG100	Ofw Karl Strohecker	Li-2	Russia	01.36	10	E
1./NJG100	Unknown	Il-4	Russia	02.04	-	E
1./NJG100	Unknown	Il-4	Russia	02.14	-	E

14-15 June 1944

2./NJG4	Uffz Kurt Leitner	Four-eng	Cambrai	01.00	1	W
7./NJG2	Uffz Hans Werthner	Halifax	North France		5*	W
2./NJG4	Hptm Josef Krahforst	Four-eng	Cambrai	01.07	4	W
8./NJG5	Oblt Werner Hopf	Lancaster	20km SW Lille	02.05	7	W
8./NJG5	Oblt Werner Hopf	Lancaster	10-20km NE St.Omer	02.24	8	W
II./NJG2	Maj Paul Semrau	Four-eng	NE Fécamp	02.24	23	W

15-16 June 1944

8./NJG5	Oblt Werner Hopf	Lancaster	Douai-Denain	00.44	9	W
2./NJG4	Hptm Josef Krahforst	Four-eng	Bouchain-Douai	00.51	5	W
II./NJG4	Hptm Hubert Rauh	Lancaster	Albert area	00.51	17	W
9./NJG5	Oblt Peter Ehrhardt	Lancaster	Cambrai-Lille-Valenciennes	00.52	20	W
Stab./NJG3	Obstlt Helmut Lent	Lancaster	SW Lille	00.52	93	W
11./NJG1	Oblt Hermann Greiner	Lancaster	20-30km NE Lille	00.56	33	W
Stab./NJG3	Obstlt Helmut Lent	Lancaster	Bethune - Armentieres	00.56	94	W
St.I./NJG3	Hptm Werner Husemann	Four-eng	180° from Thérouanne	00.59	20	W
Stab./NJG3	Obstlt Helmut Lent	Lancaster	NW Bethune	01.00	95	W
St.IV./NJG1	Hptm H-W Schnaufer	Lancaster	Bethune-Arras	01.00	78	W

16-17 June 1944

1./NJG1	Uffz Oppermann	Halifax	E beacon "Brisbar"	00.53	1	W
3./NJG1	Fw Wilhelm Morlock	Halifax	NW Tilburg	01.02	5	W
3./NJG1	Hptm Heinz Strüning	Halifax	Beacon "Gorilla"	01.07	51	W
2./NJG1	Oblt Werner Baake	Halifax	Eindhoven	01.10	33	W
3./NJG1	Oblt Josef Nabrich	Halifax	Beacon "Kürfurst"	01.12	15	W
3./NJG1	Hptm Heinz Strüning	Halifax	Beacon "Gorilla"	01.13	52	W
7./NJG2	Hptm Gerhard Wagner	Halifax	Geldern	01.13	9	W
3./NJG1	Fw Wilhelm Morlock	Halifax	Beacon "Brisbar"	01.14	6	W
6./NJG1	Oblt Johannes Hager	Halifax	Duisburg	01.20	29	W

Oberstleutnant Helmut Lent (left), the leading Experte of the nightfighter force and Kommodore of NJG 3, claimed three Lancasters on the night of 15th June. These were his 99th - 101st victories, achieved with 22 rounds of ammunition in just seven minutes. Oberleutnant Hermann Greiner of 11./NJG 1 claimed his 33rd victory on the same night. He survived the war - in hospital - but Lent would be killed in an accident on 5th October. His total score, including day kills, was 113 and he had won Germany's highest decoration; the Ritterkreuz mit Eichenlaub, Schwerter und Brillianten.

3./NJG7	Lt Kissinger	Four-eng	Thiel	01.20	2	W
6./NJG1	Fw Herbert Puchner	Lancaster	5km SW Lembeek	01.38	1	W
10./NJG3	Oblt Lippert	Four-eng	Beacon "Kürfurst"	01.40	3	W
7./NJG2	Fw Heinz Misch	Halifax	20km SE Arnhem	01.40	4	W
7./NJG2	Lt Heinz Reuter	Halifax	5km W Emmerich	01.41	7	W
6./NJG1	Oblt Johannes Hager	Halifax	15km NW Arnheim	01.43	30	W
St.III./NJG2	Hptm Berthold Ney	Four-eng	Appeldoorn	01.45	7	W
St.IV./NJG1	Hptm Adolf Breves	Four-eng	Utrecht	01.46	5	W
2./NJG1	Maj Schäfer	Four-eng	Beacon "Kürfurst"	01.46	1	W
3./NJG3	Uffz Hans Schadowski	Four-eng	30km NW Duisberg	01.47	2	W
7./NJG2	Lt Hans Schäfer	Lancaster	Near Stadtlohn	01.48	9	W
3./NJG1	Oblt Josef Nabrich	Four-eng	Beacon "Kürfurst"	01.49	16	W
St.IV./NJG1	Hptm Adolf Breves	Halifax	IM14	01.50	6	W
St.III./NJG2	Oblt Alfons Köster	Four-eng	Volkel	01.50	21	W
St.IV./NJG1	Oblt Georg Fengler	Four-eng	Beacon "Kürfurst"	01.51	5	W
6./NJG1	Lt Gottfried Hanneck	Four-eng	Volkel	01.53	3	W
St.IV./NJG1	Hptm H-W Schnaufer	Four-eng	St.Trond	01.54	79	W
St.IV./NJG1	Hptm Adolf Breves	Lancaster	Seiden	01.59	7	W
6./NJG1	Oblt Johannes Hager	Halifax		02.00	31	W
6./KG51	Uffz Ottren	Four-eng	England	02.00	2	W
St.IV./NJG1	Hptm H-W Schnaufer	Halifax	St.Trond	02.04	80	W

St.III./NJG1	Hptm Martin Drewes	Lancaster	Amsterdam	02.05	41	W
6./NJG1	Uffz Gustav Sarzio	Lancaster	5km E Utrecht	02.05	2	W
2./NJG3	Fw Hugo Fuetscher	Four-eng	Rotterdam	02.08	6	W
2./NJG3	Lt Fehre	Four-eng	Beacon "Kürfurst"	02.12	3	W
6./KG51	Fw Johann Trenke	Four-eng	England	02.15	8	W
10./NJG3	Lt Walter Briegleb	Four-eng	Beacon "Kürfurst"	02.16	11	W
St.III./NJG1	Hptm Martin Drewes	Lancaster	80km W Amsterdam	02.16	42	W
2./NJG1	Hptm Paul Forster	Halifax	Duisburg	02.20	5	W
4./NJG1	Fw Peter Müller	Halifax	30km S Bergen	02.24	1	W
8./NJG2	Lt Josef Förster	Four-eng	North Sea	02.32	2	W

17-18 June 1944

1./NJG100	Ofw Helmut Dahms	R-5	Russia	22.43	9	E
1./NJG100	Lt G E Francsi	U-2	Russia	22.44	17	E
1./NJG100	Lt G E Francsi	R-5	Russia	23.23	18	E
3./NJG100	Hptm August Fischer	Li-2	Russia	01.28	3	E

18-19 June 1944

3./NJG100	Lt Eberhard Kaross	Li-2	Russia	22.14	7	E
1./NJG100	Unknown	PS-84	Russia	23.50	-	E
3./NJG100	Oblt Josef Putzkuhl	U-2	22km SW Witebsk	00.02	11	E

20-21 June 1944

3./NJG100	Uffz "Daenbler"	Il-4	Russia	22.50	2	E
1./NJG100	Lt G E Francsi	Boston	Russia	00.27	19	E

21-22 June 1944

4./NJG3	Lt Hermann Leube	Four-eng	Eindhoven	01.06	15	W
7./NJG2	Ogfr H. Mackle	Lancaster	Volkel	01.10	1	W
4./NJG1	Oblt Fritz Lau	Lancaster	12km SW Eindhoven	01.12	11	W
1./NJG1	Hptm E-W Modrow	Lancaster	Beacon "Gorilla"	01.12	23	W
3./NJG1	Hptm Heinz Strüning	Lancaster	Venlo	01.13	53	W
6./NJG1	Hptm E-W Bonin.von	Lancaster	18km ESE Turnhout	01.14	37	W
St.IV./NJG1	Oblt Georg Fengler	Four-eng	St.Trond	01.16	6	W
3./NJG1	Oblt Josef Nabrich	Lancaster	E beacon "Brisam"	01.16	17	W
3./NJG1	Hptm Heinz Strüning	B-17	N Maastricht	01.17	54	W
2./NJG3	Uffz Weiner	Four-eng	Mönchengladbach	01.17	1	W
7./NJG2	Uffz Hans Werthner	Lancaster	Near Uden	01.18	6	W
St.IV./NJG1	Oblt Georg Fengler	Four-eng	St.Trond	01.20	7	W
7./NJG1	Ogfr Werner Kasmann	Four-eng	Near Eindhoven	01.20	1	W
2./NJG1	Oblt Werner Baake	Lancaster	W Aachen	01.22	34	W
4./NJG1	Gfr Reinecke	Four-eng	Venlo	01.22	1	W
11./NJG1	Lt Otto Teschner	Four-eng	Köln	01.22	3	W
2./NJG3	Uffz Weiner	Four-eng	NW Mönchengladbach	01.22	2	W
1./NJG1	Hptm E-W Modrow	Lancaster	Beacon "Gorilla"	01.24	24	W
7./NJG2	Uffz Hans Werthner	Lancaster	NMear Dinslaken	01.24	7	W
8./NJG2	Fw Kötter	Lancaster	4km N Gelsenkirchen	01.25	1	W
St.IV./NJG1	Hptm H-W Schnaufer	Lancaster	Heeze-Mijel	01.25	81	W
I./NJG1	Lt Richard Heuke	Four-eng	40km NE Venlo	01.26	1	W
3./NJG3	Oblt Heinz Ferger	Lancaster	Venlo	01.27	14	W
3./NJG3	Uffz G Schneider	Four-eng	N Aachen	01.27	3	W

One of the successful He219 pilots was Hauptmann Ernst-Wilhelm Modrow of I./NJG 1. He survived the war with the Ritterkreuz and 34 combat victories. A former transport pilot, he trained as a nightfighter pilot in late 1943, claimed his first successes in March 1944 and was quickly promoted to command 1 Staffel. During the spring and early summer of 1944 his score mounted rapidly and his 23rd - 26th victories came on 21/22nd June, four Lancasters falling to the powerful armament of the big He219 'Uhu' (Owl).

St.IV./NJG1	Hptm H-W Schnaufer	Lancaster	2km S Meeuven	01.30	82	W
3./NJG3	Oblt Heinz Ferger	Lancaster	Venlo	01.32	15	W
12./NJG3	Oblt Herbert Koch	Lancaster	Venlo	01.32	12	W
7./NJG2	Lt Hans Schäfer	Lancaster	W Köln	01.33	10	W
St.IV./NJG1	Hptm Adolf Breves	Lancaster	Venlo area	01.35	8	W
St.IV./NJG1	Oblt Georg Fengler	Four-eng	St.Trond	01.35	8	W
3./NJG1	Fw Wilhelm Morlock	Lancaster	Venlo	01.35	7	W
St.IV./NJG1	Hptm H-W Schnaufer	Lancaster	Aachen	01.36	83	W
3./NJG3	Uffz G Schneider	Four-eng	Mönchengladbach	01.38	4	W
1./NJG1	Hptm E-W Modrow	Lancaster	NNW Duisburg	01.39	25	W
3./NJG1	Hptm Heinz Strüning	Lancaster	W Cologne	01.39	55	W
2./NJG1	Oblt Werner Baake	Lancaster	W Aachen	01.46	35	W
1./NJG7	Uffz Nischler	Four-eng	Goch-Nijmegen	01.58	1	W
3./NJG1	Oblt Josef Nabrich	Lancaster	Beacon "Bisam"	01.59	18	W
1./NJG1	Hptm E-W Modrow	Lancaster	N Deelen	02.01	26	W
3./NJG3	Uffz Gunther Schlomberg	Four-eng	Arendonk - Hechtel	02.03	1	W
St.IV./NJG1	Hptm H-W Schnaufer	Lancaster	6km S Hamomt	02.04	84	W
8./NJG2	Uffz Egon Engling	Lancaster	Near Turnhout	02.07	3	W
St.IV./NJG1	Hptm Adolf Breves	Lancaster	Breda-Tilburg	02.08	9	W
6./NJG1	Uffz Koppelmayer	Lancaster	14km W Hertogenbosch	02.09	1	W
3./NJG1	Fw Wilhelm Morlock	Lancaster	Venlo	02.09	8	W
6./NJG1	Lt Gottfried Hanneck	Lancaster	St.Trond	02.12	4	W
12./NJG3	Ofw Heinrich Kuseling	Four-eng	Eindhoven	02.12	1	W
St.IV./NJG1	Hptm Adolf Breves	Lancaster	Rosendaal	02.14	10	W
St.I./NJG3	Lt Manfred Scheunpflug	Four-eng	N Antwerp	02.16	2	W
12./NJG3	Ofw Heinrich Kuseling	Four-eng	Beacon "Kurfürst"	02.17	2	W
7./NJG2	Lt Hans Schäfer	Lancaster	Breda-Tilburg	02.17	11	W
1./NJG4	Oblt Walter Riedlberger	Lancaster	In sea	02.24	7	W
3./NJG1	Hptm Heinz Strüning	Four-eng	Beacon "Hamster"	02.30	56	W
St.III./NJG1	Hptm Martin Drewes	Lancaster	NW Schouwen	02.32	43	W
St.III./NJG1	Hptm Martin Drewes	Lancaster	15km NW Schouwen	02.36	44	W

22-23 June 1944

7./NJG1	Oblt Walter Prues	Lancaster	LM-5		6*	W
9./NJG5	Lt Gerhard Wagner	Halifax	25km S Amiens	00.28	10	W
St.I./NJG2	Hptm Gerhard Raht	Four-eng	SE Compiegne	00.37	34	W
2./NJG3	Fw Herbert Stähler	Four-eng	FF1-1	01.01	1	W
3./NJG3	Oblt Heinz Ferger	Four-eng	S St.Quentin	01.14	16	W
3./NJG3	Oblt Heinz Ferger	Four-eng	Amiens	01.15	17	W
2./NJG3	Fw Loll	Four-eng	Amiens - Breteuil	01.19	1	W
9./NJG5	Lt Gerhard Wagner	Halifax	NE Senlis	01.20	11	W
3./NJG3	Ofw Giesecke	Four-eng	Beacon "Carl"	01.21	2	W
1./NJG4	Oblt Walter Riedlberger	Lancaster	Amiens	01.23	8	W
2./NJG3	Hptm Huck	Four-eng	Nr. Abbeville	01.32	1	W

23-24 June 1944

St,.NJG2	Oblt Johann Werth	Lancaster			11*	W
1./NJG4	Uffz Konrad Beyer	Four-eng	S Calais	00.28	7	W
2./NJG4	Hptm Josef Krahforst	Four-eng	St.Omer - Ypres	01.10	6	W
8./NJG5	Ofw Herbert Altner	Four-eng	SE Calais	01.11	15	W
St.I./NJG3	Hptm Werner Husemann	Four-eng	St.Omer - Ypres	01.15	21	W
2./NJG3	Fw Herbert Stähler	Four-eng	NNW Saintes	01.18	2	W
St.I./NJG3	Lt Manfred Scheunpflug	Four-eng	St.Omer - Ypres	01.20	3	W

24-25 June 1944

1./NJG2	Oblt Adolf Kaiser	B-17	Blangy-sur-Bresle	00.12	10	W
II./NJG2	Maj Paul Semrau	Lancaster	St.Valery	00.14	24	W
2./NJG2	Oblt Heinz Rökker	Lancaster	S Dieppe	00.16	33	W
II./NJG2	Maj Paul Semrau	Lancaster	Dieppe	00.17	25	W
2./NJG3	Hptm Huck	Four-eng	Dieppe - Abbeville	00.24	2	W
3./NJG3	Uffz Hans Schadowski	Four-eng	Nr. St.Pol	00.24	3	W
Stab./NJG5	Maj Walter Borchers	Four-eng	QD	00.25	15	W
Stab./NJG3	Obstlt Helmut Lent	Four-eng	QE-PE	00.25	96	W
3./NJG3	Uffz Hans Schadowski	Four-eng	Nr. St.Pol	00.25	4	W
2./NJG4	Hptm Josef Krahforst	Four-eng	QD	00.27	7	W
6./NJG2	Lt Erich Jung	Stirling	Nr. Amiens	00.28	4	W
3./NJG3	Uffz G Schneider	Four-eng	SW Dieppe	00.30	5	W
3./NJG4	Oblt Kadleik	Four-eng	Montreuil	00.32	2	W
2./NJG3	Uffz Alois Vogler	Four-eng	Nr. St.Pol	00.32	1	W
St.III./NJG5	Hptm Paul Zorner	Lancaster	E Boulogne-sur-Mer	00.32	53	W
3./NJG3	Uffz Hans Schadowski	Four-eng	St.Pol	00.33	5	W
6./NJG2	Lt Erich Jung	Lancaster	Nr. Amiens	00.34	5	W
3./NJG4	Oblt Rolf Klages	Four-eng	Abbeville	00.34	1	W
3./NJG3	Uffz G Schneider	Four-eng	Dieppe	00.35	6	W
6./NJG2	Lt Martin Tschiersch	Lancaster	PB	00.35	5	W
7./NJG4	Oblt Richard Delakowitz	Four-eng	Abbeville - Doullens	00.38	7	W
5./NJG2	Hptm H-H Hissbach	Lancaster	N Abbeville	00.38	17	W
6./NJG2	Hptm Kamsties	Lancaster	Nr. Dieppe	00.38	2	W
2./NJG3	Hptm Huck	Four-eng	Dieppe - Abbeville	00.39	3	W
6./NJG2	Lt Lothar Jarsch	Lancaster	Amiens-Berck-Fécamp	00.40	1	W
Stab./NJG3	Obstlt Helmut Lent	Four-eng	QE-PE	00.40	97	W

Leutnant Ernst-Ewald Hittler (right) flew with 3./NJG 1, claiming his first victory in April. His fifth and last, a Mosquito, came on 25th June. (left) Oberleutnant Wilhelm Johnen with his 8./NJG 6 crew. Note the curiously 'painted-out' swastika. They claimed their 20th victory on 26th June 1944

3./NJG3	Lt Götze	Four-eng	S St.Pol	00.41	1	W
9./NJG4	Hptm Kurt Fladrich	Lancaster	Abbeville - Doullens	00.44	13	W
8./NJG5	Uffz Rudolf Irmisch	Lancaster	between Amiens & Dieppe	00.44	1	W
6./NJG2	Lt Lothar Jarsch	Lancaster	Amiens-Berck-Fécamp	00.44	2	W
3./NJG3	Ofw Gierscke	Four-eng	Abbeville	00.45	3	W
9./NJG5	Lt Gerhard Wagner	Lancaster	NW Le Treport	00.45	12	W
Stab./NJG5	Maj Walter Borchers	Four-eng	NB-3	00.48	16	W
2./NJG3	FhjFw Herbert Stähler	Four-eng	Sea, NW Calais	00.48	3	W
St.III./NJG5	Hptm Paul Zorner	Lancaster	N-W Etaples	00.50	54	W
3./NJG3	Gfr Kruschyna	Four-eng	Somme Estuary	00.55	1	W
3./NJG1	Lt Ernst-Ewald Hittler	Mosquito	30km NW Berlin	01.37	5	W
1./NJG5	Oblt E-G Drünkler	Stirling	10km W Berck-sur-Mer	03.45	16	W

25-26 June 1944

1./NJG100	Ofw Helmut Dahms	Boston	Russia	22.11	10	E
1./NJG100	Lt G E Francsi	U-2	Russia	22.36	20	E
3./NJG100	Uffz Hausschild	Boston	S Bobruisk	22.51	2	E
1./NJG100	Lt G E Francsi	U-2	30km NE Pinsk	22.57	21	E
3./NJG100	Oblt Josef Putzkuhl	U-2	21km W Pinsk	23.07	12	E
St.III./NJG6	Oblt Wolfgang Knieling	Wellington	Plattensee	23.50	3	W
St.III./NJG6	Oblt Wolfgang Knieling	Wellington	W Budapest	23.59	4	W
8./NJG6	Oblt Wilhelm Johnen	Wellington	SW Plattensee	00.22	20	W
St.III./NJG6	Oblt Wolfgang Knieling	Wellington	N Kaposvár	00.29	5	W
7./NJG6	Oblt Josef Kraft	Halifax	Tamási	00.40	29	W
1./NJG100	Lt G E Francsi	U-2	NE Pinsk	01.38	22	E

26-27 June 1944

3./NJG100	Uffz Hausschild	Il-2	SE Baranov	20.23	3	E
4./NJG100	Fw Haberkern	DB-3	Russia	23.13	1	E
4./NJG100	Fw Haberkern	DB-3	Russia	23.43	2	E
4./NJG100	Lt Richter	PS-84	Russia	23.43	6	E
4./NJG100	Uffz Gutschmidt	PS-84	Russia	00.44	2	E

27-28 June 1944

3./NJG100	Uffz Hausschild	Boston	N Pinsk	22.16	4	E

4./NJG100	Lt Klaus Scheer	DB-3	Russia	22.37	14	E
3./NJG100	Uffz Hausschild	Boston	SSE Pinsk	23.51	5	E
1./NJG100	Fw Rudi Duding	B-25	Russia	00.27	4	E
1./NJG100	Lt G E Francsi	B-24	N Baranovichi	00.33	23	W
2./NJG2	Ofw Johannes Strassner	Lancaster	St. Valéry	00.36	1	W
1./NJG100	Lt G E Francsi	B-24	N Baranovichi	00.45	24	W
Stab./NJG3	Obstlt Helmut Lent	Four-eng	Off Dunkirk	01.10	98	W
4./NJG4	Oblt Jakob Schaus	Four-eng	Off Ostend	01.25	18	W
St.III./NJG4	Hptm Hans-Karl Kamp	Four-eng	N Mailly-le-Camp	01.49	21	W
4./KG51	Lt Markan	Four-eng		02.27	1	W
5./KG51	Lt Friedrich	Stirling		02.30	1	W
4./KG51	Uffz Walther	Lancaster		02.31	1	W
St.I./NJG5	Maj Werner Hoffmann	Halifax	NW Reims	03.50	32	W
3./NJG3	Oblt Heinz Ferger	Mosquito	7km SE Roulers	10.30	18	W

28-29 June 1944

3./NJG100	Uffz Hausschild	Boston	NE Pinsk	23.41	6	O
1./NJG100	Lt G E Francsi	Il-4	Russia	00.12	25	E
II./NJG2	Maj Paul Semrau	Halifax	N Rouen	00.31	26	W
6./NJG2	Lt Lothar Jarsch	Halifax	Breteuil	00.33	3	W
6./NJG2	Obst Günther Radusch	Halifax	Compiègne	00.37	59	W
2./NJG5	Uffz Lichert	Four-eng	Compiègne	00.40	1	W
3./NJG3	Uffz G Schneider	Four-eng	N Paris	00.40	7	W
1./NJG6	Ofhr F-W Kister	Lancaster	W Juvincourt	00.44	1	W
6./NJG2	Obst Gunther Radusch	Halifax	Soissons	00.44	60	W
II./NJG2	Maj Paul Semrau	Halifax	Etrepagny	00.45	27	W
II./NJG2	Maj Paul Semrau	Halifax	Etrepagny	00.45	28	W
1./NJG5	Lt Joachim Plass	Halifax	Chateau-Thierry	00.47	2	W
1./NJG5	Lt Joachim Plass	Halifax	Craonne/Rheims	00.53	3	W
3./NJG3	Uffz G Schneider	Four-eng	N Paris	00.55	8	W
2./NJG3	Hptm Huck	Four-eng	Mantes - Houdan	00.56	4	W
3./NJG3	Uffz G Schneider	Four-eng	N Paris	00.57	9	W
St.I./NJG5	Maj Werner Hoffmann	Halifax	NW Reims	00.58	33	W
2./NJG3	Hptm Huck	Four-eng	Mantes - Houdan	00.59	5	W
II./NJG2	Maj Paul Semrau	Halifax	Soissons	01.00	29	W
6./NJG2	Fw Gajewski	Halifax	Soissons - Fisnes	01.06	4	W
1./NJG5	Oblt E-G Drünkler	Lancaster	Château-Thierry	01.08	17	W
11./NJG6	Fw Hermann Scherer	B-24	SW Brunn	02.17	1	W
St.IV./NJG6	Maj Herbert Lütje	Wellington	N Stockerau	02.23	38	W
4./NJG4	Oblt Jakob Schaus	Four-eng	Grand Quevilly	03.10	19	W
4./NJG4	Oblt Jakob Schaus	Four-eng	Port Audemer	03.20	20	W

29-30 June 1944

3./NJG100	Oblt Josef Pützkuhl	Il-2	NE Pinsk	22.39	13	E
3./NJG100	Oblt Josef Pützkuhl	Il-4	NE Pinsk	23.10	14	E
St.I./NJG100	Hptm Gunter Bertram	Il-4	Russia	23.15	18	E
1./NJG100	Lt G E Francsi	Boston	Russia	23.29	26	E
7./NJG6	Oblt Helmuth Schulte	Wellington	Mariazell	00.30	15	W
3./NJG100	Oblt Josef Putzkuhl	Boston	NW Pinsk	00.47	15	E
3./NJG100	Ofw Walter Scheinbein	Il-2	Russia	02.07	3	E
4./NJG100	Uffz Kollrep	Boston	Russia	02.15	6	E

Hauptmann Heinz-Horst Hissbach (right) of 5./NJG 2 claimed two Lancasters on the night of 30th June to bring his tally to 23. His career had been varied. First, with I./NJG 2 flying intruder missions over England, then to the Mediterranean area, operating over Malta and North Africa before moving on to interdiction missions over Biscay with 15./KG 40. He would later become Gruppenkommandeur of II./NJG 2 and take part in the last great Luftwaffe intruder mission against England to bring his score to 34.

30 June-1 July 1944

1./NJG100	Unknown	R-5	Russia	23.28	-	E
3./NJG100	Oblt Josef Pützkuhl	B-25	NNE Pinsk	23.35	16	E
3./NJG100	Oblt Josef Pützkuhl	Il-4	N Pinsk	23.48	17	E
1./NJG100	Lt Hermann Weihe	Il-2	Russia	00.17	1	E
4./NJG100	Oblt Gunter Bertram	PS-84	Baranowicze	00.25	19	E
1./NJG100	Lt G E Francsi	Il-4	Russia	00.28	27	E
3./NJG100	Oblt Josef Pützkuhl	Il-2	20km SE Pinsk	00.43	18	E
4./NJG4	Hptm Fritz Söthe	Four-eng	E Blois	01.08	15	W
Stab./NJG2	Obst Günther Radusch	Four-eng	St Leu d'Esserent	01.12	61	W
Stab./NJG2	Obst Günther Radusch	Four-eng	St Leu d'Esserent	01.14	62	W
St.III./NJG5	Hptm Paul Zorner	Lancaster	15km SW Bourges	01.22	55	W
4./NJG4	Oblt Jakob Schaus	Four-eng	NNW Bourges	01.23	21	W
5./NJG2	Uffz Rudolf Morenz	Lancaster	E Blois	01.24	1	W
5./NJG2	Hptm H-H Hissbach	Lancaster	between Blois & Vendome	01.25	18	W
3./NJG3	Hptm Banschein	Four-eng	NNE Vendome	01.26	1	W
II./NJG2	Maj Eckhoff	Four-eng	La Ferte St.Cyr	01.26	1	W
4./NJG4	Hptm Fritz Söthe	Four-eng	Nr. Bourges	01.27	16	W
4./NJG4	Oblt Jakob Schaus	Four-eng	E Blois	01.35	22	W
6./NJG2	Oblt Franz Brinkhaus	Four-eng	Chateaudun	01.36	11	W
3./NJG3	Uffz Hans Schadowski	Four-eng	between Vendome & Blois	01.36	6	W
5./NJG2	Hptm H-H Hissbach	Lancaster	30km NW Chateaudun	01.38	19	W
3./NJG4	Lt Tennenhorst.von	Four-eng	S Chateaudun	01.38	1	W
St.I./NJG2	Hptm Gerhard Raht	Lancaster	Bailleau	01.44	35	W

1-2 July 1944

St.I./NJG100	Hptm Kurt Bonow	Il-4	Russia	22.13	7	E
3./NJG100	Oblt Josef Putzkuhl	Il-4	23km ESE Pinsk	00.36	19	E
2./NJG1	Oblt Joachim Fincke	Mosquito	Beacon "Gorilla"	01.38	1	W

2-3 July 1944

6./KG51	Fw Johann Trenke	E/A	England	00.14	9	W
6./NJG101	Oblt Hans Krause	Wellington	S Sombor	00.25	11	W
11./NJG6	Fw Hermann Scherer	B-24	W Bucharest	01.08	2	W

3-4 July 1944

6./NJG101	Oblt Hans Krause	Halifax	Sombor	22.26	12	W
6./NJG101	Oblt Hans Krause	Halifax	Sombor	22.52	13	W
1./NJG100	Fw Rudi Duding	Il-4	Russia	23.05	5	E
St.III./NJG6	Hptm Leopold Fellerer	Halifax	10km NE Kaposvár	01.50	34	W
6./NJG101	Lt Heinz Bock	Halifax	Sombor	02.15	2	W

4-5 July 1944

1./NJG100	Lt G E Francsi	Il-4	Russia	23.01	28	E
1./NJG100	Fw Rudi Duding	Il-4	Russia	23.08	6	E
1./NJG100	Lt G E Francsi	Il-4	Russia	23.08	29	E
1./NJG100	Fw Rudi Duding	B-24	Russia	23.15	7	W
1./NJG100	Lt G E Francsi	Il-4	Russia	23.16	30	E
St.I./NJG100	Hptm Kurt Bonow	Il-4	15km NW Vilna	23.20	8	E
Stab./NJG2	Ofw Willi Glitz	Four-eng	Dieppe - Yvetot	00.38	4	W
Stab./NJG2	Ofw Willi Glitz	Four-eng	Dieppe - Yvetot	00.43	5	W
Stab./NJG2	Ofw Willi Glitz	Four-eng	Dieppe - Yvetot	00.44	6	W
2./NJG2	Uffz Friedrich Frick	Lancaster	Mantes	01.08	1	W
4./NJG4	Hptm Fritz Söthe	Four-eng	Chartres	01.14	17	W
3./JG301	Uffz Gratz	Four-eng	Senarpont	01.15	1	W
2./JG301	Lt Willi Esche	Lancaster	Cayeux-sur-Mer	01.16	1	W
1./JG301	Fw Martin Schulze	Four-eng	France	01.17	1	W
Stab./NJG2	Oblt Johann Werth	Four-eng	TA-TB-UA	01.18	12	W
1./JG301	Fw Hans Engfer	Lancaster	TA-TB-UA	01.19	1	W
3./NJG3	Uffz G Schneider	Four-eng	HF	01.20	10	W
5./NJG2	Ofw H Breitenfelder	Four-eng	Fontainebleau	01.23	1	W
4./NJG4	Uffz Kurt Stockbauer	Four-eng	AE-BE	01.26	1	W
5./NJG4	Fw Georg Matt	Four-eng	BE	01.27	2	W
6./NJG2	Lt Erich Jung	Four-eng	AE	01.28	6	W
1./JG301	Fw Martin Schulze	Four-eng	W Paris	01.28	2	W
6./NJG2	Lt Erich Jung	Four-eng	AE	01.30	7	W
1./JG301	Fw Gerhard Koch	Lancaster	25km N St Leu d'Esserent	01.30	3	W
2./NJG2	Ogfr Johannes Strassner	Halifax	E Chartres	01.30	2	W
6./NJG2	Uffz Schulz	Four-eng	Rambouillet	01.32	1	W
1./JG301	Oblt Ernst Fischer	Lancaster	St Leu d'Esserent	01.35	1	W
6./NJG2	Lt Fischer	Four-eng	AE	01.36	1	W
6./NJG2	Lt Erich Jung	Four-eng	N Paris	01.36	8	W
2./NJG2	Ogfr Johannes Strassner	Halifax	Chartres	01.36	3	W
9./NJG5	Ofw Dreyer	Four-eng	N Paris	01.37	1	W
1./JG301	Oblt Ernst Fischer	Lancaster	N St Leu d'Esserent	01.39	2	W
1./NJG5	Fw Ernst Reitmeyer	Four-eng	Orléans	01.39	13	W
St.I./NJG3	Lt Manfred Scheunpflug	Four-eng	Abbeville - Crecy	01.40	4	W
1./NJG2	Oblt Adolf Kaiser	Lancaster	70km N Chateaudun	01.41	11	W
1./NJG5	Fw Ernst Reitmeyer	Four-eng	S Orléans	01.42	14	W
St.I./NJG5	Oblt Breitfeld	Four-eng	NW Creil	01.45	1	W
6./NJG2	Lt Lothar Jarsch	Four-eng	Dreux area	01.48	4	W
2./NJG5	Uffz Lichert	Four-eng	Near Gournay	01.48	2	W
3./NJG3	Uffz Gunther Schlomberg	Four-eng	Beauvais	01.49	2	W
Stab./NJG5	Maj Walter Borchers	Four-eng	Amiens	01.50	17	W
3./SKG10	Ofw Rahlhammer	Four-eng	Nr. La Longpré	01.50	1	W

A Do217 of an unidentified unit shows the curious 'wave mirror' camouflage scheme adopted by many night units in 1944.

1./JG301	Uffz Hans Zarm	Four-eng	6km W St Leu d'Esserent	01.50	1	W
1./JG301	Fw Georg Scharein	Lancaster	20km N St Leu d'Esserent	01.52	1	W
8./NJG5	Fw Blank	Four-eng	Rouen-Beauvais	01.53	1	W
4./NJG4	Hptm Fritz Söthe	Four-eng	Nr. Artenay	01.54	18	W
5./NJG2	Hptm H-H Hissbach	Lancaster	S Abbeville	01.55	20	W
2./NJG5	Uffz Lichert	Four-eng	Neufchâtel	01.55	3	W
2./JG301	Lt Willi Esche	Lancaster	Nr. Abbeville	01.58	2	W
1./NJG5	Lt Joachim Plass	Four-eng	N St. Valéry	01.59	4	W
9./NJG5	Lt Gerhard Wagner	Four-eng	N Aumale	02.00	13	W
3./NJG4	Oblt Rolf Klages	Lancaster	Aumale	02.03	2	W
11./NJG1	Hptm Hermann Greiner	Four-eng	Dieppe	02.05	34	W
1./NJG2	Oblt Adolf Kaiser	Lancaster	40km N Caen	02.06	12	W

5-6 July 1944

3./NJG100	Oblt Josef Pützkuhl	Il-4	10km SW Pinsk	22.13	20	E
3./NJG100	Oblt Josef Pützkuhl	Il-4	40km W Pinsk	22.22	21	E
3./NJG100	Oblt Josef Pützkuhl	Il-4	60km WNW Pinsk	23.00	22	E
3./NJG100	Oblt Josef Pützkuhl	Il-4	15km SW Pinsk	00.04	23	E
3./NJG100	Oblt Josef Pützkuhl	Il-4	15km W Pinsk	00.20	24	E
1./NJG100	Oblt Hans Schmid	Il-4	Russia	00.24	2	E
3./NJG3	Uffz G Schneider	Four-eng	Blanc Nez	00.32	11	W
St.I./NJG3	Hptm Werner Husemann	Four-eng	Dover Straits	00.35	22	W
1./JG301	Fw Gerhard Koch	Four-eng	Calais	00.38	4	W
1./NJG4	Oblt Walter Riedlberger	Lancaster	Goodwin Sands	00.41	9	W
3./NJG3	Uffz Gunther Schlomberg	Four-eng	NE Channel	00.43	3	W
2./JG301	Fw Hans Müller	Halifax	Channel off Griz Nez	00.44	1	W
1./JG301	Fw Gerhard Koch	Four-eng	Dunkirk	00.45	5	W
3./NJG100	Oblt Josef Pützkuhl	Il-4	65km ESE Pinsk	00.53	25	E
7./NJG1	Fw Heinrich Lahmann	Halifax	Waddenzee	01.21	1	W

2./NJG1	Ofw Josef Strohlein	B-17	Nr. Rotterdam	01.42	2	W
6./KG51	Uffz Illner	Twin-eng	Caen	02.38	1	W
2./JG301	Lt Willi Esche	Four-eng	10km E Dieppe	03.56	3	W

6-7 July 1944

1./NJG100	Oblt Hans Schmid	Il-4	Russia	00.17	3	E
1./NJG100	Oblt Hans Schmid	Il-4	Russia	00.43	4	E
7./NJG6	Oblt Helmuth Schulte	Wellington	S St.Polten	01.10	16	W
6./NJG101	Fw Brinkhaus	Wellington	N S. Pölten	01.15	1	W
8./NJG6	Oblt Wilhelm Johnen	Wellington	40km S Fels-am-Wagram	01.17	21	W
7./NJG6	Ofw Heinz Säwert	Wellington	S St.Polten	01.17	15	W
7./NJG6	Oblt Josef Kraft	Wellington	S St.Polten	01.20	30	W
8./NJG6	Oblt Wilhelm Johnen	Wellington	40km S Fels-am-Wagram	01.25	22	W
9./NJG6	Oblt Rudolf Thun	B-24	W St.Polten	01.25	2	W
7./NJG6	Uffz Willi Lauterbach	Wellington	W Vienna	01.26	2	W
St.III./NJG6	Oblt Wolfgang Knieling	Wellington	S St.Polten	01.27	6	W
7./NJG6	Oblt Josef Kraft	Wellington	S St.Polten	01.27	31	W
St.III./NJG6	Oblt Wolfgang Knieling	Four-eng	Ratten	01.34	7	W
St.III./NJG6	Hptm Leopold Fellerer	Lancaster	E Marburg	01.45	35	W
9./NJG6	Oblt Rudolf Thun	B-24	E Graz	01.45	3	W
7./NJG6	Uffz Willi Lauterbach	Wellington	N Marburg	01.47	3	W
St.III./NJG6	Hptm Leopold Fellerer	Wellington	N Laibach	02.05	36	W

7-8 July 1944

4./NJG100	Uffz Kollrep	B-25	Russia	23.08	7	E
4./NJG100	Lt Klaus Scheer	B-25	Russia	23.10	15	E
3./NJG100	Hptm August Fischer	Il-4	Russia	23.40	4	E
6./NJG2	Oblt Franz Brinkhaus	Lancaster	QD	00.54	12	W
II./NJG2	Lt Siegfried Elsasser	Four-eng	QB-AC	00.59	1	W
St.I./NJG2	Hptm Gerhard Raht	Lancaster	AB-AC	01.04	36	W
Stab./NJG2	Obstlt . Günther Radusch	Lancaster	TB-89	01.05	63	W
II./NJG2	Maj Paul Semrau	Four-eng	AD	01.07	30	W
Stab./NJG2	Ofw Willi Glitz	Lancaster	QB-UB-AC	01.10	7	W
St.I./NJG2	Hptm Gerhard Raht	Lancaster	AE	01.11	37	W
I./JG301	Hptm Wilhelm Burghardt	Lancaster	St Leu d'Esserent	01.12	1	W
II./NJG2	Lt Siegfried Elsasser	Four-eng	AC-AD	01.12	2	W
5./NJG2	Fw Hans Kurz	Lancaster	NW Paris	01.12	4	W
II./NJG2	Maj Paul Semrau	Four-eng	AD-BD-BE	01.12	31	W
6./NJG2	Lt Erich Jung	B-17	10km W Creil	01.13	9	W
2./NJG5	Hptm Alex Ressugier	Lancaster	S Gisors	01.15	2	W
I./JG301	Lt Max Kreil	Lancaster	SG	01.16	1	W
6./NJG2	Lt Martin Tschiersch	Four-eng	QE	01.16	6	W
1./JG301	Fw Hans Engfer	Lancaster	TD	01.20	2	W
I./JG301	Hptm Wilhelm Burghardt	Lancaster	UE	01.21	2	W
2./JG301	Hptm Erich Wegener	Four-eng	TD	01.21	1	W
1./NJG4	Uffz Konrad Beyer	Four-eng	N Paris	01.23	8	W
1./NJG5	Uffz Helmut Burghardt	Lancaster	SD-2	01.25	1	W
1./JG301	FhjFw Viktor Emanuel	Lancaster	UE	01.25	2	W
2./NJG2	Fw Heinrich Tetzlaff	Lancaster	SC	01.25	1	W
1./NJG5	Uffz E-A Schulze-Bergcamen	Four-eng	LE	01.26	1	W
3./JG301	Fw Manfred Gromoll	Lancaster	UD-4	01.26	1	W

On the night of 7th July 1944, the Wilde Sau pilots of JG 301 were sent to face not only the British, but also the Americans. A formation of 9th Air Force B-26 Marauders had been sent to make precision attacks on a chateau near Amiens. The American crews had been briefed to make timed attacks, one after the other, and the Wilde Sau pilots, together with several 'conventional' nightfighter crews, slaughtered the Marauders.

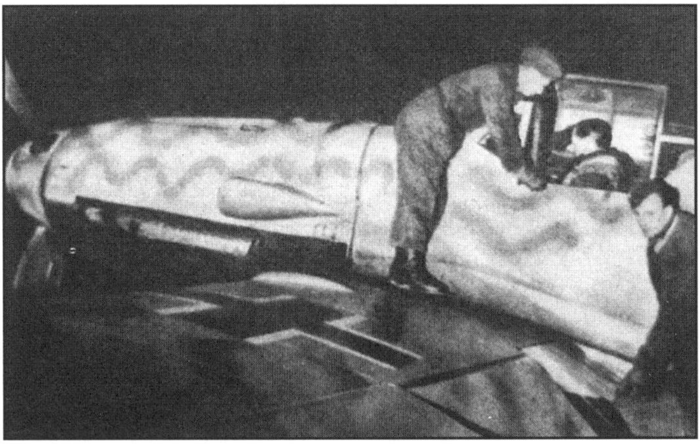

6./NJG2	Lt Martin Tschiersch	Four-eng	UE	01.26	7	W
2./NJG4	Uffz Hans Saalberg	Four-eng	TE	01.27	1	W
3./JG301	Fw Joachim Hähnel	Lancaster	QC-8	01.28	1	W
St.I./NJG2	Hptm Gerhard Raht	Lancaster	TD-TE	01.29	38	W
St.I./NJG4	Maj Gunter Vörwinkel	Four-eng	TC-TD	01.30	2	W
St.I./NJG5	Maj Werner Hoffmann	Lancaster	S Gisors	01.31	34	W
1./NJG5	Fw Ernst Reitmeyer	Lancaster	TD	01.31	15	W
6./NJG4	Hptm Hans Autenrieth	Lancaster	SD	01.32	22	W
1./NJG4	Uffz Konrad Beyer	Four-eng	SE	01.32	9	W
4./NJG4	Lt Frithjof Fensch	Four-eng	TD-8	01.32	6	W
II./NJG4	Hptm Hubert Rauh	Four-eng	SC-99	01.32	18	W
3./JG301	Lt Alfred Hiller	Lancaster	SD-4	01.33	1	W
6./NJG2	Lt Erich Jung	Lancaster	50km N Creil	01.35	10	W
3./NJG4	Oblt Kadleik	Lancaster	RD-1	01.35	3	W
5./NJG2	Ofw Robert Luddeke	Lancaster	N Rouen	01.35	9	W
2./JG301	Lt Willi Esche	Lancaster	Off Cayeux	01.38	4	W
1./NJG5	Lt Joachim Plass	Lancaster	Neufchâtel	01.38	5	W
6./NJG2	Lt Martin Tschiersch	Four-eng	Nr. Gisors	01.38	8	W
I./JG301	Lt Brinkmann	Four-eng	Senarpont	01.39	4	W
1./JG301	Fw Herbert Bower	Four-eng	Off Cayeux	01.40	1	W
St.I./NJG5	Maj Werner Hoffmann	Lancaster	3km N Foucarmont	01.40	35	W
1./JG301	Fw Hans Engfer	Four-eng	SE Troyes	01.42	3	W
3./NJG4	Oblt Kadleik	Lancaster	Nr. Cayeux	01.42	4	W
2./NJG5	Uffz Lichert	Four-eng	Douchy-les-Mines	01.42	4	W
St.I./JG301	Lt Brinkmann	Four-eng	St Quentin	01.43	5	W
2./JG301	Uffz Klaus Jacobi	Lancaster	Senarpont	01.43	1	W
1./JG301	Fw Herbert Bower	Four-eng	St. Valéry	01.44	2	W
5./NJG2	Hptm H-H Hissbach	Lancaster	Dieppe	01.45	21	W
1./NJG2	Ofw Rudolf Winn	Lancaster	Paris	01.45	3	W
3./JG301	Fw Manfred Gromoll	Lancaster	S Cambrai	01.48	2	W
1./NJG10	Oblt Krause	Mosquito	Near Brandenburg	01.48	1	W
St.I./NJG5	Maj Werner Hoffmann	Lancaster	St. Quentin	01.50	36	W
3./NJG3	Hptm Eduard Schröder	Four-eng	between Lille & Douai	01.56	12	W
1./JG301	Fw Georg Scharein	B-26	Senarpont	02.14	2	W
3./JG301	Hptm Helmut Sühr	B-26	St Quentin	02.22	4	W

3./JG301	Uffz Paul Werner	B-26	SE Dieppe	02.28	1	W
2./JG301	Ofw Heinz Günther	B-26	Amiens	02.37	1	W
9./NJG5	Lt Wolf	B-26	NW Amiens	02.37	1	W
1./JG301	Fw Georg Scharein	B-26	Doullens	02.44	3	W
4./NJG4	Lt Frithjof Fensch	B-26	Senarpont	02.46	7	W
3./JG301	Ofw Josef Löffler	Lancaster	Senarpont	02.46	1	W
3./NJG3	Hptm Eduard Schröder	Twin-eng	Amiens	02.47	13	W
Stab./NJG4	Maj Wolfgang Thimmig	B-26	Amiens	02.47	21	W
9./NJG5	Lt Gerhard Wagner	B-26	NW Amiens	02.47	14	W
3./NJG3	Hptm Banschein	Twin-eng	Amiens	02.50	2	W
1./JG301	Fw Gerhard Koch	Four-eng	Senarpont	02.50	6	W
Stab./NJG4	Maj Wolfgang Thimmig	B-26	Senarpont	02.55	22	W
Stab./NJG4	Maj Wolfgang Thimmig	B-26	Seine Bay	02.59	23	W
1./JG301	Fw Martin Schülze	B-26	Pont-Rémy	03.05	3	W
4./NJG4	Lt Frithjof Fensch	Twin-eng	St.Arnoult	03.09	8	W

8-9 July 1944

3./NJG100	Hptm August Fischer	Il-4	Russia	22.55	5	E
1./NJG100	Lt G E Francsi	Il-4	Russia	24.00	31	E
1./NJG100	Ofw Rudi Duding	Il-4	Russia	00.03	8	E
3./NJG100	Hptm August Fischer	Il-4	Russia	00.36	6	E
1./NJG100	Ofw Rudi Duding	tbr-7	Russia	00.41	9	E
3./NJG100	Hptm August Fischer	Il-4	Russia	00.52	7	E
3./NJG100	Hptm August Fischer	Il-4	Russia	01.16	8	E
7./NJG1	Oblt Daborer	Wellington	over Sea	02.40	1	W

9-10 July 1944

3./NJG100	Lt Eberhard Kaross	B-25	Russia	23.14	8	E
4./NJG100	Oblt Gunter Bertram	DB-3	SW Kobrin	23.16	20	E
1./NJG100	Fw Heinz De Fries	Il-4	Russia	23.23	3	E
1./NJG100	Ofw Helmut Dahms	Il-4	Russia	00.14	11	E
3./NJG100	Lt Truber	B-25	Russia	00.15	4	E

10-11 July 1944

1./NJG100	Oblt Waldemar Rathke	Il-4	Russia	23.14	15	E
2./NJG1	Maj Hans Karlowski	Mosquito	Off Terschelling	03.18	5	W

12-13 July 1944

8./NJG2	Fw Kleibauer	Stirling	St.Armand	00.54	3	W
1./NJG5	Fw Ernst Reitmeyer	Lancaster	Ligny-en-Barrois	01.59	16	W
7./NJG1	Uffz Wilsch	Wellington	Off Ameland	02.03	1	W
1./NJG5	Fw Ernst Reitmeyer	Lancaster	Joinville	02.08	17	W
2./NJG5	Uffz Lichert	Four-eng	BL-6	02.09	5	W
St.I./NJG5	Maj Werner Hoffmann	Lancaster	10km W Neafchateau	02.12	37	W
2./NJG5	Uffz Hengsbach	Lancaster	Rolampont	02.17	3	W
1./NJG5	Fw Ernst Reitmeyer	Lancaster	Rolampont	02.17	18	W
1./NJG5	Oblt E-G Drünkler	Lancaster	NNE Chaumont	02.18	18	W
St.I./NJG5	Maj Werner Hoffmann	Lancaster	10km SE Joinville	02.18	38	W

14-15 July 1944

4./NJG100	Oblt Gunter Bertram	PS-84	Tomashev	22.27	21	E

On 14th July, Oberleutnant Joachim Fincke of 2./NJG 1 was killed in action at Venlo in He219 G9+LK. He had possibly previously been attacked over Belgium by Lieutenant Armour R.N. of the Fighter Iinterception Unit, who claimed a Do217.

(S.Hall)

4./NJG100	Oblt Gunter Bertram	PS-84	Jitomir	22.42	22	E
1./NJG100	Lt August Kayser	PS-84	Russia	22.49	3	E
3./NJG100	Lt Eberhard Kaross	Il-2	Russia	23.42	9	E
1./NJG100	Lt G E Francsi	Il-4	Russia	23.48	32	E
3./NJG100	Lt Eberhard Kaross	Il-2	Russia	00.52	10	E
8./NJG4	Ofw Reinhard Kollak	Four-eng	Bar-le-Duc	01.54	44	W
2./NJG5	Uffz Lichert	Four-eng	N Pancey	02.00	6	W
8./NJG4	Ofw Reinhard Kollak	Four-eng	E St.Dizier	02.05	45	W
1./NJG5	Oblt E-G Drünkler	Lancaster	W Chaumont	02.08	19	W
St.I./NJG5	Maj Werner Hoffmann	Lancaster	5km E Thonnance	02.10	39	W
1./NJG5	Oblt E-G Drünkler	Lancaster	2km SE Bar sur Seine	02.17	20	W
St.I./NJG5	Oblt Breitfeld	Four-eng	N Chaumont	02.24	2	W

15-16 July 1944

3./NJG100	Uffz Hausschild	Boston	Russia	23.23	7	E
1./NJG100	Fw Heinz De Fries	Il-2	N Brest Litovk	23.28	5	E
1./NJG100	Fw Heinz De Fries	Il-4	N Shask	23.28	4	E
1./NJG100	Fw Rudi Duding	Il-4	Brest Litovsk	23.53	10	E
Stab./NJG5	Maj Walter Borchers	Four-eng	Châlons-sur-Marne	01.35	18	W
2./NJG4	Hptm Josef Krahforst	Four-eng	Nr. Chateau Thierry	01.43	8	W
8./NJG3	Lt Dräger	Four-eng	Off Danish coast	02.10	1	W

18-19 July 1944

8./NJG5	Oblt Werner Hopf	Halifax	Amiens	00.32	10	W
8./NJG5	Oblt Werner Hopf	Halifax	Amiens	00.34	11	W
3./NJG3	Uffz G Schneider	Four-eng	Off Dieppe	00.46	12	W
8./NJG5	Ofw Herbert Altner	Lancaster	Ermenonville	01.01	16	W
9./NJG4	Hptm Kurt Fladrich	Four-eng	Ermenonville	01.04	14	W
1./NJG4	Ofw Rücker	Four-eng	Valenciennes	01.05	1	W
3./NJG4	Lt Grunder	Four-eng	Chateau - Thierry	01.06	1	W
2./NJG4	Hptm Josef Krahforst	Four-eng	Valenciennes	01.06	9	W
II./NJG2	Maj Paul Semrau	Four-eng	E Paris	01.10	32	W

8./NJG5	Ofw Herbert Altner	Lancaster	Coulommirers	01.11	17	W
5./NJG2	Hptm H-H Hissbach	Four-eng	Jouarre	01.12	22	W
Stab./NJG2	Ofw Willi Glitz	Four-eng	NE Paris	01.13	8	W
8./NJG4	Fw Gunter Zimmermann	Four-eng	Chateau - Thierry	01.17	1	W
8./NJG5	Ofw Herbert Altner	Lancaster	Romilly	01.21	18	W
8./NJG5	Ofw Herbert Altner	Lancaster	Saulieu	01.25	19	W
8./NJG4	Ofw Reinhard Kollak	Four-eng	Romilly-sur-Seine	01.30	46	W
8./NJG4	Fw Gunter Zimmermann	Four-eng	Mailly-le-Camp	01.33	2	W
8./NJG5	Ofw Herbert Altner	Lancaster	W Dijon	01.34	20	W
Stab./NJG3	Obstlt Helmut Lent	Four-eng	SE Reims	01.36	99	W
1./NJG5	Oblt E-G Drünkler	Lancaster	5km NW Sommevoire	01.37	21	W
1./NJG5	Uffz E-A Schulze-Bergcamen	Lancaster	Semur-en-Auxoise	01.37	1	W
Stab./NJG3	Obstlt Helmut Lent	Four-eng	50km SE Reims	01.41	100	W
Stab./NJG2	Ofw Willi Glitz	Four-eng	Mailly-le-Camp	01.43	9	W
1./NJG5	Oblt E-G Drünkler	Lancaster	2km NW Conde en Barre	01.45	22	W
10./NJG3	Fw Richard Richter	Four-eng	Venlo	01.46	1	W
8./NJG4	Ofw Reinhard Kollak	Four-eng	area S Chalons	01.49	47	W
4./NJG4	Lt Frithjof Fensch	Four-eng	Pargny-sur-Saulx	01.51	9	W
8./NJG4	Ofw Reinhard Kollak	Four-eng	Bergères-lé-Vertu	01.55	48	W
St.II./NJG4	Hptm Hubert Rauh	Four-eng	Vitry-le-Francois	01.55	19	W
3./NJG1	Hptm Heinz Strüning	Mosquito	50km W Berlin	01.55	57*	w
1./NJG10	Uffz Wittmann	Mosquito	Gardelegen-Berlin	01.58	1	W
5./NJG2	Hptm H-H Hissbach	Four-eng	Epernay	01.58	23	W
8./NJG4	Ofw Reinhard Kollak	Four-eng	St.Dizier	02.05	49	W
St.I./NJG2	Hptm Gerhard Raht	Lancaster	Coulommiers	02.07	39	W
St.II./NJG2	Hptm Berger	Four-eng	Bergères-lé-Vertu	02.10	1	W
St.I./NJG2	Hptm Gerhard Raht	Lancaster	Gien	02.17	40	W
10./NJG3	Lt Walter Briegleb	Lancaster	NW Amsterdam	02.20	12	W
2./NJG5	Uffz Lichert	Four-eng	St.Dizier	03.27	7	W
2./NJG5	Uffz Lichert	Four-eng	St.Dizier	03.27	8	W

20-21 July 1944

4./NJG100	Oblt Gunter Bertram	Boston	Lwow	22.55	23	E
3./NJG3	Hptm Eduard Schröder	Four-eng	Brugge	01.08	14	W
4./NJG3	Ogfr Gerhard Wartenberger	Four-eng	SW Feindhoren	01.09	1	W
St.III./NJG1	Hptm Martin Drewes	Lancaster	Apeldoorn	01.11	45	W
4./NJG3	Uffz Bruno Rupp	Lancaster	W Mönchengladbach	01.14	8	W
6./NJG1	Uffz Gustav Sarzio	Halifax	12km W Roermond	01.14	3	W
10./NJG3	Lt Walter Briegleb	Halifax	5km W Venray	01.15	13	W
12./NJG3	Fw Klaus Möller	Four-eng	N Krefeld	01.16	6	W
8./NJG1	Oblt Dietrich Schmidt	Lancaster	50km W Duisberg	01.16	30	W
Sch./NJG1	Hptm Fritz Buchholtz.von	Four-eng	LM-LN	01.17	2	W
8./NJG2	Fw Allmeier	Four-eng	30km NW Duisberg	01.20	1	W
St.III./NJG1	Hptm Martin Drewes	Lancaster	W Oldenzaal	01.20	46	W
2./NJG4	Hptm Josef Krahforst	Four-eng	Nr. Dunkirk	01.24	10	W
11./NJG1	Hptm Hermann Greiner	Lancaster	LN-KN	01.28	35	W
12./NJG1	Ofw Karl-Heinz Scherfling	E/A		01.30	33*	W
8./NJG1	Lt Rolf Ebhardt	Halifax	50km W Münster	01.31	1	W
10./NJG3	Lt Walter Briegleb	Lancaster	20km NE Tilburg	01.33	14	W
St.I./NJG3	Maj Werner Husemann	Four-eng	NW Venlo	01.34	23	W
11./NJG1	Hptm Hermann Greiner	Lancaster	KL	01.37	36	W

Another Experte fell on the night of 20th July. Oberfeldwebel Karl-Heinz Scherfling of 12./NJG 1 had, in April 1941, destroyed the first four-engined bomber to fall to a nightfighter crew. By July 1944 his score stood at 33 and he had been awarded the Ritterkreuz. On his last mission he was caught near Mol in Belgium by a British intruder and shot down. Only his Bordfunker survived. His opponent was probably Pilot Officer Vandenburghe of 141 Squadron, who in fact claimed a 'Bf110 damaged'. The only other claims - by Wing Commander N.B.R.Bromley and Flight Lieutenant Fifield both of 169 Squadron - were far distant from Scherfling's crash.

St.IV./NJG1	Hptm H-W Schnaufer	Lancaster	1.5km North of Boxtel	01.40	85	W
8./NJG1	Oblt Dietrich Schmidt	Lancaster	30km NW Volkel	01.41	31	W
IV./NJG3	Lt Bruno Heilig	Four-eng	KK-KL	01.43	6	W
8./NJG2	Lt Josef Förster	Four-eng	between Tilburg & Vucht	01.45	3	W
12./NJG3	Ofw Rudolf Mangelsdorf	Four-eng	W Breda	01.45	5	W
7./NJG2	Fw Heinz Misch	Halifax	between Tilburg & Vucht	01.45	5	W
6./NJG1	Oblt Gottfried Hanneck	Lancaster	2km N Dordrecht	01.48	5	W
3./NJG10	Oblt Berger	Four-eng	N Rotterdam	01.50	1	W
8./NJG2	Uffz Gross	Four-eng	N Rotterdam	01.50	1	W
3./NJG10	Hptm H-M Hadeball	Lancaster	Schelde estuary	01.51	32	W
St.IV./NJG1	Hptm H-W Schnaufer	Lancaster	8km N Boxtel	01.51	86	W
1./NJG1	Hptm E-W Modrow	Lancaster	40 km SW beacon "Biber"	01.57	27	W
8./NJG2	Uffz Krieger	Four-eng	N Einhoven	02.03	1	W
9./NJG5	Lt Gerhard Wagner	Four-eng	50km NW Lille	02.05	15	W
Stab./NJG3	Obstlt Helmut Lent	Lancaster	Sea,Nr. Deal	02.07	101	W
8./NJG2	Lt Josef Forster	Four-eng	SW Rotterdam	02.09	4	W
1./NJG4	Fw Konrad Beyer	Four-eng	Courtrai	02.10	10	W
12./NJG1	Uffz Erwin Egeler	Halifax	50km NW Hook/Holland	02.16	1	W
Stab./NJG3	Obstlt Helmut Lent	Lancaster	Sea, nr. Dover	02.16	102	W
6./NJG1	Uffz Gustav Sarzio	Lancaster	20km W Katwijk-aan-Zee	02.16	4	W

21-22 July 1944

St.II./NJG6	Lt Peter Spoden	Wellington	NW Neunkirchen	23.23	10	W
7./NJG6	Oblt Josef Kraft	Wellington	W Brod	23.40	32	W
St.III./NJG6	Hptm Leopold Fellerer	Boston	W Brod	23.42	37	W
7./NJG6	Oblt Helmuth Schulte	Wellington	Pardubitz	00.00	17	W
6./NJG101	Oblt Hans Krause	Wellington	Grussbach	00.06	14	W

23-24 July 1944

2./NJG4	Ogfr Horst Pietzschke	Four-eng	off Fécamp	01.39	1	W
9./NJG3	Lt Heinz-Rudolf Molowitz	Four-eng	W Westerland	01.45	1	W
9./NJG3	Lt Heinz-Rudolf Molowitz	Four-eng	W Sylt	02.00	2	W
3./NJG1	Fw Wilhelm Morlock	Lancaster	SP-2	02.10	9	W

24-25 July 1944

Stab./NJG2	Ofw Willi Glitz	Lancaster	BC-CC	23.56	10	W
II./NJG2	Maj Paul Semrau	Four-eng	NW Orléans	00.01	33	W

II./NJG2	Maj Paul Semrau	Four-eng	Orléans	00.11	34	W
6./NJG2	Lt Erich Jung	Four-eng	SE Orléans	00.12	11	W
St.II./NJG4	Hptm Hubert Rauh	Lancaster	NW Orléans	00.17	20	W
3./NJG102	Lt Kehrberg	DB-3	nr Tilsit	00.32	2	E
2./NJG2	Ogfr Johannes Strassner	Four-eng	S Strasbourg	00.50	4	W
Stab./NJG3	Obstlt Helmut Lent	Four-eng	Pforzheim	01.14	103	W
3./NJG7	Lt Vollhardt	Four-eng	20km SW Heschbrunn	01.49	1	W
1./NJG7	Fw Reinhold Schwemmer	Four-eng	N Stuttgart	01.50	1	W
3./NJG6	Uffz Benno Kratz	Lancaster	Mössingen	01.56	5	W
3./NJG7	Ofw "Gunter" Scheidel	Four-eng	Beacon "Christa"	02.11	1	W
7./NJG2	Fw Heinz Misch	Lancaster	100-200km NW Stuttgart	02.24	6	W
11./NJG5	Uffz Bühlmann	Four-eng	80km WNW Stuttgart	02.25	1	W
St.III./NJG5	Hptm Paul Zorner	Lancaster	W Strasbourg	02.25	56	W
St.I./NJG5	Oblt Breitfeld	Four-eng	Nancy	02.31	3	W
St.III./NJG5	Hptm Paul Zorner	Lancaster	NW Strasbourg	02.35	57	W
St.III./NJG5	Hptm Paul Zorner	Lancaster	30km NW St.Dizier	02.54	58	W
1./NJG5	Uffz Helmut Burghardt	Lancaster	Lisieux	04.01	2	W

25-26 July 1944

3./NJG100	Oblt Josef Pützkuhl	Il-4	10km E Warka	22.27	26	E
1./NJG100	Unknown	Il-4	Russia	22.28	-	E
1./NJG100	Lt August Kayser	Il-4	Russia	22.50	4	E
5./NJG4	Oblt Helmuth Schulte	Four-eng	Chateauneuf	00.07	18	W
1./NJG2	Oblt Adolf Kaiser	Four-eng	15km W Gien	00.38	13	W
5./NJG4	Fw Burke	Four-eng	Morvan	00.40	1	W
5./NJG4	Oblt Helmuth Schulte	Four-eng	St.Sauveur	00.41	19	W
1./NJG6	Lt Rose	Four-eng	N Frankfurt-a-M	01.16	1	W
St.IV./NJG1	Hptm Adolf Breves	Four-eng		01.58	11	W
2./NJG6	Oblt Martin Becker	Lancaster	W Stuttgart	02.10	34	W
3./NJG4	Lt Erhard Duttmann	Four-eng	W Paris	02.48	1	W
2./NJG2	Oblt Heinz Rökker	Four-eng	NE Romilly	03.21	34	W
2./NJG2	Oblt Heinz Rökker	Lancaster	NW Châteaudun	04.38	35	W

26-27 July 1944

1./NJG100	Fw Heinz De Fries	Il-4	Pultuske	23.15	6	E
1./NJG102	Oblt "Rahdanz"	DB-3	Torunygen	00.07	1	E
NJG102	Oblt Rossbach	DB-3	Russia	00.07	1	E
1./NJG102	Oblt "Rahdanz"	DB-3	Torunygen	00.10	2	E
1./NJG102	Uffz Hans Keller	DB-3	Russia	00.10	1	E
2./NJG2	Lt Otto Huchler	Lancaster	Haren - Lathen	00.15	1	W
3./NJG102	Lt Kehrberg	PS-84	E-Tislit	00.15	3	E
3./NJG102	Lt Kehrberg	DB-3	E Tilsit	00.20	4	E
3./NJG102	Lt Kehrberg	DB-3	NE Target	00.25	5	E
1./NJG102	Uffz Goldmann	PS-84	N Jurburg	00.26	1	E
2./NJG2	Lt Otto Huchler	Lancaster	Rodenwerder	00.45	2	W

27-28 July 1944

8./NJG6	Ofw Josef Federle	E/A		22.52	1	E
4./NJG100	Uffz Rohde	Boston	Russia	23.22	1	E
4./NJG100	Uffz Rohde	Boston	Russia	23.36	2	E
1./NJG102	Oblt Eggeling	DB-3	Russia	00.25	1	E

Oberleutnant Fritz Lau (right) flew first with III./NJG 1 in 1943, rising to command 9 Staffel, then transferred to lead 4./NJG 1 in 1944. On the night of 28th July he claimed two four-enginged bombers destroyed, these probably his 15th and 16th victories. Awarded the Ritterkreuz on 28th April 1945, he is believed to have survived the war with a score of 28 confirmed victories, including at least one Mosquito. On 28th July Leutnant Bruno Heilig (left) of 4./NJG 3 claimed his 7th and 8th victories. He had no further success before his death in 1945.

28-29 July 1944

2./NJG2	Ogfr Johannes Strassner	Halifax	4km E Chateaudun	00.01	5	W
2./NJG2	Oblt Heinz Rökker	Four-eng	Orléans	00.14	36	W
4./NJG4	Lt Frithjof Fensch	Four-eng	EJ	00.44	10	W
St.I./NJG2	Hptm Gerhard Raht	Lancaster	EG-FG	00.45	41	W
5./NJG2	Hptm H-H Hissbach	Four-eng	DF-DG	00.46	24	W
7./NJG2	Lt Heinz Reuter	Four-eng	between Toul & Nancy	01.00	8	W
8./NJG2	Lt Josef Förster	Four-eng	Nr. Bouxwiller	01.05	5	W
II./NJG4	Hptm Hubert Rauh	Four-eng	Chaumont	01.07	21	W
1./NJG5	Oblt E-G Drünkler	Lancaster	5km N Mirecourt	01.09	23	W
IV./NJG3	Lt Bruno Heilig	Four-eng	Niedermendig	01.09	7	W
8./NJG1	Oblt Dietrich Schmidt	Halifax	W Hamburg	01.11	32	W
IV./NJG3	Lt Bruno Heilig	Four-eng	Niedermendig	01.13	8	W
7./NJG2	Fw Hans Werthner	Four-eng		01.15	8	W
10./NJG3	Fw Richard Richter	Four-eng	20km NE Hamburg	01.16	2	W
2./NJG2	Oblt Heinz Rökker	Lancaster	50km NE Chaumont	01.17	37	W
5./NJG5	Ogfr Boy	Lancaster	Colmar-Breisach	01.18	1	W
12./NJG3	Fw Klaus Möller	Four-eng	40km W Hamburg	01.18	7	W
St.II./NJG3	Oblt Fritz Brandt	Four-eng	Nr. Epinal	01.20	3	W
5./NJG6	Ogfr Eberle (Bs Griese)	Mosquito	W Strasbourg	01.20	1	W
8./NJG5	Oblt Werner Hopf	Lancaster	W Strasbourg	01.22	12	W
8./NJG1	Lt Rolf Ebhardt	Four-eng	50km NW Hamburg	01.23	2	W
5./NJG2	Hptm H-H Hissbach	Four-eng	Chaumont	01.23	25	W
2./NJG6	Oblt Martin Becker	Lancaster	N Luneville	01.25	35	W
2./NJG5	Uffz Lichert	Four-eng	Rambervillers	01.25	9	W
2./NJG1	Lt Otto Fries	Four-eng	100km SSW beacon "Kuli"	01.26	12	W
8./NJG2	Uffz "Kurt" Boldt	Lancaster	E Strasbourg	01.27	1	W
4./NJG1	Lt Richard Heuke	Four-eng	W Colmar	01.27	2	W
2./NJG5	Uffz Metzer	Mosquito	Rambervillers	01.27	1	W
2./NJG3	Fw Herbert Stähler	Four-eng	NW Hamburg	01.27	4	W
4./NJG3	Uffz Bruno Rupp	Halifax	W Strassbourg	01.28	9	W
2./NJG1	Maj Schäfer	Four-eng	20km SSW beacon "Kuli"	01.28	2	W
8./NJG1	Lt Rolf Ebhardt	Four-eng	100km NW Hamburg	01.29	3	W
2./NJG6	Ogfr Walter Swoboda	Four-eng	St.Die area	01.30	1	W
2./NJG6	Oblt Martin Becker	Lancaster	NE Luneville	01.35	36	W

3./NJG7	Lt Roppe	Four-eng	La Chaux	01.35	1	W
8./NJG1	Lt Rolf Ebhardt	Four-eng	50km E Heligoland	01.37	4	W
6./NJG1	Oblt Gottfried Hanneck	Four-eng	SW Strasbourg	01.37	6	W
8./NJG1	Oblt Dietrich Schmidt	Halifax	N Bruchsal	01.37	33	W
3./NJG3	Hptm Eduard Schröder	Four-eng	Sea, off Heligoland	01.37	15	W
St.IV./NJG1	Hptm H-W Schnaufer	Lancaster	50km W Stuttgart	01.38	87	W
8./NJG2	Uffz Egon Engling	Four-eng	Vosges du Nord	01.39	4	W
St.III./NJG101	Oblt Ernst Schüren	Lancaster	Phalsbourg	01.40	3	W
8./NJG2	Uffz Günther Schmidt	Four-eng	AR-AS-BR-BS	01.43	3	W
2./NJG6	Oblt Martin Becker	Lancaster	W Stuttgart	01.45	37	W
St.III./NJG2	Hptm Berthold Ney	Four-eng	SW Kahlsruhe	01.45	8	W
5./NJG5	Fw Resl	Four-eng	S Strasbourg	01.45	1	W
4./NJG3	Ogfr Gerhard Wartenberger	Four-eng	50-100km SW Stuttgart	01.45	2	W
6./NJG1	Oblt Johannes Hager	Four-eng	20km W Stuttgart	01.45	32	W
8./NJG2	Lt Josef Förster	Four-eng	Hagenau - Maisch	01.46	6	W
2./NJG1	Lt Otto Fries	Four-eng	W Stuttgart	01.46	13	W
St.III./NJG101	Oblt Ernst Schüren	Lancaster	Phalsbourg	01.46	4	W
St.I./NJG3	Maj Werner Husemann	Four-eng	Bitsche - Pirmasens	01.47	24	W
3./NJG6	Uffz Benno Kratz	Lancaster	Sindelfingen	01.47	6	W
Stab./NJG3	Obstlt Helmut Lent	Four-eng	Nr. Bouxwiller	01.47	104	W
III./NJG1	Maj Höfele	Lancaster	2km W St Wendell	01.48	1	W
4./NJG1	Oblt Fritz Lau	Four-eng	W Stuttgart	01.48	12	W
8./NJG2	Hptm Piwarz	Four-eng	Nr. Stuttgart	01.50	3	W
8./NJG1	Oblt Dietrich Schmidt	Four-eng	10km NW Heligoland	01.50	34	W
4./NJG1	Lt Hans Schmitz	Four-eng	20km W Stuttgart	01.50	6	W
St.IV./NJG1	Hptm H-W Schnaufer	Lancaster	30km W Stuttgart	01.50	88	W
St.III./NJG101	Oblt Ernst Schüren	Lancaster	Phalsbourg	01.50	5	W
9./NJG3	Uffz Heinz Misch	Four-eng	10-20km NE Helgoland	01.51	7	W
12./NJG1	Oblt H-H Augenstein	Four-eng	15km W Stuttgart	01.52	42	W
2./NJG1	Maj Hans Karlowski	Four-eng	SW Stuttgart	01.53	11	W
Stab./NJG3	Ofw Martin Kramer	Four-eng	10-30km NW Helgoland	01.53	1	W
2./NJG4	Hptm Josef Krahforst	Four-eng	S Hagenau	01.54	11	W
4./NJG1	Oblt Karl-Frank Stephan	Four-eng	SW Stuttgart	01.54	1	W
12./NJG3	Uffz "Kurt" Trautmann	Four-eng	Nr. Husum	01.54	1	W
6./NJG2	Hptm Kamsties	Four-eng	30km W Stuttgart	01.55	3	W
./NJG3	Obstlt Helmut Lent	Four-eng	Nr. Chateau-Salins	01.57	105	W
St.IV./NJG1	Hptm H-W Schnaufer	Lancaster	20km W Stuttgart	01.57	89	W
Stab./NJG5	Maj Walter Borchers	Four-eng	NW Stuttgart	01.58	19	W
11./NJG1	Uffz Grosser	Four-eng	Nr. Metz	01.58	2	W
6./NJG1	Oblt Johannes Hager	Four-eng	Nr. Strasbourg	01.58	33	W
5./NJG2	Hptm H-H Hissbach	Four-eng	40km W Stuttgart	01.58	26	W
8./NJG2	Uffz Egon Engling	Lancaster	Pforzheim area	01.59	5	W
11./NJG1	Uffz Grosser	Four-eng	Nr. Metz	01.59	3	W
4./NJG1	Oblt Fritz Lau	Four-eng	20km W Stuttgart	01.59	13	W
Stab./NJG3	Ofw Martin Kramer	Four-eng	40-70km NW Heligoland	02.00	2	W
2./NJG6	Oblt Martin Becker	Lancaster	Weissenburg	02.04	38	W
III./NJG1	Maj Höfele	Lancaster	North Sea	02.06	2	W
6./NJG1	Fw Robert Koch	Four-eng	30km SW Stuttgart	02.06	1	W
1./NJG6	Fw Günther Bahr	Lancaster	30-50km from Stuttgart	02.08	18	W
12./NJG1	Oblt H-H Augenstein	Four-eng	Nr. Stuttgart	02.11	43	W

On the night of 6th August, Hauptmann Helmut Bergmann, Kapitän of 8./NJG 4, took his Bf110 out over the Normandy beachhead, seeking his 37th night victory. He and his crew disappeared and have never been found. They had fallen to a 604 Squadron Mosquito, flown by Flight Lieutenant John Surman, who also claimed a Do217 this night.

2./NJG4	Uffz Altmann	Four-eng	SE Stuttgart	02.14	5	W
1./NJG101	Fw Scheler	Lancaster	Heidenheim	02.15	1	W
10./NJG3	Lt Walter Briegleb	Halifax	Normeny	02.16	15	W

29-30 July 1944

2./NJG2	Oblt Heinz Rökker	Halifax	Orleans	00.14	38	W

1-2 August 1944

8./NJG6	Ofw Josef Federle	Wellington	S Szorgrád	02.46	2	W

4-5 August 1944

3./NJG100	Hptm August Fischer	Halifax		00.04	9	W
1./NJG5	Oblt E-G Drünkler	Four-eng	BK1	01.04	24	W
3./NJG100	Fw Helmut Konter	Lancaster	80km ESE Krakow	01.19	14	W
3./NJG100	Fw Helmut Konter	Lancaster	65km E Krakow	01.46	15	W

5-6 August 1944

2./NJG5	Uffz Lichert	Halifax	DL-71	02.10	10	W

6-7 August 1944

Stab./NJG2	Ofw Willi Glitz	Mosquito	BJ-BI	23.40	11	W
6./NJG2	Oblt Erich Jung	P-38		02.31	12	W

7-8 August 1944

2./NJG2	Oblt Heinz Rökker	Lancaster	NE Le Havre	23.29	39	W
2./NJG2	Oblt Heinz Rökker	Lancaster	NE Le Havre	23.35	40	W
4./NJG4	Uffz Rolf Koch	Four-eng	Nr. Yvetot	23.47	1	W
II./NJG4	Hptm Hubert Rauh	Four-eng	Le Havre - Lisieux	23.50	22	W
4./NJG4	Lt Frithjof Fensch	Four-eng	TA-9	23.52	11	W
4./NJG4	Lt Frithjof Fensch	Four-eng	TA-3	24.00	12	W
2./NJG2	Oblt Heinz Rökker	Lancaster	NE Le Havre	00.06	41	W
2./NJG2	Ofw Johannes Strassner	Lancaster	NE Le Havre	00.06	6	W
2./NJG2	Ofhr. Johannes Naskreut	Lancaster	RA	00.12	3	W
2./NJG2	Ofhr. Johannes Naskreut	Lancaster	RA	00.14	4	W
9./NJG6	Lt Strohbach	B-24	10km S Szekesfehervar	01.04	1	W
9./NJG6	Oblt Rudolf Thun	Wellington	15km W Szekesfehervar	01.05	4	W

7./NJG6	Ofw Heinz Säwert	B-24	Szombathely	01.10	16	W
7./NJG6	Oblt Josef Kraft	Wellington	Kormend	01.27	33	W

8-9 August 1944

3./NJG4	Oblt Kadleik	Lancaster	PK-7	00.36	5	W
3./NJG4	Lt Erhard Duttmann	Four-eng	JG-3	02.29	2	W
1./NJG5	Oblt E-G Drünkler	Lancaster	GJ	03.10	25	W

9-10 August 1944

St.IV./NJG6	Maj Herbert Lütje	Wellington	Rumania	23.00	39*	W
11./NJG6	Hptm Bauer	Halifax	W Bucharest		1*	W
11./NJG6	Hptm Bauer	B-24	W Bucharest		2*	W
8./NJG6	Fw Eckart Ulmer	B-25	Russia	00.03	1	E
2./NJG100	Ofw "Karl" Maisch	Lancaster	NW Lom	00.20	7	W
8./NJG6	Fw Eckart Ulmer	B-25		00.46	2	E

10-11 August 1944

6./NJG2	Fw Oberheide	Four-eng	HH-JJ	00.49	1	W
8./NJG6	Ofw Josef Federle	B-25		01.02	3	E
8./NJG6	Oblt Wilhelm Johnen	B-25		02.08	23	E

12-13 August 1944

2./NJG4	Hptm Kranz	Four-eng	RN	23.46	1	W
11./NJG1	Lt Hans Gienanth.von	Halifax	Le Culot	23.48	4	W
12./NJG1	Uffz Werner Huhn	Four-eng	Le Culot	23.50	1	W
12./NJG1	Fw Erwin Egeler	Four-eng	Near Morbach	23.58	2	W
8./NJG2	Lt Josef Förster	Four-eng	FT-GT	24.00	7	W
9./NJG3	Lt Hans Raum	Four-eng	N Bremerhaven	00.10	10	W
3./NJG1	Oblt Josef Nabrich	Four-eng	40km W Mainz	00.14	19	W
4./NJG3	Oblt Hermann Leube	Four-eng	E Bremen	00.18	16	W
6./NJG1	Uffz Gustav Sarzio	Halifax	33km N Kaiserslautern	00.21	5	W
8./NJG2	Lt Horst-Rudiger Blüme	Four-eng	GA-FA	00.25	2	W
8./NJG3	Oblt Fritz Brinkmann	Four-eng	NW Bramsche	00.26	6	W
1./NJG1	Uffz Franz Frankenhauser	Four-eng	NW Hanover	00.26	1	W
11./NJG1	Hptm Hermann Greiner	Lancaster	W Frankfurt-am-Main	00.26	37	W
St.II./NJG6	Lt Peter Spoden	Lancaster	40km SW Darmstadt	00.26	11	W
3./NJG3	Uffz G Schneider	Four-eng	GA	00.27	13	W
8./NJG1	Uffz Pahler	Halifax	80km NW Braunschweig	00.29	5	W
St.IV./NJG1	Hptm Adolf Breves	Four-eng	Rieneth	00.30	12	W
3./NJG7	Lt Brünner	Four-eng	EU-EH	00.30	1	W
9./NJG3	Fw Walter Rohlfing	Four-eng	20km N Celle	00.30	10	W
11./NJG1	Lt Otto Teschner	Four-eng	RQ	00.30	4	W
12./NJG1	Lt Kurt Matzak	Halifax	SE Russelsheim	00.31	8	W
8./NJG5	Oblt Werner Hopf	Halifax	100km W Darmstadt	00.40	13	W
1./NJG3	Oblt Herbert Koch	Four-eng	Vechta	00.40	13	W
4./NJG1	Fw Peter Müller	Four-eng	Varrelbusch	00.43	2	W
9./NJG2	Lt Joachim Fiedler	Four-eng	ES	00.46	1	W
St.I./NJG3	Maj Werner Husemann	Four-eng	Sulingen - Wagerfeld	00.47	25	W
8./NJG1	Oblt Dietrich Schmidt	Lancaster	50km S Braunschweig	00.47	35	W
8./NJG2	Lt Josef Förster	Four-eng	FQ-FR	00.48	8	W
St.IV./NJG1	Hptm H-W Schnaufer	Lancaster	Wasserliesch	00.48	90	W

Dietrich Schmidt joined the nightfighter force in late 1941, going to 8./NJG 1. He was a relatively 'slow starter', his first success coming on the night of 24th March 1943. Three months later he became Staffelkäpitan and, within a year, had brought his score to thirty. On the night of 12 August he claimed his 35th, a Lancaster, and would add five more victories to his tally before the armistice. He was awarded the Ritterkreuz on 27th July 1944.

8./NJG5	Oblt Werner Hopf	Lancaster	120km W Darmstadt	00.49	14	W
6./NJG1	Lt Barthel	Lancaster	3km S Furstenau	00.52	1	W
10./NJG3	Fw Richard Richter	Four-eng	Remagen area	00.53	3	W
6./NJG1	Uffz Koppelmayer	Four-eng	50km S Oldenburg	00.55	2	W
10./NJG3	Fw Richard Richter	Four-eng	Nr. Bremen	01.00	4	W
8./NJG2	Uffz "Kurt" Boldt	Four-eng	CP-CQ	01.02	2	W
4./NJG3	Oblt Hermann Leube	Four-eng	Nr. Cuxhaven	01.03	17	W
6./NJG1	Fw Robert Koch	Four-eng	N Enschelde	01.05	2	W
St.IV./NJG1	Hptm H-W Schnaufer	Lancaster	Gouvy	01.05	91	W
St.IV./NJG1	Hptm H-W Schnaufer	Lancaster	3km W Mons	01.09	92	W
St.IV./NJG1	Hptm H-W Schnaufer	Lancaster	Werbomomt	01.09	93	W
9./NJG3	Hptm Banschein	Four-eng	NW Vechta	01.15	3	W
8./NJG3	Oblt Fritz Brinkmann	Four-eng	Nr. Senden	01.15	7	W
7./NJG2	Fw Heinz Misch	Halifax	40km N Norderney	01.15	8	W
11./NJG1	Lt Otto Teschner	Four-eng	10km S St.Trond	01.15	5	W
3./NJG4	Uffz Friedrich Grunder	Four-eng	Ath	01.33	2	W
8./NJG2	Uffz Gunther Schmidt	Four-eng	BR-BS	01.45	4	W
2./NJG4	Uffz Wolfrum Möckel	Four-eng	Havelange	01.50	5	W
1./NJG3	Oblt Herbert Koch	B-17	Osnabruck	01.59	14*	W

14-15 August 1944

1./NJG100	Lt G E Francsi	Lancaster	Warsaw	00.28	33	W
1./NJG100	Lt G E Francsi	Lancaster	Warsaw	01.38	34	W
1./NJG100	Lt G E Francsi	Lancaster	Warsaw	01.54	35	W

15-16 August 1944

8./NJG6	Oblt Wilhelm Johnen	B-25		22.59	24	E
7./JG300	Fw Keller	Mosquito	Nr. Berlin	23.59	2	W
8./JG300	Lt Emil Kriegbaum	Mosquito	SW Berlin	00.00	2	W

16-17 August 1944

1./NJG3	Oblt Herbert Koch	Lancaster	KA-QU	23.52	15*	W
1./NJG100	Lt G E Francsi	Halifax	Opochna	23.54	36	W
3./NJG3	Ofw Giesecke	Four-eng	North Sea	00.04	4	W
8./NJG3	Oblt Fritz Brinkmann	Four-eng	NNE Kiel	00.06	8	W
Stab./NJG3	Ofw Martin Kramer	Four-eng	SE Kiel	00.15	3	W

4./NJG1	Oblt Fritz Lau	Four-eng	W Kiel	00.28	14	W
12./NJG3	Oblt Langnickel	Four-eng	North Sea	00.37	1	W
1./NJG3	Oblt Herbert Koch	Lancaster	St. Peter-Ördìng	00.38	16	W
St.I./NJG3	Maj Werner Husemann	Four-eng		00.40	26	W
1./NJG100	Lt G E Francsi	Halifax	Bochnia	01.18	37	W
1./NJG100	Lt G E Francsi	Lancaster	Tarnów	01.50	38	W
12./NJG3	Fw Klaus Moller	Four-eng	Zeeland	01.55	8	W
1./NJG100	Ofw Helmut Dahms	Lancaster	Maribor	01.57	12	W
12./NJG3	Fw Klaus Moller	Four-eng	North Sea	02.15	9	W
1./NJG100	Lt G E Francsi	Lancaster	Przymysl	02.18	39	W
1./NJG100	Ofw Helmut Dahms	Lancaster	Krakow	02.27	13	W

18-19 August 1944

7./NJG1	Oblt Hanspeter Schickling	Lancaster	GJ-3	01.34	1	W
1./NJG1	Hptm E-W Modrow	Lancaster	Zoutkamp	02.09	28	W

20-21 August 1944

St.IV./NJG6	Maj Herbert Lutje	Wellington		23.05	40	W
7./NJG6	Oblt Josef Kraft	Wellington	E Steyr	23.10	34	W
2./NJG6	Ofw Ludwig Schmidt	B-24	Mittrerkirchen	23.16	2	W
7./NJG6	Oblt Josef Kraft	Wellington		23.18	35	W
7./NJG6	Ofw Heinz Säwert	Wellington	E Steyr	23.20	17	W
St.III./NJG6	Hptm Leopold Fellerer	B-24	N Laibach	23.50	38	W
6./NJG101	Oblt Hans Krause	Wellington		23.59	15	W

21-22 August 1944

8./NJG6	Oblt Wilhelm Johnen	B-25		22.42	25	E
II ./NJG101	Lt Hans-Joachim Beck	Lancaster		23.07	3	W
7./NJG6	Oblt Josef Kraft	Wellington	Komarom	23.14	36	W
8./NJG6	Fw Eckart Ulmer	Halifax		23.35	3	W
St.IV./NJG6	Maj Herbert Lutje	Il-4		01.05	41	E

22-23 August 1944

8./NJG6	Oblt Wilhelm Johnen	Halifax	13km S Szombor	22.23	26	W
7./NJG6	Oblt Josef Kraft	Wellington	Gyongos	23.14	37	W
7./NJG6	Oblt Josef Kraft	B-24	Tizsafured	23.23	38	W

23-24 August 1944

1./NJG10	Hptm Friedrich-Karl Muller	Mosquito	Eindhoven	23.25	24	W
4./NJG100	Lt Klaus Scheer	R-5		02.03	16	E
4./NJG100	Oblt Gunter Bertram	Il-4		02.33	24	E
4./NJG100	Oblt Gunter Bertram	Boston		02.57	25	E

24-25 August 1944

4./NJG100	Lt Richter	U-2		22.52	7	E
St.I./NJG102	Ofw Alfred Rahner	B-25		23.08	1	E
4./NJG100	Lt Klaus Scheer	U-2		01.35	17	E

25-26 August 1944

3./JG301	Hptm Helmut Suhr	P-61	France	02.22	5*	W

Flying and fighting at night was never easy. Fähnenjunker-Feldwebel Hermann-Walter Stähler of 2./NJG 3 had been successful in four air combats. On 23rd August 1944, he was taking off from Nordhorn when a searchlight blinded him. He crashed into a building from an altitude of 60 feet and was badly injured together with his crew.
(S.Hall)

25-26 August 1944

10./JG300	Lt Kurt Welter	Mosquito	Nr. Nienburg	00.38	18	W
8./NJG2	Uffz "Kurt" Boldt	Four-eng	Ludwigshafen	00.40	3	W
I./NJG101	Unknown	Four-eng	Karlstuhl	00.50	-	W
2./NJG6	Hptm Martin Becker	Lancaster	NW Russelheim	00.50	39	W
4./NJG6	Fw Helmut Bunje	Halifax	SW Russelheim	00.58	3	W
2./NJG101	Ofw Leuenz	Four-eng	Russelscheim	01.00	1	W
1./NJG6	Uffz Rechenberg	Lancaster	SE Darmstadt	01.00	1	W
8./NJG2	Uffz Egon Engling	Lancaster	S Darmstadt	01.02	6	W
2./NJG2	Lt Johannes Strassner	Lancaster	S Mainz	01.05	7	W
2./NJG6	Hptm Martin Becker	Lancaster	WSW Russelheim	01.08	40	W
9./NJG3	Uffz Rudnik	Four-eng	E Bingen	01.08	2	W
7./NJG3	Lt Achim Woeste	Four-eng	60km NW Mannheim	01.08	10	W
St.II./NJG6	Lt Peter Spoden	Lancaster	35km NW Russelheim	01.12	12	W
1./NJG6	Fw Günther Bahr	Halifax	W Russelheim	01.15	19	W
2./NJG6	Hptm Martin Becker	Halifax	W Russelheim	01.15	41	W
St.II./NJG6	Oblt Helmuth Schulte	Lancaster	70km W Russelheim	01.15	20	W
7./NJG2	Ofw Walter Exner	Four-eng	Wiesbaden	01.20	3	W
8./NJG2	Lt Josef Förster	Four-eng	Near Wittlich	01.24	9	W
5./NJG2	Hptm H-H Hissbach	Four-eng	Darmstadt	01.24	27	W
6./NJG6	Hptm Ernst Wallner	Lancaster	Russelheim	01.27	3	W
3./NJG10	Maj Harold Lange	Four-eng	Darmstadt	01.31	3	W
8./NJG2	Uffz Egon Engling	Lancaster	30km E Darmstadt	01.32	7	W
7./NJG2	Uffz Willmann	Four-eng	Darmstadt	01.34	1	W
Stab./NJG2	Hptm Gerhard Raht	Lancaster		01.38	42	W
6./NJG6	Hptm Ernst Wallner	Halifax	Sulzbach	01.38	4	W
8./NJG2	Lt Josef Förster	Four-eng	Near Dinant	01.40	10	W
4./NJG3	Oblt Hermann Leube	Four-eng	ESE Kaiserslautern	01.53	18	W

26-27 August 1944

4./NJG100	Lt Klaus Scheer	DB-3		21.11	18	E
4./NJG100	Lt Klaus Scheer	Boston		21.26	19	E
10./NJG5	Oblt Rolf Bussmann	Il-4	10km NE Skirsmenen	21.32	23	E
1./NJG102	Unknown	B-25		21.33	-	E
St.IV./NJG5	Hptm Rudolf Altendorf	Il-4		21.42	20	E

The FW190A series, which had entered service during the summer of 1941, had become the 'workhorse' of the Luftwaffe day fighter units. From mid-1943 it had also proved its worth with the Wilde Sau units.

4./NJG100	Lt Klaus Scheer	DB-3		21.52	20	E
8./NJG6	Oblt Wilhelm Johnen	Halifax	SE Baja	22.00	27	W
1./NJG102	Fw Fritz Kohnlein	E/A	near Tilsit	22.00	1	E
1./NJG102	Lt Oberle	DB-3	Near Tilsit	22.01	1	E
8./NJG6	Oblt Wilhelm Johnen	B-25		22.30	28	E
8./NJG6	Ofw Josef Federle	Halifax	S Budapest	22.34	4	W
10./NJG3	Lt Walter Briegleb	Lancaster	30km SW Kiel	23.03	16	W
3./NJG3	Lt Werner Götze	Four-eng	N Schleswig	23.05	2	W
2./NJG3	Uffz Weiner	Four-eng	WSW Schleswig	23.06	3	W
12./NJG3	Ofw Rudolf Mangelsdorf	Four-eng	W Kiel	23.07	6	W
1./NJG3	Fw G Schneider	Four-eng	N Schleswig	23.12	14	W
1./NJG3	Fw G Schneider	Four-eng	N Schleswig	23.17	15	W
12./NJG3	Uffz Koch	Four-eng	50km NNW Kiel	23.21	1	W
3./NJG3	Uffz Hans Schadowski	Four-eng	N Schleswig	23.24	7	W
3./NJG3	Uffz Hans Schadowski	Four-eng	N Schleswig	23.29	8	W
1./NJG3	Fw G Schneider	Four-eng	Bredstedt	23.35	16	W
12./NJG5	Fw Freischer	Four-eng	Samso Island	01.11	1	W
8./NJG3	Oblt Fritz Brinkmann	Lancaster	Samso Island	02.59	9	W
8./NJG3	Oblt Fritz Brinkmann	Lancaster	Samso Island	03.09	10	W
III./NJG1	Maj Höfele	Lancaster	W Hörnum	03.15	3	W
4./NJG3	Uffz Bruno Rupp	Lancaster	Samso Island	03.20	10	W
4./NJG3	Uffz Gerhard Wartenberger	Four-eng	ESE beacon "Herzog"	03.27	3	W
8./NJG3	Oblt Fritz Brinkmann	Lancaster	10km NW Hörnum	03.38	11	W

27-28 August 1944

8./NJG6	Oblt Wilhelm Johnen	Halifax	Friedrichskoog	21.50	29	W
1./NJG100	Ofw Rudi Duding	Halifax		02.33	11	W

28-29 August 1944

8./NJG6	Oblt Wilhelm Johnen	Halifax	N Mohacs	22.12	30	W

One of the lesser-known units was NJG 7, whose pilots were not among the most successful of the Nachtjagd. Oberfeldwebel Schmitz of 3 Staffel (right) was credited with shooting down two four-engined bombers on the night of 29th August 1944 but, as far as is known, these were his only recorded successes. (S.Hall) Oberleutnant Rudolf Szardenings (left) flanked by his 5./NJG 3 crew claimed his ninth victory on 30th August. He would gain one more

29-30 August 1944

6./NJG3	Hptm Decker	Four-eng	SW Göttingen	23.30	1	W
4./NJG3	Uffz Bruno Rupp	Lancaster	Nr. Tyboron-Den	00.01	11	W
3./NJG7	Ofw Schmitz	Four-eng	LT	00.05	1	W
2./NJG7	Fw Heller	Four-eng	LG	00.09	1	W
1./NJG7	Lt Appel	Four-eng	LT-JT	00.10	1	W
3./NJG7	Ofw Schmitz	Four-eng	LT	00.10	2	W
St.II./NJG3	Oblt Fritz Brandt	Lancaster	LU	00.12	4	W
St.II./NJG3	Oblt Fritz Brandt	Four-eng	LA	00.22	5	W
St.I./NJG7	Lt Ludger Krimphone	Four-eng	NG-LG	00.25	1	W
3./NJG3	Ofw Gierscke	Four-eng	NC-ND-OD	00.31	5	W
3./NJG3	Ogfr Walter Schuster	Four-eng	LF-LG	00.45	1	W
1./NJG3	Fw G Schneider	Four-eng	KG	00.45	17	W
St.I./NJG3	Maj Werner Husemann	Lancaster	LH	00.48	27	W
1./NJG3	Fw G Schneider	Four-eng	LG	00.53	18	W
1./NJG102	Oblt Heinrich Barth	Lancaster	4km S Jesau	01.17	2	W
St.I./NJG3	Maj Werner Husemann	Lancaster	OK-PK	01.19	28	W
10./NJG5	Oblt Rolf Bussmann	Lancaster	W Kranz	01.21	24	W
9./NJG5	Lt Otto Keller	Lancaster	SA	01.23	11	W
11./NJG5	Uffz Bauer	Four-eng	Königsberg	01.25	1	W
1./NJG102	Lt Oberle	Lancaster	1km W Hellgenheim	01.31	2	W
1./NJG5	Oblt E-G Drünkler	Lancaster	18km NW Königsberg	01.32	26	W
12./NJG5	Fw Freischer	Four-eng	E Königsberg	01.33	2	W
St.I./NJG5	Maj Werner Hoffmann	Lancaster	25km NW Königsberg	01.37	40	W
9./NJG5	Oblt Breitfeld	Lancaster	E edge Königsberg	01.38	4	W
St.IV./NJG5	Hptm Rudolf Altendorf	Lancaster	Königsberg	01.39	21	W
9./NJG5	Oblt Breitfeld	Lancaster	W Königsberg	01.42	5	W
2./NJG5	Oblt Peter Ehrhardt	Lancaster	S Königsberg	01.42	21	W
1./NJG102	Fw Schafer	Lancaster	RU-9	01.44	1	W
9./NJG5	Oblt Breitfeld	Lancaster	W edge Königsberg	01.48	6	W
1./NJG102	Fw Schäfer	Lancaster	Gross Haferbeck	01.48	2	W
11./NJG5	Uffz Georg Bargel	Four-eng	WNW Königsberg	01.50	3	W
11./NJG5	Hptm Gustav Tham	Four-eng	W Königsberg	01.50	11	W
1./NJG102	Lt Oberle	Lancaster	Langeland	01.53	3	W

1./NJG5	Hptm Alex Ressugier	Lancaster	Königsberg	01.53	3	W
E./NJG2	Oblt Scholl	Lancaster	SW Pielenwisken	01.53	1	W
E./NJG2	Oblt Scholl	Lancaster	E Stettin	01.53	2	W
12./NJG5	Fw Otto Hiller	Lancaster	BB1	01.54	1	W
1./NJG102	Oblt Heinrich Barth	Lancaster		01.55	3	W
1./NJG5	Oblt E-G Drünkler	Lancaster	30km NW Pillau	01.56	27	W
12./NJG5	Fw Otto Hiller	Lancaster	BB-1	01.58	2	W
St.IV./NJG5	Hptm Rudolf Altendorf	Lancaster	BJ-1	01.59	22	W
10./JG300	Lt Kurt Welter	Lancaster	CJ-BJ	02.05	19	W
Stab./NJG3	Ofw Martin Kramer	Four-eng	FJ-FK	02.10	4	W
10./JG300	Lt Kurt Welter	Lancaster	BJ-CJ	02.10	20	W
10./JG300	Lt Kurt Welter	Lancaster	BJ-CJ	02.12	21	W
10./JG300	Lt Kurt Welter	Lancaster	BJ-CJ	02.15	22	W
4./NJG3	Oblt Hermann Leube	Four-eng	NA	03.30	19	W
1./NJG1	Hptm E-W Modrow	Four-eng	LT	03.51	29	W
5./NJG3	Oblt Rudolf Szardenings	Four-eng		03.58	9	W
12./NJG3	Hptm Huck	Lancaster		04.10	6	W

Chapter Eleven

Overwhelming Odds

September 1944 - January 1945

The relentless pressure of the continuing Allied air attacks was now taking its toll on the now-beleaguered Luftwaffe defences. Although the aircraft production continued, having been dispersed away from large factory complexes, the standard of crew training had greatly deteriorated; with the daylight skies over Germany now ruled by the USAAF long-range fighters, there was no 'safe area' for fledgling aircrews to fly training sorties. Of particular importance was the worsening fuel situation. Serviceable aircraft could not be employed against attacks and, more and more, those with fuel would be allocated to the most successful pilots.

1-2 September 1944

6./NJG101	Ofw Ludwig	Halifax	Friedrichshafen	22.20	5	W
1./NJG5	Oblt E-G Drünkler	DB-3	30km SE Georgensburg	01.56	28	E
8./NJG6	Ofw Josef Federle	Halifax	S. Pálanka	02.08	5	W
8./NJG6	Ofw Josef Federle	Halifax	S. Mohács	02.25	6	W
8./NJG6	Ofw Josef Federle	Halifax	Vinkovca	02.47	7	W

10-11 September 1944

6./NJG101	Ofw Ludwig	Halifax	Hassberge Area	22.44	6	W
8./NJG6	Oblt Wilhelm Johnen	Halifax	W Szeged	23.38	31	W
6./NJG101	Ofw Ludwig	Halifax	Kaufunder Wald	03.10	7	W
4./NJG6	Lt Lange	Lancaster	10km W Darmstedt	03.59	1	W

11-12 September 1944

3./NJG3	Uffz Hans Schadowski	Lancaster	OA	22.43	9	W
3./NJG3	Uffz Hans Schadowski	Lancaster	OB	22.44	10	W
8./NJG3	Oblt Fritz Brinkmann	Stirling	W Westerland	22.52	12	W
3./NJG3	Lt Wolf Gutter	Four-eng	Nr. Copenhagen	22.53	4	W
10./JG300	Lt Kurt Welter	Mosquito	FG-4	22.58	23	W
3./NJG3	Lt Wolf Gutter	Four-eng	S Copenhagen	23.00	5	W
3./NJG7	Ofw Schmitt	Four-eng	NA-OA	23.37	3	W
7./NJG101	Uffz Weismann	Four-eng	Hirson area	00.09	1	W
4./NJG6	Fw Helmut Bunje	Lancaster	15km SSW Darmstadt	00.10	4	W
4./NJG6	Lt Lange	Lancaster	50km W Darmstadt	00.10	2	W
2./NJG4	Hptm Kranz	Four-eng	E Alzey	00.12	2	W
St.II./NJG6	Lt Peter Spoden	Lancaster	20km NW Darmstedt	00.12	13	W
1./NJG4	Oblt Walter Riedlberger	Four-eng	Darmstadt	00.12	10	W
II./NJG6	Oblt Kopp	Four-eng	30km WNW Darmstadt	00.13	1*	W
1./NJG6	Ofw Clemens Ostrowitzki	Lancaster	WSW Darnstadt	00.13	1	W
St.I./NJG6	Hptm Gerhard Friedrich	Lancaster	S Darmstadt	00.14	19	W
1./NJG4	Lt "..I.....adt"	Four-eng	Darmstadt	00.17	-	W
3./NJG4	Uffz Friedrich Grunder	Four-eng	Nr. Worms	00.18	3	W
St.II./NJG6	Lt Peter Spoden	Halifax	25km WNW Darmstadt	00.18	14	W
II./NJG6	Ofhr Junghans	Four-eng	10km S Darmstadt	00.20	1*	W

1./NJG11	Oblt Krause	Lancaster	Nr. Darmstadt	00.20	2	W
9./NJG4	Fw Gottfried Kurzwernhart	Four-eng	Bad Kreuznach	00.20	2	W
9./NJG4	Hptm Kurt Fladrich	Four-eng	Hunsrück	00.21	15	W
2./NJG6	Ofw Ludwig Schmidt	Halifax	70km W Darmstadt	00.21	3	W
3./NJG6	Uffz Benno Kratz	B-17	NW Darmstadt	00.25	7	W
12./NJG3	Fw Klaus Möller	Four-eng	SW Rostock	00.25	10	W

12-13 September 1944

2./NJG6	Fw Siegmann	Lancaster	Hagenau	22.40	1	W
7./NJG101	Uffz Karl Ludy	B-17	UR-AR	22.44	1	W
2./NJG6	Hptm Martin Becker	Lancaster	SW Frankfurt-a-M	22.45	42	W
1./NJG6	Fw Günther Bahr	B-17	NW Mannheim	22.46	20	W
St.I./NJG6	Hptm Gerhard Friedrich	Lancaster	NW beacon "Christa"	22.46	20	W
6./NJG6	Lt Otto Drähne	Lancaster	20km SE Mannheim	22.47	2	W
1./NJG11	Lt Josef Wolfsberger	Lancaster	W Frankfurt-am-Main	22.50	2	W
2./NJG6	Hptm Martin Becker	Lancaster	SSW Frankfurt-am-Main	22.53	43	W
2./NJG6	Ofw Ludwig Schmidt	Lancaster	E Mannheim	22.55	4	W
1./NJG11	Uffz Hans Dürscheidt	Lancaster	NW Frankfurt-am-Main	23.02	1	W
1./NJG11	Uffz Ulrich	Four-eng	N Frankfurt-am-Main	23.04	1	W
I./NJG11	Hptm Friedrich-Karl Müller	Lancaster	NW Frankfurt-am-Main	23.06	25	W
St.IV./NJG1	Hptm H-W Schnaufer	Lancaster	SW Frankfurt-am-Main	23.07	94	W
5./NJG6	Hptm Fritz Griese	Lancaster	20km NW Frankfurt-a-M	23.08	1	W
4./NJG4	Oblt Jakob Schaus	Four-eng	NE Zell	23.10	23	W
10./JG300	Lt Kurt Welter	Mosquito	Hopsten - Bramsche	23.10	24	W
10./JG300	Ofw Baltenschat	Mosquito	E Berlin	23.13	1	W
1./NJG11	Ofw Günther Migge	Lancaster	Frankfurt-am-Main	23.15	5	W
St.I./NJG6	Oblt Wilhelm Engel	Lancaster	E Frankfurt-am-Main	23.16	15	W
3./NJG4	Oblt Kadleik	Four-eng	Katzenelbogen	23.16	6	W
5./NJG6	Uffz Heider	Lancaster	20km E Frankfurt-am-M	23.20	1	W
1./NJG101	Oblt Bernard Schlimpen	Four-eng	NW Frankfurt-am-Main	23.20	1	W
1./NJG11	Lt Gustav Mohr	Lancaster	SW Wiesbaden	23.21	1	W
10./JG300	Ofw Baltenschat	Mosquito	SW Berlin	23.25	2	W
1./NJG6	Ofw Kurt Karsten	Lancaster	SW Stuttgart	22.35	6*	W
10./JG300	Lt Kurt Welter	Mosquito	Tangerhutte	23.25	25	W
9./NJG4	Lt Hromadnik	Four-eng	area S Wittlich	23.35	6	W

13-14 September 1944

8./NJG6	Oblt Wilhelm Johnen	Halifax	SE Baja	21.30	32	W
I./NJG10	Ofw Egbert Jaacks	Mosquito	SE Berlin	23.00	3	W
10./JG300	Lt Karl Mitterdorfer	Mosquito	Braunschweig	23.15	1	W

14-15 September 1944

2./NJG100	Ofw "Karl" Maisch	PS-84		21.39	8	E
2./NJG100	Ofw "Karl" Maisch	PS-84		21.39	9	E
St.II./NJG101	Lt Hans-Joachim Beck	DB-3		21.54	4	E
7./NJG6	Oblt Josef Kraft	Il-4	20km N Budapest	22.00	39	E
St.II./NJG101	Lt Hans-Joachim Beck	DB-3	Budapest	22.01	5	E
6./NJG101	fhj Edwin Brochhaus	DB-3	15km S Matrageb	22.11	1	E
6./NJG101	Oblt Hans Krause	DB-3	Budapest	22.23	16	E

Two Luftwaffe crews who came to grief on 17/18th September were those of Unteroffiziers Heinz Gesse of 12./NJG 1 (centre) and Walter Sjuts of 11./NJG 1, both caught near Arnhem and shot down by Flight Lieutenant A.J.Owen of 85 Squadron (right). There were no survivors from either crew.

15-16 September 1944

9./NJG6	Lt Resiger	DB-3	60km NE Debrecen	21.55	1	E
4./NJG101	Oblt Karl Dorscheln	DB-3	NE Debrecen	21.59	1	E
10./JG300	Fw Reichenbach	Mosquito	7km NW Wittenburg	01.59	1	W
10./JG300	Lt Kurt Welter	Mosquito	25km S Berlin	02.03	26	W
10./JG300	Lt Kurt Welter	Mosquito	5km N Aachmer	02.30	27	W

16-17 September 1944

2(F)./Nacht	Uffz Liebenau	Twin-eng		22.17	1	W

17-18 September 1944

9./NJG6	Lt Resiger	DB-3		21.34	2	E
6./NJG101	fhj Ludwig	DB-3		21.41	8	E
St.II./NJG101	Oblt Hans-Joachim Beck	DB-3		22.26	6	E

18-19 September 1944

St.II./NJG101	Oblt Hans-Joachim Beck	DB-3		21.00	7	E
7./NJG6	Oblt Josef Kraft	PS-84	Tokaj	21.40	40	E
6./NJG101	Lt Wilhelm Dietz	DB-3		21.44	1	E
3./NJG3	Uffz Hans Schadowski	Lancaster	QQ	22.30	11	W
3./NJG3	Uffz Hans Schadowski	Lancaster	AP	22.32	12	W
10./JG300	Lt Kurt Welter	Mosquito	N Wittenberg	23.05	28	W

19-20 September 1944

St.III./NJG6	Hptm Leopold Fellerer	DC-3		20.26	39	E
7./NJG6	Oblt Josef Kraft	PS-84	Nyiregyhaza	21.04	41	E
7./NJG6	Oblt Josef Kraft	PS-84	Nyiregyhaza	21.20	42	E
4./NJG101	Oblt Karl Dörscheln	DB-3		21.37	2	E
4./NJG101	Oblt Karl Dörscheln	DB-3		21.55	3	E
4./NJG101	Oblt Karl Dörscheln	DB-3		22.05	4	E
I./NJG11	Fw Wilhelm Jansen	Lancaster	Mönchengladbach	22.53	1	W
6./NJG101	Hptm Hans Krause	DB-3		22.53	17	E
11./NJG1	Hptm Hermann Greiner	Lancaster	WSW Mönchengladbach	22.55	38	W
IV./NJG1	Fw Erwin Egeler	Lancaster	WSW Mönchengladbach	22.58	3	W
4./NJG1	Oblt Fritz Lau	Lancaster	WNW Mönchengladbach	22.59	15	W
6./NJG101	Hptm Hans Krause	DB-3		23.00	18	E
11./NJG1	Hptm Hermann Greiner	Lancaster	WSW Mönchengladbach	23.02	39	W
5./NJG2	Hptm H-H Hissbach	Lancaster	WNW Mönchengladbach	23.05	28	W

IV./NJG1	Lt Friedrich Forstener	Lancaster	WNW Mönchengladbach	23.11	1	W
4./NJG1	Oblt Fritz Lau	Lancaster	WNW Mönchengladbach	23.11	16	W
5./NJG4	Unknown	E/A		01.14	-	W

20-21 September 1944

9./NJG6	Oblt Rudolf Thun	B-25	NE Debrecen	21.46	5	E
4./NJG101	Oblt Karl Dörscheln	DB-3		22.25	5	E
7./NJG6	Oblt Josef Kraft	Wellington	Szombathely	23.08	43	W
7./NJG6	Oblt Josef Kraft	Wellington	Szombathely	23.12	44	W
10./NJG6	Fw Josef Brunner	Wellington	N Neusielder lake	23.23	6	W
St.IV./NJG6	Maj Herbert Lütje	Wellington	NE Neusieder See	23.30	42	W
4./NJG100	Oblt Gunter Bertram	DB-3		23.46	26	E

21-22 September 1944

8./NJG6	Oblt Wilhelm Johnen	Halifax	W Senta	20.32	33	W

22-23 September 1944

1./NJG100	Lt G E Francsi	Il-4		01.10	40	E

23-24 September 1944

4./NJG2	Oblt Lothar Jarsch	Lancaster	SE Mönchengladbach	22.24	5	W
4./NJG2	Oblt Lothar Jarsch	Lancaster	SE Mönchengladbach	22.29	6	W
1./NJG1	Hptm E-W Modrow	Lancaster	45km WNW Düsseldorf	22.40	30	W
11./NJG1	Fw Anton Weissensel	Four-eng	Maastricht	22.43	1	W
4./NJG2	Ofw Wolfram Möckel	Lancaster	NW Mönchengladbach	22.51	3	W
5./NJG2	Ofw Heinrich Breitenfelder	Lancaster	NW Münster	22.53	2	W
St.IV./NJG1	Hptm H-W Schnaufer	Lancaster	W Münster	22.53	95	W
St.IV./NJG1	Hptm H-W Schnaufer	Lancaster	W Münster	23.10	96	W
1./NJG1	Hptm E-W Modrow	Lancaster	75km WNW Düsseldorf	23.11	31	W
St.IV./NJG1	Hptm H-W Schnaufer	Lancaster	W Münster	23.15	97	W
11./NJG1	Hptm Hermann Greiner	Lancaster	WSW Munster	23.16	40	W
IV./NJG1	Lt Otto Teschner	Lancaster	N Nijmegen	23.23	6	W
St.IV./NJG1	Hptm H-W Schnaufer	Lancaster	W Münster	23.25	98	W
8./NJG1	Lt Theodor Adamski	Lancaster	30km W Den Helder	23.27	1	W
8./NJG1	Oblt Dietrich Schmidt	Lancaster	S edge of Zuiderzee	23.31	36	W

28-29 September 1944

St.II./NJG3	Hptm H-H Müller	Stirling	LP	00.27	10	W

2-3 October 1944

12./NJG3	Fw Klaus Möller	Wellington	AQ	21.43	11	W

3-4 October 1944

10./NJG3	Oblt Walter Briegleb	Wellington	St. Avold	23.13	17	W

4-5 October 1944

5./NJG3	Oblt August Gyory	Four-eng	Werl	20.25	1	W
6./NJG3	Hptm Decker	Four-eng	Attendorn	21.30	2	W
5./NJG3	Oblt August Gyory	Four-eng		22.05	2	W
5./NJG3	Oblt Rudolf Szardenings	Four-eng	Ludenscheid	22.05	10*	W

On the night of 6th October Hauptmann Erich Frey of Stab II./NJG 1 (far right) was reported missing with his crew over Belgium. They were probably shot down by Flying Officer R.H.Finlayson of 409 Squadron. Three nights later, on 9/10th October, Hauptmann Heinz-Wolfgang Schnaufer, Kommandeur of IV./NJG 1, - pictured here with his crew Bordfunker Leutnant Fritz Rumpelhardt (left) and Bordschütze Oberfeldwebel Wilhelm Gänsler (right) - claimed two Halifaxes two bring his score to an astonishing 100.

5-6 October 1944

10./JG300	Lt Karl Mitterndorf	Mosquito	Nr. Berlin	20.30	3	W
2(E)./NJSt	Uffz Günter	Mosquito	Berlin	20.32	1	W

6-7 October 1944

12./NJG3	Lt Hermann Stock	Lancaster	In sea UP-94	19.43	7	W
10./JG300	Uffz Schmid	Mosquito	Berlin	20.30	1	W
10./JG300	Lt Kurt Welter	Mosquito	Berlin	20.30	29	W
12./NJG3	Uffz Carl-Hans Hurth	Four-eng	W Bremen	20.36	2	W
10./JG300	Lt Georg Czypionka	Mosquito	Berlin	20.40	1	W
10./NJG3	Oblt Walter Briegleb	Lancaster	5km N Leer	20.44	18	W
6./NJG3	Hptm Decker	Stirling	LH-59	09.23	3	W

7-8 October 1944

2./NJG100	Ofw "Karl" Maisch	DB-3		19.22	10	E
2./NJG100	Ofw "Karl" Maisch	DB-3		19.45	11	E
2./NJG100	Ofw "Karl" Maisch	DB-3		20.16	12	E
2./NJG100	Ofw "Karl" Maisch	DB-3		20.40	13	E
9./NJG102	Oblt Heinrich Barth	B-25	SW Plumkenau	20.43	4	E

9-10 October 1944

St.IV./NJG1	Hptm H-W Schnaufer	Halifax	S Bochum	20.32	99	W
IV./NJG1	Oblt Georg Fengler	Lancaster	SW Bochum	20.35	9	W
3./NJG1	Fw Wilhelm Morlock	Halifax	Münster	20.45	10	W
1./NJG102	Fw Heidsick	B-25	QE-6	20.55	1	E
St.IV./NJG1	Hptm H-W Schnaufer	Halifax	S Münster	20.55	100	W

13-14 October 1944

7./NJG6	Oblt Josef Kraft	B-24	Tamasi	20.36	45	W
7./NJG6	Oblt Josef Kraft	Wellington	SE Barcs	20.46	46	W

6./NJG101	Hptm Hans Krause	Wellington		21.01	19	W
2./NJG100	Ofw Gemünd	PS-84		22.48	1	E
1./NJG100	Ofw Rudi Duding	Li-2		00.42	12	E
1./NJG10	Uffz Hans Durscheidt	Mosquito	N Duisberg	02.11	2	W
7./NJG2	Lt Arnold Doring	Four-eng	Kevlaer	03.43	2	W

15-16 October 1944

Stab./NJG5	Maj Walter Borchers	PS-84		19.23	20	E
2./NJG100	Ofw "Karl" Maisch	Li-2	NE Tarnow	19.37	14	E
8./NJG3	Oblt Fritz Brinkmann	Halifax	German Bight	20.25	13	W
10./NJG3	Uffz Werner Kroschke	Four-eng	German Bight	20.26	1	W
6./NJG3	Hptm Decker	Lancaster		20.50	4	W
5./NJG3	Oblt August Gyory	Four-eng	LT-67	21.02	3	W
St.II./NJG3	Oblt Fritz Brandt	Halifax	HU-56	21.06	6	W
6./NJG3	Hptm Decker	Lancaster		22.40	5	W

16-17 October 1944

2./NJG100	Ofw "Karl" Maisch	B-24	30km NE Tarnow	20.34	15	W
2./NJG100	Ofw "Karl" Maisch	B-24	30km NW Tarnow	20.46	16	W

17-18 October 1944

1./NJG5	Oblt E-G Drünkler	PS-84		17.57	29	E
St.IV./NJG5	Lt Wilhelm Seuss	Li-2		18.07	8	E

19-20 October 1944

1./NJG6	Lt Clemens Ostrowitzki	Lancaster	Tubingen	20.41	2	W
1./NJG6	Lt Clemens Ostrowitzki	Lancaster	SE Stuttgart	20.50	3	W
2./NJG101	Fw "Lucas"	Lancaster	SW Würzberg	21.00	1	W
1./NJG6	Lt Clemens Ostrowitzki	Lancaster	SE Stuttgart	21.30	4	W
2./NJG2	Hptm Heinz Rökker	Lancaster	Pirmasens	21.48	42	W

20-21 October 1944

7./NJG6	Oblt Josef Kraft	Wellington	Szombathely	21.43	47	W
7./NJG6	Oblt Josef Kraft	B-24	Szombathely	22.10	48	W
7./NJG6	Oblt Josef Kraft	Wellington	Szombathely	22.15	49	W
St.III./NJG6	Hptm Leopold Fellerer	B-24	Szombathely	22.21	40	W

21 October 1944

6./NJG101	Uffz "Ragmonville"	P-51		14.25	1	W
6./NJG101	Uffz "Ragmonville"	P-51		14.25	2	W

21-22 October 1944

6./NJG101	Hptm Hans Krause	Wellington		20.36	20	W

23-24 October 1944

Stab./NJG5	Maj Walter Borchers	DB-3		18.09	21	E
1./NJG5	Oblt E-G Drünkler	PS-84		18.12	30	E
St.IV./NJG5	Unknown	DB-3	ENE Gumbinnen	18.52	-	E
2./NJG100	Ofw "Karl" Maisch	Li-2	E Poprad	20.37	17	E
2./NJG100	Ofw "Karl" Maisch	Li-2	E Poprad	21.07	18	E

As 1945 approached, the Nachtjagd continued to fight, although with less spectacular success than it had previously attained. Hauptmann Ernst-Georg Drünkler (far right) had joined NJG 5 late in 1942 claimed his 29th and 30th victories in October 1944 and would end the war with 46 confirmed and the Ritterkreuz. His Kommodore, Major Walter Borchers, had 10 victories as a Zerstörer pilot before joining the nightfighters. On 23rd October he brought his score to 31, receiving the Ritterkreuz six days later. On 3/4th March 1945, with a score of 59, he was shot down and killed by Wing Commander Gibb of 239 Squadron.

25-26 October 1944

3./NJG5	Fw Ernst Reitmeyer	PS-84		17.38	19	E

26-27 October 1944

10./NJG3	Uffz Werner Kroschke	Halifax	S Heligoland	19.42	2	W
6./NJG101	Hptm Hans Krause	DB-3	Budapest	22.47	21	E

27-28 October 1944

10./JG300	Lt Kurt Welter	Mosquito	N Berlin	01.20	30	W

29-30 October 1944

1./NJG100	Ofw Helmut Dahms	Li-2		20.51	14	E
10./NJG3	Uffz Werner Kroschke	Wellington	Near Borkum	21.00	3	W

1-2 November 1944

5./NJG2	Unknown	Four-eng	SW Duisburg	20.32	-	W
4./NJG2	Unknown	Four-eng	LN-KN	20.40	-	W
IV./NJG1	Unknown	Four-eng	KN	20.41	-	W
6./NJG1	Unknown	Four-eng	S Venlo	20.42	-	W
IV./NJG1	Unknown	Four-eng	KM-KN	20.46	-	W
6./NJG2	Unknown	Four-eng	KN-KO-LO-LN	20.46	-	W
IV./NJG1	Unknown	Four-eng	LM	20.47	-	W
IV./NJG1	Unknown	Four-eng	NM	20.58	-	W

2-3 November 1944

IV./NJG1	Unknown	Four-eng	Düsseldorf	19.13	-	W
II ./NJG2	Maj Paul Semrau	Four-eng	Ruhr area	19.15	35	W
II./NJG2	Maj Paul Semrau	Four-eng	Ruhr area	19.17	36	W
6./NJG1	Lt Barthel	Four-eng	W Düsseldorf	19.19	2	W
II./NJG2	Maj Paul Semrau	Four-eng	W Düsseldorf	19.21	37	W
6./NJG1	Ofw Burger	Four-eng	W Düsseldorf	19.22	1	W
3./NJG1	Ofw Wilhelm Morlock	Four-eng	S Düsseldorf	19.25	11	W
6./NJG1	Ofw Josef Strohlein	Four-eng	W Dusseldorf	19.25	3	W
3./NJG1	Oblt Ernst Runzel	Four-eng	S Rheydt	19.26	1	W
3./NJG1	Ofw Wilhelm Morlock	Four-eng	S Düsseldorf	19.27	12	W
3./NJG1	Ofw Wilhelm Morlock	Four-eng	S Düsseldorf	19.28	13	W
IV./NJG1	Unknown	Four-eng	S Düsseldorf	19.30	-	W

6./NJG2	Unknown	Four-eng	LO-8-MO-3	19.30	-	W
3./NJG1	Ofw Wilhelm Morlock	Four-eng	S Düsseldorf	19.30	14	W
2./NJG1	Uffz Werner Wollenhaupt	Halifax	Ruhr area	19.30	1	W
I./NJG11	Fw Wilhelm Jansen	Lancaster	S Düsseldorf	19.31	2	W
3./NJG1	Ofw Wilhelm Morlock	Four-eng	S Düsseldorf	19.31	15	W
3./NJG1	Ofw Wilhelm Morlock	Four-eng	S Düsseldorf	19.33	16	W
3./NJG1	Oblt Thurner	Four-eng	S Düsseldorf	19.33	1	W
6./NJG1	Oblt Johannes Hager	Four-eng	Mönchengladbach	19.34	34	W
6./NJG1	Uffz Janicke	Four-eng	25km SE Düsseldorf	19.34	1	W
4./NJG1	Hptm Fritz Lau	Four-eng	SW Düsseldorf	19.34	17	W
II./NJG1	Unknown	Four-eng		19.35	-	W
5./NJG2	Unknown	Four-eng	NW Düsseldorf	19.35	-	W
6./NJG2	Oblt Erich Jung	Lancaster	10km NW Düsseldorf	19.35	13	W
IV./NJG1	Lt Hans Gienanth.von	Four-eng	Hechtel area	19.36	5	W
4./NJG1	Hptm Fritz Lau	Four-eng	SW Düsseldorf	19.36	18	W
2./NJG11	Lt Gustav Mohr	Lancaster	Düsseldorf-Maas	19.36	2	W
5./NJG2	Unknown	Four-eng	SW Düsseldorf	19.37	-	W
3./NJG1	Ofw Wilhelm Morlock	Four-eng	S Düsseldorf	19.37	17	W
6./NJG2	Oblt Erich Jung	Lancaster	NW Dusseldorf	19.38	14	W
6./NJG1	Oblt Johannes Hager	Four-eng	W Mönchengladbach	19.40	35	W
2./NJG11	Fw Schermutzki	Lancaster	SW Düsseldorf	19.40	1	W
4./NJG1	Hptm Fritz Lau	Four-eng	W the Maas	19.41	19	W
4./NJG2	Unknown	Four-eng	NK-NL-OK	19.42	-	W
6./NJG2	Oblt Erich Jung	Lancaster	25km NW Düsseldorf	19.42	15	W
IV./NJG1	Unknown	Four-eng	MK-NK	19.43	-	W
4./NJG2	Unknown	Four-eng	NK-NL-OK-OL	19.44	-	W
2./NJG11	Lt Stitz	Lancaster	Ruhr area	19.44	1	W
6./NJG2	Unknown	Four-eng	Hechtel-Hasselt	19.46	-	W
III./NJG2	Unknown	Four-eng	Verviers-Eupen	19.46	-	W
II./NJG1	Unknown	Four-eng	Hasselt-St. Trond	19.57	-	W

4-5 November 1944

11./NJG1	Hptm Hermann Greiner	Halifax	W Essen	19.21	41	W
11./NJG3	Oblt Josef Förster	Four-eng	JP-KP	19.29	11	W
2./NJG2	Hptm Heinz Rökker	Lancaster	N Dortmund	19.31	43	W
11./NJG1	Hptm Hermann Greiner	Halifax	W Essen	19.35	42	W
I./NJG1	Hptm Werner Baake	Halifax	8km NW Mettingen	19.36	36	W
2./NJG2	Hptm Heinz Rökker	Lancaster	N Dortmund	19.36	44	W
6./NJG2	Oblt Erich Jung	Lancaster	KO-KP-LO-LP	19.37	16	W
4./NJG1	Ofhr Ramsauer	Halifax	S Essen	19.37	1	W
6./NJG2	Unknown	Four-eng	LO	19.40	-	W
6./NJG2	Oblt Erich Jung	Lancaster	KO-KP-LO-LP	19.44	17	W
6./NJG2	Unknown	Four-eng	MO-LO	19.45	-	W
6./NJG1	Uffz Janicke	B-17	NE Düsseldorf	19.45	2	W
1./NJG10	Oblt Krause	Lancaster	S Essen	19.45	3	W
IV./NJG1	Unknown (Matzak?)	Four-eng	NO	19.48	-	W
6./NJG1	Lt Barthel	Halifax		19.50	3	W
11./NJG3	Lt Horst-Rudiger Blume	Lancaster		19.50	3	W
1./NJG2	Uffz Helmut Calgeer	Four-eng	LP	19.50	1	W
IV./NJG1	Unknown	Four-eng		19.52	-	W

By late 1944 the standard Luftwaffe night interceptors were the Messerschmitt Bf110G-4 and the Junkers Ju88G-6. Both had their performance lowered by the drag of the large radar arrays, but they still outperformed every Allied bomber except for the Mosquito. The range of the Bf110 had been enhanced by the addition of underwing fuel tanks. Note the twin upward-firing 20mm cannon in the mid-fuselage of the Ju88. These were sighted by a periscope and employed low-illumination ammunition to make RAF aircrews believe that they were being engaged by Flak.

8./NJG1	Fw Karl Neumann	Halifax	Duisburg	19.52	1	W
5./NJG2	Fw Kubetz	Halifax	SW Essen	19.54	1	W
5./NJG2	Fw Kubetz	Mosquito	SW Essen	19.55	2	W
4./NJG1	Fw Peter Muller	Halifax	SW Essen	19.55	3	W
4./NJG2	Unknown	Four-eng	KO-LO	19.58	-	W
2./NJG2	Hptm Heinz Rökker	Halifax	N Geldern	20.00	45	W
4./NJG2	Unknown	Four-eng	MN-MO-NN-NO	20.01	-	W
8./NJG1	Fw Karl Neumann	Halifax	Duisburg	20.01	2	W
5./NJG2	Fw Rudolf Morenz	Four-eng	NO-NN	20.04	2	W
3./NJG1	Lt Jurgen Prietze	Four-eng	Bonn	20.04	1	W
2./NJG2	Hptm Heinz Rökker	Halifax	N Geldern	20.06	46	W
3./NJG1	Lt Jurgen Prietze	Four-eng	Bonn	20.10	2	W
IV./NJG1	Lt Heinz Rolland	Four-eng	OK-PK	20.13	13	W
IV./NJG1	Lt Hans Gienanth.von	Halifax	Recklinghausen	20.26	5	W
12./NJG1	Oblt H-H Augenstein	Lancaster			44	W
12./NJG1	Oblt H-H Augenstein	Lancaster			45	W
12./NJG1	Oblt H-H Augenstein	Lancaster			46	W

5-6 November 1944

6./NJG101	Hptm Hans Krause	B-24	S Klagenfurt	23.22	22	W
6./NJG101	Hptm Hans Krause	B-24	S Klagenfurt	23.31	23	W

6-7 November 1944

I./NJG1	Hptm Werner Baake	Lancaster	12km SE Doetinchem	19.23	37	W
1./NJG1	Hptm E-W Modrow	Lancaster	GP	19.24	32	W
IV./NJG4	Hptm H-W Schnaufer	Lancaster	KP-LP	19.26	101	W
2./NJG1	Lt Otto Fries	Lancaster	W Rheine	19.27	14	W
11./NJG3	Oblt Josef Förster	Four-eng		19.28	12	W
1./NJG1	Hptm E-W Modrow	Lancaster	GP	19.28	33	W
IV./NJG4	Hptm H-W Schnaufer	Lancaster	KP-LP	19.34	102	W
3./NJG1	Lt Jurgen Prietze	Lancaster	Handorf	19.37	3	W
2./NJG1	Lt Otto Fries	Lancaster	W Rheine	19.38	15	W
IV./NJG4	Hptm H-W Schnaufer	Lancaster	KP-IP	19.41	103	W
2./NJG1	Lt Otto Fries	Lancaster	W Rheine	19.43	16	W
10./NJG3	Uffz Heinrich Herr	Lancaster		21.40	2	W

11-12 November 1944

9./NJG2	Hptm Dietrich Kornblum	Four-eng	NW Hamburg	19.03	1	W
St.III./NJG2	Hptm Heinz Ferger	Four-eng	NW Hamburg	19.09	19	W
7./NJG2	Oblt Walter Briegleb	Four-eng		19.41	19	W
7./NJG2	Oblt Dietrich	Four-eng		1941	1	W
St.III./NJG2	Hptm Heinz Ferger	Four-eng		19.45	20	W

17-18 November 1944

6./NJG101	Hptm Hans Krause	Halifax	Cape Ancona	21.57	24	W
6./NJG101	Hptm Hans Krause	Halifax	Cape Ancona	22.03	25	W

21-22 November 1944

II./NJG4	Hptm Hubert Rauh	Four-eng	Ruhr area	18.45	23	W
II./NJG4	Hptm Hubert Rauh	Four-eng	Ruhr area	19.02	24	W
11./NJG1	Oblt Arthur Turowski	Four-eng	Ruhr area	19.03	1	W
IV./NJG4	Hptm H-W Schnaufer	Halifax	Ruhr area	19.05	104	W
IV./NJG4	Hptm H-W Schnaufer	Halifax	Ruhr area	19.07	105	W
4./NJG1	Hptm Fritz Lau	Four-eng	SW Bochum	19.10	20	W
4./NJG1	Hptm Fritz Lau	Four-eng	Radevormwald	19.11	21	W
2./NJG1	Uffz Werner Wollenhaupt	Four-eng	Ruhr area	19.11	2	W
9./NJG2	Oblt Erich Jung	Lancaster	Ruhr area	19.13	18	W
11./NJG6	Fw Emil Weinmann	Lancaster	Spessart	19.26	1	W
1./NJG4	Fw Konrad Beyer	Four-eng	Gross Unstadt	19.30	11	W
1./NJG1	Fw Buer	Four-eng		19.30	1	W
11./NJG6	Fw Emil Weinmann	Lancaster	W Aschaffenburg	19.34	2	W

22-23 November 1944

St.II./NJG101	Oblt Karl Dörscheln	B-24	Szombathely	19.59	6	W
6./NJG101	Hptm Hans Krause	Wellington	Szombathely	20.01	26	W
St.II./NJG101	Oblt Karl Dörscheln	Wellington	Szombathely	20.05	7	W
7./NJG6	Uffz Stender	B-24	N Szombathely	20.06	1	W
7./NJG6	Oblt Helmut Bruder	Wellington	N Szombathely	20.13	1	W
II ./NJG101	Oblt Jeschke	Wellington	N Szombathely	20.14	2	W
6./NJG101	Hptm Hans Krause	B-24	W Szombathely	20.17	27	W
7./NJG6	Ofhr Tangelst	B-24	Zalaegerszeg	20.21	1	W
7./NJG4	Unknown	E/A		23.48	-	W

27-28 November 1944

9./NJG3	Hptm Banschein	Four-eng		19.14	4	W

28-29 November 1944

7./NJG4	Unknown	E/A		00.49	-	W
8./NJG1	Lt Rolf Ebhardt	Lancaster	between Essen & Aachen	06.01	5	W

29-30 November 1944

3./NJG3	Hptm Eduard Schroder	Four-eng		20.43	16	W
1./NJG3	Oblt Herbert Koch	Lancaster	Skaggerak	23.49	17*	W

2-3 December 1944

7./NJG2	Oblt Walter Briegleb	Wellington	6km N Baltrum	21.35	20	W
3./NJG3	Ofw Giesecke	Stirling		04.19	6	W

Hubert Rauh (left) was a pre-war officer, who joined the nightfighter force in 1942. In late 1944 he was leading II./NJG 4. On 21/22nd November he claimed his 23rd and 24th victories and his total would reach 31 before the surrender. With Hauptmann Dietrich Kornblum there is a mystery. He was officially reported lost on 26/27th November in the explosion of his sole victory. There is no RAF loss to relate to this. However the victory listings show a credit for him on 11/12th November, when there certainly were RAF losses in the specified area. It is possible that an error of date has taken place.

4-5 December 1944

6./NJG6	Hptm Ernst Wallner	Lancaster	SSE Heilbronn	19.30	5	W
2./NJG11	Lt Gustav Mohr	Lancaster	Heilbronn-Karlsruhe	19.33	3	W
St.II./NJG6	Lt Peter Spoden	Lancaster	S Heilbronn	19.33	15	W
4./NJG6	Fw Helmut Bunje	Lancaster	NW Heilbronn	19.34	5	W
I./NJG11	Hptm Friedrich-Karl Müller	Lancaster	Worth-Lautersbach	19.35	26	W
2./NJG11	Fw Wilhelm Jansen	Lancaster	NW Heilbronn-Karlsruhe	19.37	3	W
St.II./NJG6	Lt Peter Spoden	Lancaster	SW Heilbronn	19.38	16	W
St.II./NJG6	Lt Peter Spoden	Lancaster	SW Heilbronn	19.42	17	W
I./NJG11	Hptm Friedrich-Karl Müller	Lancaster	SW Kahlsruhe	19.43	27	W
2./NJG11	Lt Gustav Mohr	Four-eng	NW Heilbronn	19.49	4	W

6-7 December 1944

IV./NJG1	Lt Hans Gienanth.von	Halifax	Quakenbruck	19.23	7	W
IV./NJG1	Ofw Gerhard Graefe	Halifax	GP-GQ	20.06	2	W
8./NJG1	Hptm Dietrich Schmidt	Lancaster	NW Twente	20.14	37	W
2./NJG11	Ofw Schuster	Lancaster	Giessen	20.15	1	W
St.IV./NJG6	Fw Gunther Liersch	Lancaster	Giessen	20.16	3	W
2./NJG11	Uffz Ulrich	Lancaster	Giessen	20.18	2	W
St.IV./NJG6	Fw Gunther Liersch	Lancaster	NW suburb Giessen	20.19	4	W
11./NJG6	FhjFw Erichsen	Lancaster	Giessen	20.20	1	W
2./NJG11	Fw Loos	Lancaster	Nr. Giessen	20.20	1	W
12./NJG3	Lt Heinz Reuter	Four-eng		20.20	9	W
St.IV./NJG6	Fw Gunther Liersch	Lancaster	NW Giessen	20.22	5	W
10./NJG6	Ofw Peter Tiedemann	Lancaster	25km N Giessen	20.23	2	W
11./NJG6	Fw Emil Weinmann	Lancaster	WSW Giessen	20.24	3	W
7./NJG4	Unknown	E/A		21.54	-	W
II./NJG2	Fw Rudolf Morenz	Halifax			3	W
II./NJG2	Lt Martin Tschiersch	Lancaster			9	W
II./NJG2	Lt Martin Tschiersch	Lancaster			10	W
12./NJG3	Oblt Alfons Köster	Halifax			22*	W
3./NJG101	Oblt Aicher	Four-eng	GG-6	05.17	1	W

9-10 December 1944

4./NJG11	Fw Reichenbach	Mosquito	Nr. Berlin	22.02	2	W

11-12 December 1944

II./NJG11	Oblt Kurt Welter	Mosquito			31	W

12-13 December 1944

3./NJG10	Hptm Gustav Tham	Lancaster	Essen	19.50	12	W
8./NJG1	Uffz Scherl	Mosquito	E Hagen	19.59	1	W
IV./NJG4	Hptm H-W Schnaufer	Lancaster		20.00	106	W
3./NJG10	Hptm H-M Hadeball	Lancaster		01.51	33	W
2./NJG11	Lt Gustav Mohr	Lancaster			5	W

14-15 December 1944

St.I./NJG5	Maj Werner Hoffmann	DB-3		16.54	41	E
3./NJG5	Fw Ernst Reitmeyer	PS-84		17.11	20	E
3./NJG5	Fw Ernst Reitmeyer	PS-84	E Libau	18.12	21	E
I./NJG100	Lt G E Francsi	Il-4			41	E

15-16 December 1944

7./NJG1	Uffz Kichling	Four-eng	S Stuttgart	22.33	1	W
3./NJG3	Hptm Eduard Schröder	Four-eng	LD-19	03.54	17	W

17-18 December 1944

3./NJG6	Uffz Walter Ludwig	Lancaster	CA-4	19.40	1	W
4./NJG5	Unknown	Lancaster	CU-63	19.45	-	W
4./NJG1	Hptm Fritz Lau	Mosquito			22	W
4./NJG1	Hptm Dietrich Schmidt	Four-eng	W Duisburg	06.36	38	W
4./NJG1	Lt Walcher	Four-eng	NW Duisburg	06.37	1	W
4./NJG1	Uffz Reinecke	Mosquito	W Duisburg	06.48	2	W

18-19 December 1944

2./NJG5	Oblt Peter Ehrhardt	Lancaster	NE Gotenhafen	21.58	22	O
2./NJG5	Uffz Hengsbach	Lancaster	SE Danzig (Gdansk)	22.05	4	O
2./NJG5	Oblt Peter Ehrhardt	Lancaster	N Gotenhafen	22.10	23	O
3./NJG5	Hptm Walter Engel	Lancaster	N Gotenhafen	22.11	1	O
II./NJG1	Hptm Adolf Breves	Mosquito	2km N Dusseldorf airport	22.39	13	W

20-21 December 1944

St.I./NJG5	Maj Werner Hoffmann	PS-84	E Libau	17.19	42	E
St.I./NJG5	Maj Werner Hoffmann	DB-3	E Libau	18.09	43	E
St.I./NJG5	Maj Werner Hoffmann	DB-3	E Libau	18.28	44	E
3./NJG5	Fw Ernst Reitmeyer	DB-3		18.31	22	E
3./NJG5	Hptm Walter Engel	DB-3	E Libau	18.40	2	E
5./NJG5	Lt Boy	Il-2		20.00	2	E
I./NJG100	Oblt Gunter Bertram	Il-4			27	E
I./NJG100	Oblt Gunter Bertram	Il-4			28	E
I./NJG100	Oblt Gunter Bertram	Il-4			29	E
I./NJG100	Ofw Helmut Dahms	Il-4			15	E
I./NJG100	Ofw Helmut Dahms	Il-4			16	E
St.I./NJG100	Lt G E Francsi	Li-2			42	E
3./NJG5	Hptm Walter Engel	DB-3	E Memel	00.51	3	E
3./NJG5	Hptm Walter Engel	DB-3	NE Memel	01.01	4	E
3./NJG5	Hptm Walter Engel	DB-3	40km E Memel	01.07	5	E

21-22 December 1944

5./NJG5	Oblt Werner Rapp	Li-2	RC-48	20.15	11	E

Oberleutnant Hermann Leube of 4./NJG 3 (left) claimed his first victory on 2/3rd August 1943 and his 19th and last on 29/30th August 1944. On 27th December he was shot down and killed near Benthuizen, Holland (S.Hall). Werner Hoffmann (right) fought in France with ZG 2 during 1940 and after a time as an instructor, joined the nightfighter force. He survived the war with 51 night victories, three of them on the night of 20th December.

I./NJG100	Ofw Helmut Dahms	Il-4			17	E
I./NJG100	Ofw Helmut Dahms	Il-4			18	E
I./NJG100	Ofw Rudi Duding	Il-4			13	E
I./NJG100	Ofw Karl Strohecker	Il-4			11	E

22-23 December 1944

3./NJG5	Hptm Walter Engel	PS-84	5km SW Libau	17.17	6	E
2./NJG5	Oblt Peter Ehrhardt	DB-3	Libau	17.22	24	E
3./NJG5	Hptm Walter Engel	DB-3	20km SE Libau	17.39	7	E
3./NJG5	Hptm Walter Engel	DB-3	25km E Libau	17.50	8	E
5./NJG5	Uffz Otto Hille	li-2		18.16	1	E
3./NJG5	Hptm Walter Engel	B-25	60km E Libau	18.17	9	E
5./NJG5	Uffz Otto Hille	li-2		18.37	2	E
St.II./NJG6	Oblt Peter Spoden	Halifax	Near Trier	18.59	18	W
11./NJG6	Hptm Kurt-Heinz Weigel	Halifax	SW Bingen	19.05	1	W
5./NJG5	Oblt Werner Rapp	li-2		21.00	12	E
5./NJG5	Oblt Werner Rapp	li-2		21.22	13	E
5./NJG5	Oblt Werner Rapp	li-2		22.04	14	E
II./NJG2	Lt Wolfram Mockel	Mosquito	Hasselt		4	W

23-24 December 1944

I./NJG100	Lt Eberhard Kaross	Il-4			11	E

24-25 December 1944

St.II./NJG6	Lt Kalmbach	Lancaster	30km SW Köln	18.39	1	W
I./NJG1	Oblt Werner Baake	Lancaster		18.50	38	W
III./NJG1	Lt Theodor Adamski	Lancaster		18.55	2	W
St.II./NJG6	Ofw Alfred Siewert	P-61	Luxembourg	01.50	2	W

26-27 December 1944

St.I./NJG3	Lt Manfred Scheunpflug	Four-eng	Off Danish coast	20.47	5	W
II./NJG2	Hptm Kamsties	Mosquito			4	W

27-28 December 1944

IV./NJG3	Fw Hans Werthner	Lancaster	SW of Thionville	02.00	9	W

28-29 December 1944

7./NJG2	Oblt Walter Briegleb	Beaufighter	20km NW Borkum	19.13	21	W
III./NJG2	Fw Richard Richter	Lancaster			5	W

29-30 December 1944

7./NJG2	Oblt Walter Briegleb	Beaufighter	40km NW Borkum	18.39	22	W

31 December 1944-1 January 1945

I./NJG1	Fw Sieben	Lancaster			1	W
I./NJG11	Oblt Krause	Lancaster	S Oberhausen	18.50	4	W
6./NJG1	Hptm Johannes Hager	Lancaster		19.00	36	W
6./NJG1	Hptm Johannes Hager	Lancaster		19.05	37	W
3./NJG3	Hptm Eduard Schröder	Lancaster	50kn NW Hanstholm	21.00	18	W
3./NJG3	Fw Heigl	Stirling		02.02	1	W

1-2 January 1945

2./NJG2	Hptm Heinz Rökker	Lancaster	SE Geldern	20.07	47	W
12./NJG3	Hptm Alfons Köster	Lancaster			23	W

2-3 January 1945

11./NJG6	Hptm Kurt-Heinz Weigel	Four-eng	W Stuttgart	18.44	2	W
St.IV./NJG6	Hptm Martin Becker	Lancaster	Bruschsal	18.45	44	W
11./NJG6	Hptm Kurt-Heinz Weigel	Four-eng	SE Stuttgart	18.55	3	W
St.IV./NJG6	Hptm Martin Becker	Lancaster	Luxembourg	19.32	45	W
II./NJG11	Oblt Kurt Welter	Mosquito			32	W

3 January 1945

III./NJG3	Lt Goerke	P-51			3	W

5-6 January 1945

7./NJG2	Oblt Walter Briegleb	Halifax	SW Delmenhorst	18.50	23	W
7./NJG2	Oblt Walter Briegleb	Halifax	SW Delmenhorst	19.04	24	W
I./NJG1	Hptm Werner Baake	Halifax	50km N Emden	19.05	39	W
I./NJG1	Hptm Werner Baake	Halifax	80km N Emden	19.12	40	W
11./NJG1	Hptm Hermann Greiner	Lancaster	Nr. Hanover	19.12	43	W
11./NJG1	Hptm Hermann Greiner	Lancaster	Nr. Hanover	19.15	44	W
11./NJG1	Hptm Hermann Greiner	Lancaster	Nr. Hanover	19.18	45	W
2./NJG2	Hptm Heinz Rokker	Halifax	N Nienburg	19.19	48	W
11./NJG1	Hptm Hermann Greiner	Halifax	Nr. Hanover	19.22	46	W
2./NJG2	Hptm Heinz Rökker	Lancaster	NW Hanover	19.29	49	W
II./NJG1	Lt Otto Fries	Lancaster	N Groningen	19.30	17	W
8./NJG5	Ofw Herbert Altner	Halifax		19.35	21	W
I./NJG1	Hptm Werner Baake	Halifax	50km W Hanover	19.44	41	W
12./NJG1	Oblt Kurt Matzak	Halifax	NE Munster	19.55	9	W
4./NJG1	Hptm Fritz Lau	Halifax	20km NW Hesepe	19.58	23	W
IV./NJG1	Lt Hans Gienanth.von	Halifax	E Zutphen	20.15	8	W
8./NJG1	Lt Rolf Ebhardt	Lancaster	Zuiderzee		6	W
St.III./NJG2	Hptm Heinz Ferger	Halifax			22	W

Alfons Köster (left) was one of the original intruder pilots with I./NJG 2 and eventually transferred to NJG 3. With a credited score of 26 (although 28 claims have been found), he crashed fatally on 6/7th January 1945. Otto Fries, with his Bordfunker Alfred Staffa had, in May 1944, become the first crew in history to carry out a combat ejection while flying an He219. They claimed their last victory on 5/6th January and, on 16/17th January they were again attacked and shot down by a Mosquito - Flight Lieutenant Vaughan of 85 Squadron - and ejected for a second time. They both survived the war.

I./NJG1	Uffz Heeren	Halifax			1	W
8./NJG5	Hptm Werner Hopf	Halifax			18	W
8./NJG5	Hptm Werner Hopf	Halifax			19	W
8./NJG5	Hptm Werner Hopf	Halifax			20	W
8./NJG5	Hptm Werner Hopf	Halifax			21	W
Stab./NJG2	Maj Hans Karlowski	Halifax			6	W
Stab./NJG2	Maj Hans Karlowski	Halifax			7	W
Stab./NJG2	Maj Hans Karlowski	Halifax			8	W
1./NJG1	Hptm E-W Modrow	Halifax			34	W
III./NJG2	Ofw "Klaus" Moller	Halifax			1	W
IV./NJG3	Oblt Heinz Reuter	Halifax			10	W
III./NJG2	Oblt Gunter Rogge	Halifax			5	W
I./NJG1	Ofw Josef Strohlein	Lancaster			4	W
II./NJG11	Oblt Kurt Welter	Mosquito			33	W

6-7 January 1945

3./NJG3	Hptm Eduard Schroder	Lancaster	40km NW Helsingor	19.21	19	W
St.I./NJG5	Maj Werner Hoffmann	Halifax	Danziger Bight	21.12	45	W
Stab./NJG2	Maj Paul Semrau	Halifax			38	W
Stab./NJG2	Maj Paul Semrau	Halifax			39	W
Stab./NJG2	Maj Paul Semrau	Halifax			40	W

8-9 January 1945

3./NJG6	Fw Gunther Bahr	Lancaster	W Munich	22.27	21	W
3./NJG6	Fw Gunther Bahr	Lancaster	W Munich	22.30	22	W
2./NJG6	Ofw Ludwig Schmidt	Lancaster	S Munich	22.31	5	W

3./NJG6	Fw Gunther Bahr	B-24	Munich	22.36	23	W
3./NJG6	Fw Gunther Bahr	Lancaster	WNW Munich	22.38	24	W
St.I./NJG6	Oblt Wilhelm Engel	Lancaster	E Munich	22.39	16	W

10-11 January 1945

10./NJG11	Oblt Kurt Welter	Mosquito			34	W

12-13 January 1945

3./NJG3	Ofw Hans Schadowski	Halifax	Langeland	21.05	13	W
3./NJG3	Hptm Eduard Schroder	Halifax	SW Faborg	21.08	20	W
3./NJG3	Ofw Hans Schadowski	Halifax	W Abenra	21.20	14	W
3./NJG3	Hptm Eduard Schroder	Halifax	W Rome	21.26	21	W
8./NJG3	Hptm Fritz Brinkmann	Halifax		23.00	14	W

13-14 January 1945

II./NJG4	Uffz Scherer	Mosquito		22.34	1	W
I./NJG11	Hptm Friedrich-Karl Müller	Lancaster			28	W
3./NJG3	Ofw Hans Schadowski	Lancaster	50km NW Hanstholm	00.28	15	W

14-15 January 1945

Stab./NJG2	Hptm Gerhard Raht	Lancaster		20.09	43	W
2./NJG6	Ofw Ludwig Schmidt	Lancaster	W Geissen	23.10	6	W
III./NJG1	Lt Theodor Adamski	Lancaster			3	W
Stab./NJG5	Obstlt Walter Borchers	Lancaster			22	W
Stab./NJG5	Obstlt Walter Borchers	Lancaster			23	W
Stab./NJG5	Obstlt Walter Borchers	Lancaster			24	W
III./NJG5	Hptm Mohring	Lancaster			1	W
IV./NJG5	Oblt Karl Teufert	Lancaster	Halle		1	W
I./NJG6	Lt Artus	Lancaster	S Leuna	00.12	1	W
2./NJG6	Ofw Ludwig Schmidt	Lancaster		00.36	7	W
I./NJG100	Ofw Rudi Duding	Li-2			14	E
I./NJG100	Oblt G E Francsi	Li-2			43	E
I./NJG100	Oblt Klaus Scheer	Li-2	Olyta		21	E
I./NJG100	Oblt Klaus Scheer	Li-2			22	E

15-16 January 1945

8./NJG1	Hptm Dietrich Schmidt	Lancaster	NE Frankfurt-am-Oder	21.15	39	W
III./NJG1	Fw Heinrich Lahmann	Halifax		21.56	2	W
III./NJG1	Lt Sommerau	Halifax		21.56	1	W
IV./NJG3	Lt Hans Schafer	Lancaster		22.03	12	W
III./NJG2	Ofw Rudolf Mangelsdorf	Halifax	S Hanover	22.05	7	W
III./NJG2	Ofw Rudolf Mangelsdorf	Halifax	SW Hanover	22.11	8	W
IV./NJG3	Lt Hans Schafer	Lancaster		22.14	13	W
Stab./NJG5	Obstlt Walter Borchers	Lancaster			25	W
Stab./NJG5	Obstlt Walter Borchers	Lancaster			26	W
Stab./NJG5	Obstlt Walter Borchers	Lancaster			27	W
9./NJG5	Oblt Breitfeld	Halifax			7	W
9./NJG5	Oblt Breitfeld	Halifax			8	W
IV./NJG3	Hptm Johann Dreher	Halifax			1	W
2./NJG5	Oblt Peter Ehrhardt	Lancaster			25	W

The Germans had, by early 1944, realised the threat posed by the British Mosquito intruder force. An earlier attempt to produce a 'Luftwaffe Mosquito' in the form of the Me410 had not proved a spectacular success. Therefore a new approach was tried. This was the Focke Wulf TA154, a product of Prof. Kurt Tank's design team. This prototype was placed on the strength of NJG 3 for evaluation in December 1944 but, although the aircraft showed great promise, it never went into production.

St.III./NJG2	Hptm Heinz Ferger	Lancaster			21	W
8./NJG5	Hptm Werner Hopf	Halifax			15	W
8./NJG5	Hptm Werner Hopf	Halifax			16	W
8./NJG5	Hptm Werner Hopf	Halifax			17	W
IV./NJG3	Fw Rannefeld	Halifax			3	W
I./NJG1	Fw Sieben	Lancaster			2	W
III./NJG1	Lt Sommerau	Halifax			2	W
IV./NJG3	Hptm Tesmar.von	Halifax	Duderstadt		1	W
II./NJG2	Lt Martin Tschiersch	Lancaster			11	W
St.III./NJG3	Oblt Wiethuchter	Halifax	Magdeburg		1	W

16-17 January 1945

II./NJG11	Oblt Kurt Welter	Lancaster			35	W
II./NJG11	Oblt Kurt Welter	Lancaster			36	W

19-20 January 1945

5./NJG5	Oblt Werner Rapp	Il-4		20.00	15	E
I./NJG100	Fw Hausschild	Il-4			8	E
I./NJG100	Oblt Hans Schmid	Il-4			5	E

21-22 January 1945

IV./NJG1	Obstlt Hans-Joachim Jabs	Lancaster			31	W
I./NJG100	Lt August Kayser	Li-2			5	E

28-29 January 1945

2./NJG6	Ofw Ludwig Schmidt	Lancaster	Stuttgart	23.28	8	W
3./NJG6	Fw Gunther Bahr	Lancaster	Stuttgart	23.32	25	W
3./NJG6	Fw Gunther Bahr	Lancaster	W Stuttgart	23.39	26	W

3./NJG6	Fw Gunther Bahr	Lancaster	Ludwigsberg	23.46	27	W
St.I./NJG6	Hptm Gerhard Friedrich	Lancaster	S Wildbad	23.53	21	W
3./NJG6	Fw Gunther Bahr	Lancaster	E Mannheim	23.59	28	W

Chapter Twelve

Warriors to the End

February - May 1945

With the Allied ground forces advancing from west, south and east, the Luftwaffe air defences were now in a hopeless situation. Fuel shortages had become acute and units literally began to disappear. Trained aircrews were handed rifles and Panzerfaust anti-tank weapons and were ordered to become infantry soldiers. Yet there were still some surprises in store for the Allies. *Oberleutnant* Kurt Welter of *II./NJG 11* had, in December 1944, been ordered to form *10./NJG 11*, a unit specifically established to hunt the elusive RAF Mosquitos. To this end the unit was given Me262A fighters, modified with radar equipment. Four Me262B nightfighters also arrived and, on the morning of 8th February, *Oberfeldwebel* Herbert Altner landed his 8./NJG 5 Ju88 at Burg airfield after a successful mission to be greeted by Welter. *'Any luck?'* he was asked and, when he replied that he had shot down three Halifaxes, Welter said *'How would you like to join my Me262 Staffel?'* Altner agreed and, within a few days, became the only qualified pilot to fly the Me262 two-seater in combat. But even the Me262 could not halt the inevitable.

1-2 February 1945

St.II./NJG6	Oblt Johannes Engels	Lancaster	Mannheim	19.20	2	W
St.I./NJG6	Hptm Gerhard Friedrich	Lancaster	Ludwigshafen	19.20	22	W
St.II./NJG6	Oblt Johannes Engels	Lancaster	Bruschsal	19.23	3	W
St.II./NJG6	Hptm Helmuth Schulte	Lancaster	Mannheim	19.30	21	W
St.II./NJG6	Hptm Helmuth Schulte	Lancaster	Bruschsal	19.33	22	W
St.II./NJG6	Oblt Peter Spoden	Lancaster	SW Mannheim	19.34	19	W
I./NJG2	Oblt Erich Jung	Lancaster	SW Kahlsruhe	19.43	19	W
2./NJG2	Hptm Heinz Rökker	Lancaster	N Koblenz	19.46	50	W
St.II./NJG6	Ofw Alfred Siewert	Mosquito	W Waibstadt	23.30	3	W
8./JG301	Ofw Hans Todt	U-2			2	W
St./NJG2	Maj Hans Karlowski	Lancaster			9	W
Stab./NJG2	Maj Hans Karlowski	Lancaster			9	W

2-3 February 1945

St.I./NJG6	Hptm Gerhard Friedrich	Lancaster	Karlsruhe	23.20	23	W
2./NJG6	Ofw Ludwig Schmidt	Lancaster	Karlsruhe	23.21	9	W

The jet age arrives in the night air war. (left) One of the Me262B two-seat fighters operated by 10./NJG 11 under evaluation after the war. (right) An Me262A1A single-seat fighter, showing the nose radar array.

Stab./NJG6	Maj Herbert Lütje	Lancaster	W Karlsruhe	23.24	43	W
St.I./NJG6	Hptm Gerhard Friedrich	Lancaster	Karlsruhe	23.25	24	W
St.IV./NJG6	Oblt Engelbert Heiner	Lancaster	S Karlsruhe	23.25	12	W
St.IV./NJG6	Oblt Engelbert Heiner	Lancaster	Karlsruhe	23.29	13	W
St.IV./NJG6	Oblt Engelbert Heiner	Lancaster	Karlsruhe	23.32	14	W
St.II./NJG6	Hptm Helmuth Schulte	Lancaster	Karlsruhe	23.40	23	W
3./NJG2	Oblt Breithaupt	Halifax	S Bocholt	23.45	1	W
Stab./NJG2	Hptm Gerhard Raht	Halifax	SW Mannheim	23.49	44	W
II./NJG4	Lt Wolfgang Bretschneider	Lancaster	NE Riedelbach		1	W

3-4 February 1945

2./NJG2	Hptm Heinz Rökker	Lancaster	NE Krefeld	19.31	51	W
2./NJG2	Hptm Heinz Rökker	Lancaster	N Geldern	19.51	52	W
2./NJG2	Hptm Heinz Rökker	Lancaster	N Geldern	19.56	53	W
Stab./NJG2	Hptm Gerhard Raht	Lancaster		20.00	45	W
I./NJG1	Hptm Werner Baake	Lancaster	15km NE Aachen	20.07	42	W
Stab./NJG4	Maj H-W Schnaufer	Lancaster	NE Krefeld	20.09	107	W
I./NJG1	Hptm Alex Ressugier	Lancaster	Roermond	20.30	4	W

7-8 February 1945

Stab./NJG2	Hptm Gerhard Raht	Four-eng		22.22	46	W
Stab./NJG2	Hptm Gerhard Raht	Four-eng		22.41	47	W
Stab./NJG2	Hptm Gerhard Raht	Four-eng		23.00	48	W
8./NJG5	Ofw Herbert Altner	Halifax	Stettin		22	W
8./NJG5	Ofw Herbert Altner	Halifax	Stettin		23	W
8./NJG5	Ofw Herbert Altner	Halifax	Stettin		24	W
10./NJG1	Oblt Georg Fengler	Lancaster			12	W
10./NJG1	Oblt Georg Fengler				13	W
Stab./NJG2	Hptm Gerhard Raht	Four-eng		00.08	49	W
Stab./NJG2	Hptm Gerhard Raht	Four-eng		00.16	50	W
2./NJG2	Hptm Heinz Rökker	Boston	Eindhoven airfield	00.21	54	W
Stab./NJG2	Hptm Gerhard Raht	Four-eng		00.23	51	W

8-9 February 1945

1./NJG5	Hptm E-G Drünkler	Lancaster	SW Stettin	21.03	31	W
I./NJG3	Hptm Herbert Koch	Lancaster	NE Copenhagen	21.04	18*	W
1./NJG5	Hptm E-G Drünkler	Lancaster	SW Stettin	21.08	32	W
1./NJG5	Hptm E-G Drünkler	Lancaster	S Stettiner Hafen	21.12	33	W
III./NJG2	Ofw Rudolf Mangelsdorf	Lancaster		22.03	9	W
Stab./NJG5	Obstlt Walter Borchers	Lancaster			28	W
Stab./NJG5	Obstlt Walter Borchers	Lancaster			29	W
Stab./NJG5	Obstlt Walter Borchers	Lancaster			30	W

13-14 February 1945

II./NJG5	Maj Hans Leickhardt	Lancaster			11	W
II./NJG5	Maj Hans Leickhardt	Lancaster			12	W

14-15 February 1945

Stab./NJG6	Maj Herbert Lutje	Lancaster	50km NE Frankfurt-a-M	20.27	44	W
St.I./NJG3	Maj Werner Husemann	Halifax	Seeland Island	20.41	29	W
1./NJG5	Hptm E-G Drünkler	Lancaster	NW Chemnitz	21.01	34	W

Engelbert Heiner was one of the 'mystery men' of the Nachtjagd. He had seen service with the Condor Legion in Spain as a bomber pilot and flew with 9./KG 27 in the campaigns over France, England and in Russia. Here he served with an unknown nightfighter unit attached to KG 27, achieving eleven victories and winning the Ritterkreuz. He joined IV./NJG 6 in 1944, becoming Kapitän of 12./ NJG 6, with which unit he gained four further victories (listed as 12 - 15 in the tables). No details of victories 1 - 11 are known. His demise is similarly puzzling. He is reported to have been shot down by a British intruder near Gelnhausen on 18/19th March 1945, but no British claim has been traced. Possibly his Ju88 was brought down by Warrant Officer D.Taylor of 157 Squadron, reported on 17/18th March and that an error of 24 hours has occurred in either the British or German records. Heiner and his crew were all killed.

3./NJG3	Hptm Eduard Schröder	Lancaster	Sjaeland	21.01	22	W
I./NJG3	Hptm Herbert Koch	Lancaster	40km NE Moen	21.04	19	W
St.I./NJG3	Maj Werner Husemann	Halifax	Moen Island	21.17	30	W
I./NJG3	Hptm Herbert Koch	Lancaster	75km N Bornholm	21.31	20	W
11./NJG6	Hptm Kurt-Heinz Weigel	Lancaster	Plauen	21.37	4	W
I./NJG3	Hptm Herbert Koch	Lancaster	50km E Odense	21.45	21	W
4./NJG6	Ofhr Helmut Bunje	Lancaster	W Saalfeld	21.50	6	W
St.I./NJG5	Maj Werner Hoffmann	B-17	SW Bad Kissingen	21.58	46	W
2./NJG2	Hptm Heinz Rökker	Lancaster	S Fulda	22.03	55	W
St.I./NJG5	Maj Werner Hoffmann	Lancaster	SW Bad Kissingen	22.11	47	W
St.IV./NJG6	Oblt Engelbert Heiner	Lancaster	SW Chemnitz	00.43	15	W
2./NJG6	Ofw Ludwig Schmidt	Lancaster	S Chemnitz	00.44	10	W

15 February 1945

II./NJG11	Fw Karl-Heinz Becker	P-38	5km N Spremberg	15.46	1	W

15-16 February 1945

I./NJG100	Oblt G E Francsi	Il-4			44	E
I./NJG100	Fw Hausschild	Il-4			9	E

18-19 February 1945

IV./NJG3	Hptm Johann Dreher	Lancaster			2	W
IV./NJG3	Oblt Heinz Reuter	Lancaster			11	W

20-21 February 1945

I./NJG11	Hptm Friedrich-Karl Müller	Four-eng			29	W
I./NJG11	Hptm Friedrich-Karl Müller	Four-eng			30	W
II./NJG1	Oblt Georg Fengler	Four-eng			10	W
II./NJG1	Oblt Georg Fengler	Four-eng			11	W
Stab./NJG1	Obstlt Hans-Joachim Jabs	Lancaster			32	W
Stab./NJG1	Obstlt Hans-Joachim Jabs	Lancaster			33	W
12./NJG1	Oblt Kurt Matzak	Lancaster	55km SSW Dortmund		12	W
I./NJG1	Hptm Schirrmacher	Lancaster			1	W
St.II./NJG5	Oblt Bernhard Jungmann	Four-eng			3	W
II./NJG4	Ofw Georg Matt	Four-eng			3	W
III./NJG1	Lt Stocker	Four-eng		00.55	1	W
III./NJG1	Lt Stocker	Four-eng		01.04	2	W

12./NJG1	Hptm Josef Kraft	Halifax	Duisburg	01.05	50	W
IV./NJG1	Uffz Grosser	Four-eng	Dortmund	01.06	4	W
12./NJG1	Hptm Josef Kraft	Halifax	Duisburg	01.09	51	W
4./NJG1	Hptm Dietrich Schmidt	Lancaster		01.10	40	W
II./NJG1	Hptm Adolf Breves	Lancaster	70km S Dortmund	01.11	14	W
II./NJG2	Ofw Robert Luddeke	Four-eng		01.12	10	W
II./NJG2	Ofw Robert Luddeke	Four-eng		01.14	11	W
II./NJG4	Maj Hubert Rauh	Four-eng		01.15	25	W
II./NJG1	Hptm Adolf Breves	Lancaster	50km S Dortmund	01.19	15	W
Stab./NJG4	Maj H-W Schnaufer	Lancaster	WSW Mönchengladbach	01.53	108	W
Stab./NJG4	Maj H-W Schnaufer	Lancaster	SW Roermond	01.58	109	W

21-22 February 1945

3./NJG6	Fw Günther Bahr	Halifax	SW Worms	20.34	29	W
2./NJG6	Ofw Ludwig Schmidt	Halifax	S Worms	20.34	11	W
3./NJG6	Fw Günther Bahr	Halifax	SW Worms	20.37	30	W
12./NJG1	Oblt Kurt Matzak	B-24	Venlo	20.37	10	W
2./NJG6	Ofw Ludwig Schmidt	Halifax	S Worms	20.38	12	W
3./NJG6	Fw Günther Bahr	Halifax	SW Worms	20.39	31	W
St.IV./NJG6	Hptm Martin Becker	Halifax	SW Worms	20.39	46	W
2./NJG6	Ofw Ludwig Schmidt	Halifax	SW Worms	20.39	13	W
St.II./NJG6	Oblt Peter Spoden	Halifax	30km SW Worms	20.40	20	W
3./NJG6	Fw Günther Bahr	Halifax	SW Worms	20.42	32	W
St.IV./NJG6	Hptm Martin Becker	Lancaster	NW Worms	20.42	47	W
Stab./NJG6	Maj Herbert Lütje	Halifax	SW Worms	20.43	45	W
11./NJG6	Fw Emil Weinmann	Halifax	Worms	20.43	4	W
3./NJG6	Fw Günther Bahr	Halifax	SW Worms	20.44	33	W
St.I./NJG6	Oblt Wilhelm Engel	Halifax	SW Worms	20.44	17	W
./NJG4	Maj H-W Schnaufer	Lancaster	NNW Münster	20.44	110	W
St.I./NJG6	Hptm Gerhard Friedrich	Halifax	10km SW Worms	20.45	25	W
3./NJG6	Fw Günther Bahr	Halifax	SW Worms	20.46	34	W
2./NJG2	Hptm Heinz Rökker	Lancaster	S Wageningen	20.46	56	W
St.IV./NJG6	Hptm Martin Becker	Halifax	NW Worms	20.47	48	W
Stab./NJG4	Maj H-W Schnaufer	Lancaster	SE Enschelde	20.48	111	W
11./NJG6	Fw Emil Weinmann	Halifax	W Worms	20.48	5	W
11./NJG6	Fw Emil Weinmann	Halifax	W Worms	20.49	6	W
3./NJG6	Fw Günther Bahr	Halifax	SW Worms	20.50	35	W
Stab./NJG4	Maj H-W Schnaufer	Lancaster	S Enschelde	20.51	112	W
Stab./NJG6	Maj Herbert Lütje	Halifax	Saarbrucken	20.55	46	W
Stab./NJG4	Maj H-W Schnaufer	Lancaster	E Zutphen	20.55	113	W
2./NJG2	Hptm Heinz Rökker	Lancaster	SW Hertogenbosch	20.56	57	W
Stab./NJG4	Maj H-W Schnaufer	Lancaster	Bocholt	20.58	114	W
Stab./NJG4	Maj H-W Schnaufer	Lancaster	SW Kleve	21.00	115	W
Stab./NJG4	Maj H-W Schnaufer	Lancaster	E s'Hertogenbosch	21.03	116	W
2./NJG2	Hptm Heinz Rökker	Lancaster	SE Eindhoven	21.06	58	W
2./NJG2	Hptm Heinz Rökker	Lancaster	SW Eindhoven	21.12	59	W
2./NJG2	Hptm Heinz Rökker	Lancaster	SW Eindhoven	21.13	60	W
2./NJG2	Hptm Heinz Rökker	Lancaster	SW Eindhoven	21.19	61	W
6./NJG1	Oblt Johannes Hager	Halifax		23.09	38	W
6./NJG1	Oblt Johannes Hager	Halifax		23.11	39	W
II./NJG1	Hptm Adolf Breves	Lancaster	20km NW Krefeld	23.12	16	W

On the night of 21st February 1945, Bomber Command attacked Duisburg, Worms and the Mitteland Canal, losing 34 bombers. Heinz Rökker (left), and Johannes Hager claimed six apiece, while Günther Bahr and Heinz Wolfgang Schnaufer each claimed seven. Bahr would end the war with 37 victories including one as a Zerstörer pilot. Schnaufer had raised his personal tally to 116 almost by accident. Taking off late, he ran straight into the bomber stream near Münster and shot down seven Lancasters in succession.

6./NJG1	Oblt Johannes Hager	Halifax		23.12	40	W
II./NJG1	Lt Barthel	Lancaster	10km NW Krefeld	23.15	4	W
6./NJG1	Oblt Johannes Hager	Halifax		23.17	41	W
4./NJG1	Hptm Dietrich Schmidt	Lancaster	10km W Venlo	23.18	41	W
6./NJG1	Oblt Johannes Hager	Halifax		23.19	42	W
II./NJG1	Hptm Adolf Breves	Lancaster	50km NW Krefeld	23.20	17	W
6./NJG1	Oblt Johannes Hager	Halifax		23.22	43	W
III./NJG11	Ofw Frank	Lancaster	NW Neuss	23.23	1	W
4./NJG1	Hptm Dietrich Schmidt	Lancaster	25km SW Goch	23.25	42	W
11./NJG1	Hptm Hermann Greiner	Halifax	Nr. Neuss	23.31	47	W
III./NJG11	Fw Tahler	Lancaster		23.36	1	W
11./NJG1	Hptm Hermann Greiner	Halifax	Nr. Neuss	23.40	48	W
II./NJG2	Maj Eckhoff	Lancaster			2	W
II./NJG2	Maj Eckhoff	Lancaster			3	W
II./NJG2	Maj Eckhoff	Lancaster			4	W
II./NJG2	Maj Eckhoff	Lancaster			5	W
II./NJG2	Maj Eckhoff	Lancaster			6	W
II./NJG100	Oblt Karl Gerber	Il-4			1	E
II./NJG11	Fw Reichenbach	Mosquito			3	W
II./NJG11	Oblt Kurt Welter	Mosquito			37	W
II./NJG11	Oblt Kurt Welter	Mosquito			38	W

22-23 February 1945

II./NJG11	Fw Reichenbach	Mosquito			4	W
II./NJG11	Fw Reichenbach	Mosquito			5	W
II./NJG11	Ofw Richards	Mosquito			1	W
II./NJG11	Oblt Kurt Welter	Mosquito			39	W
II./NJG11	Oblt Kurt Welter	Mosquito			40	W
II./NJG100	Oblt Karl Gerber	Il-4			2	E

23-24 February 1945

St.I./NJG6	Hptm Gerhard Friedrich	Lancaster	E Pforzheim	20.00	26	W
St.I./NJG6	Oblt Wilhelm Engel	Lancaster	Nr. Pforzheim	20.05	18	W
St.I./NJG6	Hptm Gerhard Friedrich	Lancaster	E Pforzheim	20.08	27	W
3./NJG6	Fw Günther Bahr	Lancaster	Nr. Pforzheim	20.09	36	W

On the night of 3rd March, the Luftwaffe launched 'Operation Gisela', with over 140 Ju88 crews briefed to fly a mass intruder mission over England. The German crews succeeded in shooting down 23 bombers and one Mosquito, with more than 20 other aircraft damaged. However, the Germans also sustained losses, for eight fighters failed to return, nine more crashed with the loss of three crews and a further eleven had sustained landing damage. Hauptmann Johann Dreher of 13./NJG 3 (right) was killed with his crew when he flew into the ground near Elvington and Feldwebel Heinrich Conze of 7./NJG 5 perished with his crew in a crash at Welton, Lincolnshire. Dreher, who had won the Ritterkreuz as a bomber pilot in Russia, had been credited with two victories before his death. Opposite Page: (top) Lancaster ME323 of 12 Squadron, shot down near Blyton at 01.10 hours killing Flying Officer Thomas and his crew. (lower) Halifax LV255 of 192 Squadron, crash-landed at Fulmodeston, Norfolk, by Flying Officer E.D.Roberts at 00.59 hours. Five of the crew were killed and the rest injured.

St.I./NJG6	Oblt Wilhelm Engel	Lancaster	ESE Pforzheim	20.09	19	W
St.II./NJG6	Oblt Peter Spoden	Lancaster	30km SW Pforzheim	20.11	21	W
St.I./NJG6	Hptm Gerhard Friedrich	Lancaster	SE Pforzheim	20.12	28	W
4./NJG6	Ofhr Helmut Bunje	Lancaster	W Tubingen	20.15	7	W
St.I./NJG6	Hptm Gerhard Friedrich	Lancaster	S Pforzheim	20.15	29	W
St.II./NJG6	Oblt Peter Spoden	Lancaster	30km SW Pforzheim	20.15	22	W
St.I./NJG6	Oblt Wilhelm Engel	Lancaster	SE Pforzheim	20.18	20	W
4./NJG6	Ofhr Helmut Bunje	Lancaster	S Tubingen	20.20	8	W
St.I./NJG6	Oblt Wilhelm Engel	Lancaster	SE Pforzheim	20.22	21	W
4./NJG6	Ofhr Helmut Bunje	Lancaster	SW Freudenstadt	20.30	9	W
II./NJG4	Lt Wolfgang Bretschneider	Lancaster	NE Riedelbach		2	W

24-25 February 1945

12./NJG1	Oblt Kurt Matzak	B-17	25km NW Düsseldorf	20.08	11	W
II./NJG11	Ofw Kreibaum	Lancaster	NW Krefeld	20.12	1	W
12./NJG1	Hptm Josef Kraft	Halifax	Duisburg	20.15	52	W
St.I./NJG100	Oblt G E Francsi	Il-4			45	E
II./NJG2	Maj Eckhoff	Halifax			7	W
IV./NJG3	Oblt Heinz Reuter	Halifax			12	W

25-26 February 1945

3./NJG3	Hptm Eduard Schröder	Halifax	NW Skagen	21.25	23	W

27-28 February 1945

IV./NJG3	Lt Hans Schäfer	B-24	SW Heligoland	20.35	14	W

2-3 March 1945

II./NJG11	Oblt Kurt Welter	Mosquito			41	W
II./NJG11	Oblt Kurt Welter	Mosquito			42	W
II./NJG11	Oblt Kurt Welter	Mosquito			43	W
St.III./NJG1	Maj Martin Drewes	Lancaster	W Münster	01.24	47	W

3-4 March 1945

Stab./NJG4	Maj H-W Schnaufer	Lancaster	SW Osnabrück	21.55	117	W
11./NJG1	Hptm Hermann Greiner	Lancaster	Nr. Dortmund	21.57	49	W
12./NJG1	Hptm Josef Kraft	Lancaster	W Münster	21.59	53	W
Stab./NJG4	Maj H-W Schnaufer	Lancaster	NE Münster	22.04	118	W
12./NJG1	Hptm Josef Kraft	Lancaster	W Münster	22.07	54	W
11./NJG1	Hptm Hermann Greiner	Lancaster	Nr. Dortmund	22.08	50	W
11./NJG1	Hptm Hermann Greiner	Lancaster	Nr. Dortmund	22.12	51	W
6./NJG2	Oblt Franz Brinkhaus	Hudson	NE Cambridge		13	W
IV./NJG3	Lt Arnold Döring	B-17	Oulton		3	W
IV./NJG3	Lt Arnold Döring	Lancaster	England		4	W
11./NJG3	Oblt Josef Förster	Four-eng	England		13	W
11./NJG3	Oblt Josef Förster	Four-eng	England		14	W
II./NJG2	Hptm H-H Hissbach	Four-eng	England		29	W

II./NJG2	Hptm H-H Hissbach	Four-eng	England		30	W
III./NJG2	Fw Heinz Koppe	Lancaster	England		7	W
IV./NJG3	Ofw Heinz Misch	Four-eng	England		9	W
9./NJG4	Hptm Kurt Fladrich	Lancaster	NE Cambridge	00.33	16	W
7./NJG2	Oblt Walter Briegleb	Lancaster	S Waddington	00.36	25	W
7./NJG2	Oblt Walter Briegleb	Lancaster	W Lincoln	00.56	26	W
Stab./NJG2	Hptm Gerhard Raht	Four-eng	England	01.04	52	W
9./NJG5	Lt Robert Wolf	Lancaster	Humber Estuary	01.08	18	W
Stab./NJG2	Hptm Gerhard Raht	Four-eng	England	01.26	53	W
IV./NJG3	Fw Schmidt	Halifax	Nr. Croft	02.14	1	W

4-5 March 1945

St.I./NJG3	Maj Werner Husemann	Lancaster	E Aarhus	20.29	31	W
St.I./NJG3	Maj Werner Husemann	Lancaster	Bourtanger Moor	23.32	32	W

5-6 March 1945

II./NJG4	Maj Hubert Rauh	Four-eng		20.22	26	W
12./NJG1	Oblt Kurt Matzak	Halifax		20.30	13	W
II./NJG4	Maj Hubert Rauh	Four-eng		20.40	27	W
II./NJG4	Maj Hubert Rauh	Four-eng		20.50	28	W
1./NJG5	Fw Schuster	Lancaster		21.51	1	W
1./NJG5	Hptm E-G Drünkler	Four-eng	SE Zwickau	22.05	35	W
Stab./NJG5	Obstlt Walter Borchers	Four-eng			31	W
Stab./NJG5	Obstlt Walter Borchers	Four-eng			32	W
II./NJG2	Maj Eckhoff	Four-eng			8	W
II./NJG2	Maj Eckhoff	Four-eng			9	W
II./NJG2	Maj Eckhoff	Four-eng			10	W
II./NJG2	Maj Eckhoff	Four-eng			11	W
II./NJG2	Maj Eckhoff	Four-eng			12	W
II./NJG100	Oblt Karl Gerber	Wellington			3	W
IV./NJG1	Ofw Gerhard Graefe	Four-eng			3	W
II./NJG5	Hptm Hans Leickhardt	Four-eng			13	W
III./NJG1	Oblt Sommerau	Lancaster			3	W
II./NJG100	Hptm Paul Zorner	B-24	SSE Graz	01.30	59	W

6-7 March 1945

II./NJG11	Oblt Kurt Welter	Mosquito			45	W

7-8 March 1945

I./NJG1	Hptm Werner Baake	Lancaster	20km NW Boxmeer	20.17	43	W
1./NJG4	Hptm Ludwig Meister	Lancaster	N Kassel	20.30	39	W
Stab./NJG4	Maj H-W Schnaufer	Lancaster	E Attendoorn	20.41	119	W
Stab./NJG4	Maj H-W Schnaufer	Lancaster	SSW Korbach	20.47	120	W
II./NJG4	Maj Hubert Rauh	Lancaster	Friesenhagen-Altenkirchen	20.55	29	W
II./NJG4	Maj Hubert Rauh	Lancaster		21.42	30	W
Stab./NJG4	Maj H-W Schnaufer	Lancaster	NE Helmstedt	21.56	121	W
IV./NJG3	Fw Egon Engling	Lancaster		22.10	8	W
IV./NJG3	Fw Egon Engling	Lancaster		22.11	9	W
IV./NJG3	Ofw Heinz Misch	Lancaster	SE Bremervörde	22.12	10	W
IV./NJG3	Fw Egon Engling	Lancaster		22.13	10	W
4./NJG3	Fw Bruno Rupp	Lancaster		22.15	12	W

(left) In early 1945 NJG 102 began ro re-equip with Ju88G-6 fighters. Note the camouflage. the black-outline crosses and the Schräge Musik cannon. (right) Major Paul Zorner was another 'first-to-last' Luftwaffe pilot. Originally a tranport pilot, he transferred to nightfighting in 1940 and served with NJG 3, NJG 5 and finally became Kommandeur II./NJG 100. He claimed his 59th and last combat success on 6th March 1945 and survived the war.

1./NJG5	Hptm E-G Drünkler	Lancaster	NE Erfurt	22.30	36	W
1./NJG5	Fw Burkhard	Lancaster	SSW Meinungen	22.45	1	W
8./NJG3	Hptm Fritz Brinkmann	Lancaster	Bremen	22.50	15	W
IV./NJG3	Oblt Fritz Brandt	Lancaster	Hamburg		7	W
9./NJG5	Oblt Breitfeld	Halifax			9	W
II./NJG2	Maj Eckhoff	Lancaster			13	W
II./NJG2	Maj Eckhoff	Lancaster			14	W
II./NJG2	Maj Eckhoff	Lancaster			15	W
II./NJG2	Maj Eckhoff	Lancaster			16	W
II./NJG2	Maj Eckhoff	Lancaster			17	W
10./NJG1	Oblt Georg Fengler	Lancaster			14	W
10./NJG1	Oblt Georg Fengler	Lancaster			15	W
I./NJG2	Oblt Erich Jung	Four-eng	Dessau		20	W
I./NJG2	Oblt Erich Jung	Lancaster			21	W
III./NJG5	Oblt Rehkate	Halifax			1	W
III./NJG5	Oblt Rehkate	Halifax			2	W
I./NJG1	Fw Sieben	Lancaster			3	W
I./NJG1	Oblt Thurner	Lancaster			2	W

8-9 March 1945

St.II./NJG6	Oblt Peter Spoden	Lancaster	Dessau	22.06	23	W
St.II./NJG6	Hptm Helmuth Schulte	Lancaster		22.29	24	W
St.II./NJG6	Oblt Peter Spoden	Lancaster	50km SW Dessau	22.29	24	W
I./NJG100	Ofw Rudi Duding	Il-4			16	E
I./NJG100	Oblt G E Francsi	Il-4			55	E
I./NJG100	Oblt G E Francsi	Il-4			56	E
I./NJG100	Lt Helmut Konter	Il-4			20	E
I./NJG100	Lt Hermann Weihe	Il-4			3	E

11 March 1945

I./NJG100	Fw Hausschild	Il-4			10	E

12-13 March 1945

3./NJG3	Hptm Eduard Schröder	Lancaster	W Samso	21.16	24	W
3./NJG3	Hptm Eduard Schröder	Lancaster	S Hov	21.27	25	W
3./NJG3	Hptm Eduard Schröder	Lancaster	75km N Bornholm	21.45	26	W

14-15 March 1945

Stab./NJG6	Maj Herbert Lütje	Lancaster	NW Nuremberg	21.24	47	W
St.IV./NJG6	Hptm Martin Becker	Lancaster	Bad Berka	21.53	49	W
St.IV./NJG6	Hptm Martin Becker	Lancaster	2km NE Apolda	21.59	50	W
St.IV./NJG6	Hptm Martin Becker	Lancaster	Naumberg	22.03	51	W
St.IV./NJG6	Oblt Daborer	Lancaster	SW Leuna	22.04	2	W
St.IV./NJG6	Hptm Martin Becker	Lancaster	3km S Naumburg	22.05	52	W
St.IV./NJG6	Hptm Martin Becker	Lancaster	5km S Naumburg	22.06	53	W
4./NJG6	Ofhr Helmut Bunje	Lancaster	10km W Leuna	22.10	10	W
St.IV./NJG6	Hptm Martin Becker	Lancaster	10km NE Jena	22.15	54	W
St.IV./NJG6	Hptm Martin Becker	Lancaster	8km SW Heidelheim	23.00	55	W
St.IV./NJG6	Hptm Martin Becker	Lancaster	3km W Göppingen	23.15	56	W
St.I./NJG6	Hptm Gerhard Friedrich	Lancaster		23.15	30	W
St.I./NJG6	Hptm Gerhard Friedrich	Lancaster		23.18	31	W
St.I./NJG6	Hptm Gaul	Lancaster	Nr. Pforzheim	23.21	3	W
St.I./NJG6	Hptm Gerhard Friedrich	Lancaster		23.24	32	W
St.IV./NJG6	Hptm Martin Becker	Lancaster	5km SE Baiersbronn	23.37	57	W

15-16 March 1945

Stab./NJG2	Hptm Gerhard Raht	Four-eng		20.49	54	W
2./NJG2	Hptm Heinz Rökker	Four-eng	N Düsseldorf	20.50	62	W
2./NJG2	Hptm Heinz Rökker	Four-eng	N Düsseldorf	20.52	63	W
Stab./NJG2	Hptm Gerhard Raht	Four-eng		21.00	55	W
Stab./NJG2	Hptm Gerhard Raht	Four-eng		21.09	56	W
Stab./NJG2	Hptm Gerhard Raht	Four-eng		21.14	57	W
Stab./NJG2	Hptm Gerhard Raht	Four-eng		21.20	58	W
2./NJG2	Hptm Heinz Rökker	B-25	flugplatz St.Trond	21.26	64	W
2./NJG2	Hptm Heinz Rökker	Mosquito	flugplatz St.Trond	21.34	65	W
III./NJG6	Oblt Wilhelm Johnen	Lancaster	15km SE Wurzburg	02.45	34	W

16-17 March 1945

St.I./NJG6	Hptm Gerhard Friedrich	Lancaster	7km S Echterdingen	21.03	33	W
2./NJG6	Ofw Ludwig Schmidt	Lancaster		21.07	14	W
I./NJG2	Oblt Erich Jung	Lancaster	Nuremberg	21.12	22	W
I./NJG2	Oblt Erich Jung	Lancaster	Crailsheim	21.12	23	W
I./NJG2	Oblt Erich Jung	Lancaster	Nuremberg	21.12	24	W
I./NJG2	Oblt Erich Jung	Lancaster	Nuremberg	21.12	25	W
I./NJG2	Oblt Erich Jung	Lancaster	Nuremberg	21.12	26	W
I./NJG2	Oblt Erich Jung	Lancaster	Nuremberg	21.12	27	W
I./NJG2	Oblt Erich Jung	Lancaster	W Nuremburg	21.12	28	W
4./NJG6	Ofhr Helmut Bunje	Lancaster	S Schwabisch Hall	21.14	11	W
St.I./NJG6	Hptm Gaul	Lancaster		21.14	4	W
1./NJG5	Hptm E-G Drünkler	Lancaster	Ansbach	21.15	37	W
St.II./NJG6	Hptm Helmuth Schulte	Lancaster		21.15	25	W
1./NJG5	Hptm E-G Drünkler	Lancaster	E Ansbach	21.17	38	W
11./NJG6	Fw Emil Weinmann	Lancaster	Stuttgart	21.17	7	W

Oberleutnant Martin 'Tino' Becker of 2./NJG 6 claimed seven Halifaxes on the disastrous Nuremburg raid of 30/31st March. These were his 20th - 26th victories. He surpassed this on the night of 14th March 1945 when no less than nine Lancasters were credited to him, bringing his score to 57. He would survive the war with a total of 58 night victories. Of the Nuremburg raid, he later said, "They seemed to be lining up to be shot down. I just had to stop after the seventh one, I was sick of the killing."

St.I./NJG5	Maj Werner Hoffmann	Lancaster	SW Schwabisch Hall	21.18	48	W
11./NJG6	Fw Emil Weinmann	Lancaster		21.24	8	W
4./NJG6	Ofhr Helmut Bunje	Lancaster	N Schwabisch Hall	21.26	12	W
St.I./NJG5	Maj Werner Hoffmann	Halifax	Ansbach	21.26	49	W
11./NJG6	Fw Emil Weinmann	Lancaster		21.27	9	W
St.I./NJG5	Maj Werner Hoffmann	Lancaster	E Ansbach	21.30	50	W
2./NJG6	Ofw Ludwig Schmidt	Lancaster		21.30	15	W
3./NJG6	Fw Günther Bahr	Lancaster	SW Nuremberg	21.31	37	W
1./NJG5	Fw Schuster	Lancaster	N Nuremberg	21.31	2	W
Stab./NJG6	Maj Herbert Lütje	Lancaster	SSW Nuremberg	21.33	48	W
St.IV./NJG6	Hptm Martin Becker	Lancaster	30-50km W Nuremberg	21.35	58	W
II./NJG101	Lt Hans Rasper	Lancaster	NW Nuremberg	21.35	8	W
St.II./NJG6	Oblt Johannes Engels	Lancaster	N Nuremberg	21.37	4	W
2./NJG6	Ofw Ludwig Schmidt	Lancaster	Nuremberg	21.39	16	W
4./NJG6	Lt Lange	Lancaster	WNW Nuremberg	21.40	3	W
1./NJG5	Hptm E-G Drünkler	Lancaster	20km S Ansbach	21.43	39	W
I./NJG2	Oblt Erich Jung	Lancaster	20km E Würzburg	21.43	27	W
11./NJG6	Fw Emil Weinmann	Lancaster	SW Nuremberg	21.55	10	W
10./NJG6	Fw Josef Brunner	Lancaster	E Ulm	22.06	7	W
St.I./NJG6	Hptm Gaul	Lancaster		22.30	5	W
I./NJG2	Oblt Erich Jung	Four-eng	Dessau		29	W

17-18 March 1945

III./NJG2	Fw Hommel	Lancaster	Hull	21.55	1	W
II./NJG2	Fw Rudolf Morenz	Lancaster	Pocklington		4	W
II./NJG2	Fw Rudolf Morenz	Halifax	Pocklington		5	W
II./NJG11	Oblt Kurt Welter	Mosquito			45	W
II./NJG11	Oblt Kurt Welter	Mosquito			46	W

18-19 March 1945

II./NJG2	Oblt Friedrich Ballier	Halifax			1	W
I./NJG100	Oblt G E Francsi	Il-4			48	E

I./NJG100	Oblt G E Francsi	Il-4			49	E
III./NJG1	Lt Stocker	Halifax		04.19	3	W
4./NJG1	Hptm Fritz Lau	Halifax	15km NE Dortmund	04.20	24	W
Stab./NJG6	Maj Herbert Lütje	Lancaster		04.45	49	W
St.IV./NJG6	Ofw Erichsen	Lancaster	Hanau	04.50	2	W

20-21 March 1945

St.III./NJG3	Maj Walter Barte	Lancaster	W Hamburg		20	W
Stab./NJG2	Maj Hans Karlowski	Lancaster			10	W
II./NJG2	Fw Rudolf Morenz	Lancaster			6	W
II./NJG2	Fw Rudolf Morenz	Stirling			7	W
III./NJG2	Unknown	Lancaster	Humber Estuary		-	W
6./NJG1	Oblt Johannes Hager	Four-eng		01.12	44	W
6./NJG1	Oblt Johannes Hager	Four-eng		01.23	45	W
6./NJG1	Oblt Johannes Hager	Lancaster	SW Kassel	03.01	46	W
6./NJG1	Oblt Johannes Hager	Lancaster	SW Kassel	03.04	47	W
6./NJG1	Oblt Johannes Hager	Lancaster	W Kassel	03.11	48	W
4./NJG1	Hptm Fritz Lau	Lancaster	25km W Kassel	03.28	25	W
1./NJG5	Hptm E-G Drünkler	Lancaster	20km SW Leipzig	03.38	40	W
5./NJG5	Lt Boy	Lancaster		04.00	3	W
II./NJG4	Maj Hubert Rauh	Lancaster		04.35	31	W
1./NJG5	Fw Burkhard	B-17	SSE Aschaffenburg	04.58	2	W
Stab./NJG6	Maj Herbert Lütje	Lancaster		04.59	50	W

21-22 March 1945

II./NJG11	Fw Karl-Heinz Becker	Mosquito	10km SW Berlin	21.34	2	W
II./NJG11	Fw Reichenbach	Mosquito		21.38	7	W
II./NJG11	Oblt Kurt Welter	Mosquito		21.38	47	W
II./NJG11	Fw Reichenbach	Mosquito		21.48	6	W
II./NJG11	Oblt Kurt Welter	Mosquito		22.04	48	W
I./NJG100	Oblt G E Francsi	Il-4			50	E
I./NJG100	Oblt G E Francsi	Il-4			51	E
I./NJG100	Oblt Hans Schmid	Il-4			6	E
8./NJG1	Hptm Dietrich Schmidt	Lancaster	Cologne	04.00	43	W
12./NJG1	Hptm Josef Kraft	Lancaster	Bochum	04.14	55	W

22-23 March 1945

12./NJG1	Hptm Josef Kraft	B-17		04.00	56	W
8./NJG3	Hptm Fritz Brinkmann	Stirling	NW Vlieland	05.30	16	W

24-25 March 1945

II./NJG11	Fw Karl-Heinz Becker	Mosquito	3km S Fehrbellin	21.32	3	W
II./NJG11	Fw Karl-Heinz Becker	Mosquito	3km S Fehrbellin	21.32	4	W
II./NJG11	Oblt Kurt Welter	Mosquito			49	W

25-26 March 1945

II./NJG3	Lt Hans Meissner	B-24			19	W
II./NJG3	Oblt Fehre	Mosquito		20.50	4	W

27-28 March 1945

II./NJG11	Fw Karl-Heinz Becker	Mosquito	10km W Nauen	21.38	5	W

(left) The controversial Oberleutnant Kurt Welter, Kapitän of 10./NJG 11. Officially credited with 56 night victories including 33 Mosquitos, his claim list show 59 night victories and 36 Mosquitos! (right) Me262s of 10./NJG 11 lined up at Burg airfield, March/April 1945.

II./NJG11	Fw Karl-Heinz Becker	Mosquito	10km W Nauen	21.38	6	W
I./NJG100	Lt Helmut Dahms	Il-4			19	E
I./NJG100	Ofw Rudi Duding	Il-4			15	E
I./NJG100	Oblt G E Francsi	Il-4			52	E
I./NJG100	Oblt G E Francsi	Il-4			53	E
I./NJG100	Oblt G E Francsi	Il-4			54	E
I./NJG100	Lt Helmut Konter	R-5			16	E
I./NJG100	Lt Helmut Konter	R-5			17	E
I./NJG100	Lt Helmut Konter	R-5			18	E
I./NJG100	Lt Helmut Konter	R-5			19	E
I./NJG100	Ofw Karl Strohecker	Il-4			12	E
I./NJG100	Ofw Karl Strohecker	Il-4			13	E
I./NJG100	Lt Hermann Weihe	Li-2			2	E
II./NJG11	Oblt Kurt Welter	Mosquito			50	W

30-31 March 1945

II./NJG11	Fw Karl-Heinz Becker	Mosquito		21.52	7	W
II./NJG11	Fw Reichenbach	Mosquito			8	W
II./NJG11	Oblt Kurt Welter	Mosquito			51	W
II./NJG11	Oblt Kurt Welter	Mosquito			52	W

31 March-1 April 1945

II./NJG11	Fw Karl-Heinz Becker	Mosquito	Tegel (Berlin)	21.52	8	W

3-4 April 1945

II./NJG11	Lt Herbert Altner	Mosquito	Berlin	21.40	25	W
I./NJG1	Hptm Alex Ressugier				5	W
II./NJG11	Oblt Kurt Welter	Mosquito			53	W

4-5 April 1945

3./NJG3	Hptm Eduard Schröder	Lancaster	WNW Anholt	23.02	27	W
3./NJG3	Ofw Hans Schadowski	Lancaster	NW Hanstholm	23.38	16	W

Hauptmann Alex Graf Ressugier first saw action as a JG 300Wilde Sau pilot and later transferred to I./NJG 5 and I./ NJG 1. He claimed his fifth and final night victory on the night of 3rd April 1945.

3./NJG3	Ofw Hans Schadowski	Lancaster	NW Hanstholm	23.42	17	W
3./NJG3	Hptm Eduard Schröder	Lancaster	25km W Klitmoller	23.43	28	W
II./NJG4	Ofw Georg Matt	Lancaster			4	W
3./NJG3	Hptm Eduard Schröder	Lancaster	75km W Stenbjerg	00.04	29	W

7-8 April 1945

III./NJG2	Ofw Rudolf Mangelsdorf	Lancaster	Flensburg	20.10	10	W
III./NJG11	Ofw Kreibaum	Spitfire	Diepholz	20.30	2	W

8-9 April 1945

1./NJG2	Ofw Rudolf Winn	B-17	W Hamburg	22.36	4	W
1./NJG5	Hptm E-G Drünkler	Lancaster	10km E Kolleda	22.56	41	W
St.I./NJG5	Maj Werner Hoffmann	Lancaster	Weissenfels	23.00	51	W
2./NJG5	Oblt Peter Ehrhardt	Lancaster			26	W
4./NJG3	Fw Bruno Rupp	Lancaster			13	W
4./NJG3	Fw Bruno Rupp	Lancaster			14	W
4./NJG3	Fw Bruno Rupp	Lancaster			15	W

9-10 April 1945

I./NJG100	Lt Helmut Dahms	Il-4			23	E
I./NJG100	Lt Helmut Dahms	Il-4			24	E
IV./NJG3	Lt Arnold Döring	Lancaster			5	W
III ./NJG2	Hptm Heinz Ferger	Lancaster			23	W
IV./NJG3	Hptm Tesmar.von	Lancaster			2	W
I./NJG100	Lt Hermann Weihe	Il-4			4	E

10-11 April 1945

1./NJG5	Hptm E-G Drünkler	Lancaster	NW Leipzig	22.55	42	W
1./NJG5	Hptm E-G Drünkler	Lancaster	NW Leipzig	22.57	43	W
1./NJG5	Hptm E-G Drünkler	Lancaster	E Leipzig	23.05	44	W
II./NJG11	Fw Stade.von	Mosquito			1	W
I./NJG100	Ofw Karl Strohecker	Lancaster			14	W
I./NJG100	Oblt Günter Bertram	Lancaster	Bitterfeld	22.57	30	W
I./NJG100	Hptm August Fischer	Lancaster	N Leipzig	23.02	10	W
II./NJG11	Oblt Kurt Welter	Mosquito			54	W
I./NJG3	Hptm Herbert Koch	Halifax	NE Anholt	03.53	22	W

As the advancing Allied forces took abandoned airfields, they found the aircraft of the once-proud Luftwaffe nightfighter force, impotently left, without fuel to fly and fight further...

11-12 April 1945

II./NJG11	Oblt Kurt Welter	Mosquito			55	W

13-14 April 1945

II./NJG11	Oblt Kurt Welter	Mosquito			56	W

14-15 April 1945

II./NJG11	Oblt Kurt Welter	Mosquito			57	W

15 April 1945

II./NJG11	Oblt Kurt Welter	Mosquito			58	W

16-17 April 1945

I./NJG100	Lt Helmut Dahms	Li-2			20	E
I./NJG100	Ofw Heinz De Fries	Li-2			7	E
III./NJG1	Oblt Witzleb	Mosquito			6	W
2./NJG6	Ofw Ludwig Schmidt	Halifax	10km N Gablingen	03.44	17	W

17-18 April 1945

1./NJG5	Hptm E-G Drünkler	Il-4	NW Strasbourg	23.21	45	E
1./NJG5	Hptm E-G Drünkler	Il-4	N Furstenwalde	23.47	46	E
I./NJG100	Lt Helmut Dahms	Il-4			21	E
I./NJG100	Oblt G E Francsi	Il-4			57	E
I./NJG100	Oblt G E Francsi	Il-4			58	E
I./NJG100	Oblt Klaus Scheer	Il-4			23	E
I./NJG100	Hptm Hans Schmid	Il-4			7	E
II./NJG11	Oblt Kurt Welter	Mosquito			59	W

19-20 April 1945

I./NJG100	Lt Helmut Dahms	Il-4			22	E
I./NJG100	Hptm Hans Schmid	Il-4			8	E

As Allied troops overran airfields they discovered the remains of the nightfighter force, like this Ju88G-6 equipped with SN2d.

24-25 April 1945

I./NJG100	Oblt G E Francsi	Il-4			59	E
I./NJG100	Oblt G E Francsi	Il-4			60	E
Stab./NJG5	Maj Rudolf Schoenert	Il-4	S Berlin		61	E
Stab./NJG5	Maj Rudolf Schoenert	Il-4	S Berlin		62	E
Stab./NJG5	Maj Rudolf Schoenert	Il-4	S Berlin		63	E
./NJG5	Maj Rudolf Schoenert	Il-4	S Berlin		64	E
5./NJG5	Oblt Werner Rapp	Il-4	Berlin	01.00	16	E
I./NJG3	Hptm Herbert Koch	Lancaster	W Skagen	01.21	23	W

26-27 April 1945

I./NJG2	Oblt Erich Jung	Lancaster			30	W

2-3 May 1945

IV./NJG3	Oblt Fritz Brandt	Halifax	N Hamburg		8	W